METHODS IN MOLECULAR BIOLOGY

For further volumes:
http://www.springer.com/series/7651

For over 35 years, biological scientists have come to rely on the research protocols and methodologies in the critically acclaimed *Methods in Molecular Biology* series. The series was the first to introduce the step-by-step protocols approach that has become the standard in all biomedical protocol publishing. Each protocol is provided in readily-reproducible step-by-step fashion, opening with an introductory overview, a list of the materials and reagents needed to complete the experiment, and followed by a detailed procedure that is supported with a helpful notes section offering tips and tricks of the trade as well as troubleshooting advice. These hallmark features were introduced by series editor Dr. John Walker and constitute the key ingredient in each and every volume of the *Methods in Molecular Biology* series. Tested and trusted, comprehensive and reliable, all protocols from the series are indexed in PubMed.

Phytochromes

Methods and Protocols

Edited by

Andreas Hiltbrunner

Faculty of Biology, Institute of Biology II and Signalling Research Centres BIOSS and CIBSS, University of Freiburg, Freiburg, Germany

Humana Press

Editor
Andreas Hiltbrunner
Faculty of Biology, Institute of Biology II and Signalling Research Centres BIOSS and CIBSS
University of Freiburg
Freiburg, Germany

ISSN 1064-3745 ISSN 1940-6029 (electronic)
Methods in Molecular Biology
ISBN 978-1-4939-9611-7 ISBN 978-1-4939-9612-4 (eBook)
https://doi.org/10.1007/978-1-4939-9612-4

Cover Caption: This image shows four-day-old wild-type Arabidopsis seedling grown in the dark (right) or in red light (left).

This Humana imprint is published by the registered company Springer Science+Business Media, LLC, part of Springer Nature.
The registered company address is: 233 Spring Street, New York, NY 10013, U.S.A.

Preface

Phytochromes are red/far-red light receptors in plants that play a pivotal role in regulation of growth and development and also contribute to the adaptation to changes in the ambient environment [1, 2]. Research in the field of phytochrome signaling is approaching a point, where the key components are known. Current research is focusing on the investigation of molecular mechanisms and interactions between these components, on the interaction of phytochrome signaling with other signaling pathways, including hormone and temperature signaling, and the evolutionary history of phytochromes. Thus, phytochrome research is relying more and more on methods and techniques that have not been used in traditional photobiology. This volume of *Methods in Molecular Biology* on phytochromes aims at providing the necessary background and a comprehensive collection of state-of-the art protocols for the next decade of phytochrome research. Also other research fields are expanding and researchers not familiar with phytochrome and light signaling may find that the signaling pathway they are investigating is linked to phytochrome signaling. Therefore, this volume of *Methods in Molecular Biology* also includes protocols and background information on methods and techniques traditionally used in phytochrome research to provide a guideline to researchers that are new to the field.

Phytochromes are synthesized in the inactive, red light absorbing Pr form and convert to the physiologically active far-red absorbing Pfr form by absorption of light. Reversion to the inactive Pr state can occur either by absorption of light or by a light-independent thermal relaxation process, referred to as dark reversion [3, 4]. Differential interaction of the Pr and Pfr state of phytochromes with downstream signaling components is critical for light regulation of gene expression and physiological responses controlled by phytochromes [5–8]. Chapters 1–3 describe how to test for light-dependent interaction of phytochromes and phytochrome-binding proteins using yeast two hybrid assays, in vitro pull down, and co-immunoprecipitation from stable transgenic lines. Protein modifications such as phosphorylation, sumoylation, and ubiquitination play an important role in phytochrome signaling [7–10]. The activity of phytochromes and phytochrome downstream signaling components is modulated by phosphorylation, and sumoylation negatively regulates light signaling by phyB, which is the primary phytochrome in light-grown seed plants. Furthermore, many factors involved in phytochrome signaling are targets of E3 ubiquitin ligase complexes. Chapters 4–6 provide protocols to investigate phosphorylation, sumoylation, and ubiquitination of phytochromes and other proteins. Photoconversion and dark reversion rates as well as spectral properties are important parameters that determine the activity of phytochromes [4, 11]. Chapter 7 describes the procedure to express phytochromes in the yeast *Pichia pastoris*, purify it, and characterize its spectral properties. Chapter 8 explains how in vivo spectroscopy can be used to measure Pfr levels of phytochromes in planta. Together, these approaches provide a powerful tool to characterize phytochromes and distinguish between properties intrinsic to the phytochrome molecule itself and properties of the phytochrome system depending on the cellular environment. Furthermore, both approaches provide data that can be used to build mathematical models. Chapter 9 describes the background and provides a *Python* script to predict the pool size of phytochrome in the active and inactive state and in different compartments, including cytosol, nucleus, and photobodies. Light-activated phytochromes are transported from the cytosol into the nucleus and assemble into subnuclear structures called nuclear bodies or photobodies [12]. The function of photobodies is not fully understood but they stabilize phyB in the active state. Many phyB mutants with reduced activity do not properly form photobodies. Furthermore, phytochrome downstream signaling components often co-localize with phytochromes in photobodies. Chapter 10 provides a protocol to quantitatively analyze photobodies of phytochromes and other proteins localizing to these structures. Phytochromes do not only regulate gene expression at the transcriptional level but also by alternative splicing, which can be tested as described in Chapter 11 [13–15]. Seed germination, the response to canopy shade and neighbor competition, accumulation of the chlorophyll precursor protochlorophyllide, input to the circadian clock, and induction of flowering are only few examples of physiological responses under regulation by phytochromes [16, 17]. Chapters 12–16 provide guidelines to characterize these responses. Often, phytohormones play an important role downstream of phytochromes or there is cross-talk between phytochrome and phytohormone signaling [18]. Chapter 17 describes methods to identify hormonal action and co-action in light-regulated processes. Phytochromes are ubiquitous in the plant kingdom and derive from a single phytochrome gene in the last common ancestor of plants [19]. Also in the liverwort *Marchantia*, phytochrome and many downstream signaling factors are represented by single copy genes. This lack of redundancy and the position in the evolutionary history of land plants make *Marchantia* an interesting model system. Chapter 18 provides protocols for testing basic phytochrome responses in *Marchantia*. In contrast to *Marchantia*, independent gene duplication events resulted in small phytochrome gene families in seed plants, polypod ferns, and mosses. The moss model species *Physcomitrella patens* contains seven phytochromes, which regulate gene expression in the nucleus and also have a function at the plasma membrane in photo- and polarotropism [20]. Chapters 19 and 20 describe how to analyze phytochrome-dependent phototropic and polarotropic growth in *Physcomitrella* and how to use the CRISPR/Cas9 system to efficiently generate *Physcomitrella* mutant lines. Phytochromes have a wide range of functions that evolved at different stages of the evolutionary history of phytochromes. Phylogenetic methods to infer how a specific function evolved along the branches of the species tree and to reconstruct the ancestral state are described in Chapter 21.

Once, phytochromes were called "pigment of the imagination" because they existed only as a concept to explain red/far-red reversible physiological responses [21]. Research over the previous 50 years has shown that phytochromes physically exist and not are the product of mere imagination. Detailed protocols were key to this success, but also future work to elucidate physiological functions of phytochromes and underlying molecular mechanisms will rely on carefully established protocols. Thus, I want to thank all the authors for their willingness to share their knowledge and experience and contribute to this book. Equally important, I want to thank John Walker, Series Editor, for his support and for providing advice and assistance whenever needed. Thank you!

Freiburg, Germany ***Andreas Hiltbrunner***

References

1. Li J, Li G, Wang H, Deng XW (2011) Phytochrome signaling mechanisms. Arabidopsis Book 9:e0148
2. Casal JJ (2012) Shade avoidance. Arabidopsis Book 10:e0157
3. Viczián A, Klose C, Ádám É, Nagy F (2016) New insights of red light-induced development. Plant Cell Environ 11:2457–2468
4. Mancinelli AL (1994) The physiology of phytochrome action. In: Kendrick RE, Kronenberg GHM (eds) Photomorphogenesis in plants, 2nd edn. Kluwer Academic Publishers, Dordrecht, pp 211–269
5. Pham VN, Kathare PK, Huq E (2018) Phytochromes and phytochrome interacting factors. Plant Physiol 176:1025–1038
6. Lee N, Choi G (2017) Phytochrome-interacting factor from *Arabidopsis* to liverwort. Curr Opin Plant Biol 35:54–60
7. Podolec R, Ulm R (2018) Photoreceptor-mediated regulation of the COP1/SPA E3 ubiquitin ligase. Curr Opin Plant Biol 45:18–25
8. Hoecker U (2017) The activities of the E3 ubiquitin ligase COP1/SPA, a key repressor in light signaling. Curr Opin Plant Biol 37:63–69
9. Medzihradszky M, Bindics J, Ádám É, et al (2013) Phosphorylation of phytochrome B inhibits light-induced signaling via accelerated dark reversion in *Arabidopsis*. Plant Cell 25:535–544
10. Sadanandom A, Ádám É, Orosa B, Viczián A, Klose C, Zhang C, Josse E-M, Kozma-Bognár L, Nagy F (2015) SUMOylation of phytochrome-B negatively regulates light-induced signaling in Arabidopsis thaliana. Proc Natl Acad Sci USA 112:11108–13
11. Klose C, Venezia F, Hussong A, Kircher S, Schäfer E, Fleck C (2015) Systematic analysis of how phytochrome B dimerization determines its specificity. Nat Plants 1:15090
12. Van Buskirk EK, Decker PV, Chen M (2012) Photobodies in light signaling. Plant Physiol 158:52–60
13. Shikata H, Hanada K, Ushijima T, Nakashima M, Suzuki Y, Matsushita T (2014) Phytochrome controls alternative splicing to mediate light responses in Arabidopsis. Proc Natl Acad Sci USA 111:18781–18786
14. Hartmann L, Drewe-Bo\s s P, Wie\s sner T, et al (2016) Alternative splicing substantially diversifies the transcriptome during early photomorphogenesis and correlates with the energy availability in Arabidopsis. Plant Cell 28:2715–2734
15. Wu H-P, Su Y-S, Chen H-C, Chen Y-R, Wu C-C, Lin W-D, Tu S-L (2014) Genome-wide analysis of light-regulated alternative splicing mediated by photoreceptors in Physcomitrella patens. Genome Biol 15:R10
16. Franklin KA, Quail PH (2010) Phytochrome functions in *Arabidopsis* development. J Exp Bot 61:11–24
17. Mochizuki N, Tanaka R, Grimm B, Masuda T, Moulin M, Smith AG, Tanaka A, Terry MJ (2010) The cell biology of tetrapyrroles: a life and death struggle. Trends Plant Sci 15:488–498
18. Vandenbussche F, Verbelen J-P, Van Der Straeten D (2005) Of light and length: regulation of hypocotyl growth in Arabidopsis. Bioessays 27:275–284
19. Li F-W, Melkonian M, Rothfels CJ, Villarreal JC, Stevenson DW, Graham SW, Wong GK-S, Pryer KM, Mathews S (2015) Phytochrome diversity in green plants and the origin of canonical plant phytochromes. Nat Commun 6:7852
20. Hughes J (2013) Phytochrome cytoplasmic signaling. Annu Rev Plant Biol 64:377–402
21. Sage LC (1992) Pigment of the imagination: a history of phytochrome research. Academic Press, San Diego

Contents

Contributors

MENG CHEN • *Department of Botany and Plant Sciences, Institute for Integrative Genome Biology, University of California, Riverside, CA, USA*
JAE-YONG CHO • *Department of Biotechnology and Kumho Life Science Laboratory, Chonnam National University, Gwangju, Republic of Korea*
GILTSU CHOI • *Department of Biological Sciences, KAIST, Daejeon, South Korea*
SETH J. DAVIS • *Department of Biology, University of York, York, UK*
ANNA LENA ERMERT • *Institute for Plant Physiology, Justus Liebig University, Giessen, Germany*
CHRISTIAN FLECK • *Laboratory of Systems and Synthetic Biology, Wageningen UR, Wageningen, The Netherlands; Department of Biosystems Science and Engineering, ETH Zurich, Basel, Switzerland*
TANJA GANS • *Institute for Plant Physiology, Justus Liebig University, Giessen, Germany*
PÉTER GYULA • *Agricultural Biotechnology Institute, National Agricultural Research and Innovation Centre, Gödöllő, Hungary*
YUN-JEONG HAN • *Department of Biotechnology and Kumho Life Science Laboratory, Chonnam National University, Gwangju, Republic of Korea*
WENJIN HE • *Fujian Normal University, Fuzhou, Fujian, China*
ANDREAS HILTBRUNNER • *Faculty of Biology, Institute of Biology II and Signalling Research Centres BIOSS and CIBSS, University of Freiburg, Freiburg, Germany*
JON HUGHES • *Institute for Plant Physiology, Justus Liebig University, Giessen, Germany*
ENAMUL HUQ • *Department of Molecular Biosciences, The Institute for Cellular and Molecular Biology, The University of Texas at Austin, Austin, TX, USA; Biological Laboratories, University of Texas at Austin, Austin, TX, USA*
KEISUKE INOUE • *Graduate School of Biostudies, Kyoto University, Kyoto, Japan*
SYLWIA M. KACPRZAK • *School of Biological Sciences, University of Southampton, Southampton, UK*
JEONG-IL KIM • *Department of Biotechnology and Kumho Life Science Laboratory, Chonnam National University, Gwangju, Republic of Korea*
EVA KLEMENT • *Laboratory of Proteomics Research, Biological Research Centre, Szeged, Hungary*
CORNELIA KLOSE • *Faculty of Biology, Institute of Biology II, University of Freiburg, Freiburg, Germany*
TAKAYUKI KOHCHI • *Graduate School of Biostudies, Kyoto University, Kyoto, Japan*
FAY-WEI LI • *Boyce Thompson Institute, Cornell University, Ithaca, NY, USA*
CHENTAO LIN • *Department of Molecular, Cell and Developmental Biology, University of California, Los Angeles, CA, USA*
QING LIU • *Basic Forestry and Proteome Research Center, Fujian Agriculture and Forestry University, Fuzhou, Fujian, China*
JAIME F. MARTINEZ-GARCIA • *Centre for Research in Agricultural Genomics (CRAG), CSIC-IRTA-UAB-UB, Barcelona, Spain; Institució Catalana de Recerca i Estudis Avançats (ICREA), Barcelona, Spain*

SARAH MATHEWS • *Australian National Herbarium, CSIRO National Research Collections Australia, Canberra, ACT, Australia; Department of Biological Sciences, Lousiana State University, Baton Rouge, LA, USA*
TOMONAO MATSUSHITA • *Faculty of Agriculture, Kyushu University, Fukuoka, Japan*
RYUICHI NISHIHAMA • *Graduate School of Biostudies, Kyoto University, Kyoto, Japan*
FABIEN NOGUÉ • *Institut Jean-Pierre Bourgin, INRA, AgroParisTech, CNRS, Université Paris-Saclay, Versailles, France*
RACHAEL J. OAKENFULL • *Department of Biology, University of York, York, UK*
BEATRIZ OROSA • *School of Biological and Biomedical Sciences, University of Durham, Durham, County Durham, UK*
INYUP PAIK • *Department of Molecular Biosciences, The Institute for Cellular and Molecular Biology, The University of Texas at Austin, Austin, TX, USA*
SANDI PAULIŠIĆ • *Centre for Research in Agricultural Genomics (CRAG), CSIC-IRTA-UAB-UB, Barcelona, Spain*
IRMA ROIG-VILLANOVA • *Centre for Research in Agricultural Genomics (CRAG), CSIC-IRTA-UAB-UB, Barcelona, Spain*
JAMES RONALD • *Department of Biology, University of York, York, UK*
DAVID J. SHEERIN • *Faculty of Biology, Institute of Biology II, University of Freiburg, Freiburg, Germany*
ROBERT W. SMITH • *Laboratory of Systems and Synthetic Biology, Wageningen UR, Wageningen, The Netherlands; Sir William Dunn School of Pathology, University of Oxford, Oxford, UK*
KIJONG SONG • *Department of Biological Sciences, KAIST, Daejeon, South Korea*
FABIAN STAHL • *Institute for Plant Physiology, Justus Liebig University, Giessen, Germany*
MATTHEW J. TERRY • *Biological Sciences, University of Southampton, Southampton, UK*
FILIP VANDENBUSSCHE • *Laboratory for Functional Plant Biology, Department of Biology, Ghent University, Gent, Belgium*
ANDRÁS VICZIÁN • *Biological Research Centre, Plant Biology Institute, Szeged, Hungary*
QIN WANG • *Basic Forestry and Proteome Research Center, Fujian Agriculture and Forestry University, Fuzhou, Fujian, China; Department of Molecular, Cell and Developmental Biology, University of California, Los Angeles, CA, USA*
XU WANG • *Basic Forestry and Proteome Research Center, Fujian Agriculture and Forestry University, Fuzhou, Fujian, China; Department of Molecular, Cell and Developmental Biology, University of California, Los Angeles, CA, USA*
DESIREE WILLIAMS • *Department of Botany and Plant Sciences, Institute for Integrative Genome Biology, University of California, Riverside, CA, USA*
QI XIE • *State Key Laboratory of Plant Genomics, National Center for Plant Gene Research, Institute of Genetics and Developmental Biology, Chinese Academy of Sciences, Beijing, People's Republic of China; University of Chinese Academy of Sciences, Beijing, People's Republic of China*
CHAN YUL YOO • *Department of Botany and Plant Sciences, Institute for Integrative Genome Biology, University of California, Riverside, CA, USA*
FEIFEI YU • *State Key Laboratory of Plant Genomics, National Center for Plant Gene Research, Institute of Genetics and Developmental Biology, Chinese Academy of Sciences, Beijing, People's Republic of China*
LING ZHU • *Department of Molecular Biosciences, The Institute for Cellular and Molecular Biology, The University of Texas at Austin, Austin, TX, USA*

Chapter 1

Investigation of Light-Regulated Protein–Protein Interactions Using Yeast Two-Hybrid Assays

David J. Sheerin

Abstract

Yeast two-hybrid allows for investigation of interactions between two proteins in vivo. Yeast media can be supplemented with phycocyanobilin, allowing the formation of photoactive phytochromes, and the investigation of their light-regulated interactions. Three reporter assays are described; the histidine growth selection assay for the *HIS3* reporter, and the filter lift and quantitative ONPG assays for the *LacZ* reporter. Design considerations for yeast two- and three-hybrid experiments are also covered.

Key words Yeast two-hybrid, Protein–protein interaction, Phytochrome, Light regulated interaction

1 Introduction

Yeast two-hybrid allows for detection of the interaction between two proteins of interest in vivo through activation of reporter genes. The gene coding sequence for the first protein is cloned into a vector creating a fusion to the DNA binding domain of a transcription factor, normally either *GAL4* from *S. cerevisiae* or *LexA* from *E. coli* [1]. The sequence for the second protein of interest is then cloned into a second vector creating a fusion to the transcription activation domain of a transcription factor, normally either *GAL4* from *S. cerevisiae*, or *VP16* from Herpes virus [1]. The activation and DNA binding domains do not need to originate from the same protein, but both fusions must contain a nuclear localization sequence. Importantly, each vector contains an *E. coli* origin of replication and antibiotic resistance marker (*AmpR*, *KanR*) to facilitate cloning, a yeast origin of replication (2μ), and an amino acid biosynthetic gene (*LEU2*, *TRP1*) [1]. A yeast two-hybrid reporter strain must be used, which lacks the two biosynthetic genes, *leu2* and *trp1*, and possess a reporter construct in its genome such as $URA3::GAL1_{UAS}\text{-}GAL1_{TATA}\text{-}LacZ$ present in the strain Y190 [2]. *GAL4* based reporter strains possess further mutations

Andreas Hiltbrunner (ed.), *Phytochromes: Methods and Protocols*, Methods in Molecular Biology, vol. 2026,
https://doi.org/10.1007/978-1-4939-9612-4_1,

of *GAL4* and *GAL80* and must be grown on media containing Glucose [3]. If the two proteins of interest are able to interact in vivo, then the transcriptional activation domain is recruited to the reporter gene promoter, activating its expression [1]. Due to the requirement for proteins to interact within the nucleus, yeast two-hybrid will perform best with cytoplasmic and nuclear proteins.

Phytochromes are relatively large proteins (approximately 120 kDa) with a complex tertiary structure that requires the *N*-terminus to be free to fold correctly, thus they can only be used as *C*-terminal fusions (PHYA-GAL4BD, PHYB-GAL4AD) [4]. Furthermore, they must covalently bind a chromophore to form a functional holoprotein (represented in lower case—phyA, phyB) [5]. Yeasts do not naturally produce phytochromobilin, the chromophore of land plant phytochromes. Fortunately, an alternate chromophore phycocyanobilin, purified from blue-green algae, functionally complements, and can be supplemented into yeast growth media [6, 7]. Phytochromes from other phyla can use other chromophores, most commonly phycocyanobilin and biliverdin [8]. Yeast are capable of synthesizing biliverdin, though media may need to be supplemented to have the levels required for a yeast two-hybrid assay [9]. Several of the procedures presented here are adapted from methods described in the Clontech yeast protocols handbook [10]. Alterations have been made to accommodate the behavior of phytochromes, their requirement for a chromophore, and optimal light treatments. These procedures can also be used for other light-responsive proteins, including blue-light responsive proteins such as the cryptochromes. If not working with phytochromes, or other bilin proteins, PCB should be omitted and red/far-red light treatments replaced by appropriate wavelengths. Cloning of vectors and immunoblotting should be performed using standard procedures and are not described here.

2 Materials

Prepare all solutions with either double deionized or double distilled water, and use analytical quality reagents. When working with solvents (methanol, hexane, butyl methyl ether, dimethylformamide) or β-mercaptoethanol, work within a fume hood. Do not mix waste solvents, and dispose of these and β-mercaptoethanol containing solutions according to local regulations.

2.1 Common Materials

1. Laminar flow bench.
2. 30 °C incubator.
3. Temperature controlled dark room capable of 26 °C.
4. Red (660 nm) and far-red (740 nm) LED light sources.
5. Photometer for measuring light intensities.
6. Heat blocks.

7. Spectrophotometer with disposable 1 mL cuvettes (*see* **Note 1**).
8. pH meter.
9. Microwave.
10. CSM agar part A: 40 g L^{-1} Agar. Maximum 40% of container volume. Autoclave.
11. CSM agar part B: 40 g L^{-1} D-glucose monohydrate, 0.2 g L^{-1} adenine hemisulfate, 13.4 g L^{-1} yeast minimal base, 1.2–1.3 g L^{-1} desired amino acids mixture (−Trp, −Leu −Trp). Autoclave at reduced pressure (10 PSI) for 10 min at 115 °C (*see* **Note 2**). Store at 4 °C.
12. *Z*-buffer: 60 mM Na_2HPO_4, 40 mM NaH_2PO_4, 10 mM KCl, 1 mM $MgSO_4$. Autoclave. Check that pH is 7.0.
13. β-mercaptoethanol (toxic!—work in a fume hood).
14. Sterile clear 96-well microtiter plates.

2.2 Phycocyanobilin Purification

1. Fume hood with sink, water supply, and water jet vacuum pump.
2. Rotary evaporator with condenser and water bath.
3. 1 L spherical flasks (3×).
4. Filter funnel.
5. 1 L vacuum flask.
6. 500 mL separation flasks (3×) with support stands.
7. Corvex glass centrifuge tubes, with rubber adapters (10×) (*see* **Note 3**).
8. Refrigerated super-speed centrifuge and Sorval SS-34 rotor.
9. 1% citric acid: weigh 3 g citric acid and prepare a 300 mL solution. Store at 4 °C.
10. Saturated NaCl: weigh 35 g NaCl and prepare a 100 mL solution.
11. 0.5% $NaHCO_3$: weigh 1 g $NaHCO_3$ and prepare a 200 mL solution. Store at 4 °C.
12. Methanol (toxic!).
13. Chloroform (toxic!).
14. Hexane (toxic!).
15. Butyl methyl ether (toxic!).
16. Ammonium sulfate.
17. 95% methanol–1.85% HCl: measure 9.5 mL methanol into a disposable tube and slowly add 0.5 mL concentrated 37% HCl. Prepare fresh.
18. Dimethylsulfoxide (DMSO).
19. *Spirulina* powder (*see* **Note 4**).

2.3 Yeast Transformation

1. Swinging bucket centrifuge for 50 mL tubes.
2. YPDA agar. 20 g L^{-1} D-glucose monohydrate, 0.1 g L^{-1} adenine hemisulfate, 20 g L^{-1} peptone, 10 g L^{-1} yeast extract, 20 g L^{-1} agar. Autoclave at reduced pressure (10 PSI) for 10 min at 115 °C (*see* **Note 2**).
3. CSM complete: 20 g L^{-1} D-glucose monohydrate, 0.1 g L^{-1} adenine hemisulfate, 6.7 g L^{-1} yeast minimal base, 0.32 g L^{-1} amino acids –leucine, –tryptophan (*see* **Note 2**).

Autoclave at reduced pressure (10 PSI) for 10 min at 115 °C.

4. 10× TE: 100 mM Tris base, 10 mM ethylenediaminetetraacetic acid disodium (EDTA), adjust pH to 8.0 with concentrated HCl. Autoclave.
5. 10× LiAc: 1 M lithium acetate, adjust pH to 8.0 with dilute acetic acid. Autoclave.
6. 50% PEG 4000. Autoclave.
7. 10 mg mL^{-1} single-stranded DNA: Dissolve high quality sheared DNA (Sigma D1626) in 1× TE. Heat to 95 °C for 10 min, cool on ice, then store at −20 °C.

2.4 Histidine Growth Selection Assay

1. 1 M 3-amino triazole (3-AT). Filter-sterilize and store at −20 °C.
2. Multichannel pipette.

2.5 Filter Lift Assay

1. 80 mm filter paper discs. Wrap in foil, autoclave, then dry in a 60 °C oven.
2. 20 mg mL^{-1} 5-bromo-4-chloro-3-indolyl-β-D-galactopyranoside (X-gal). Dissolve 100 mg X-gal in 5 mL *N*,*N*-dimethylformamide (DMF, toxic!), protect from light, and store at −20 °C.
3. *Z*-buffer X-gal. Mix 5 mL *Z*-buffer, 13.5 μL β-mercaptoethanol (β -ME) and 83.5 μL of 20 mg mL^{-1} X-gal. Prepare fresh.
4. Glass spreaders.

2.6 ONPG Quantitative Assay

1. Microtiter plate reader.
2. Drum roller.
3. Sterile 160 × 16 mm glass culture tubes.
4. *Z*-buffer o-nitrophenol-β-D-galactopyranoside (ONPG). Weigh ONPG using a sterilized metal spatula for a final concentration of 4 mg mL^{-1}. Dissolve in *Z*-buffer with vigorous shaking. Prepare on the day of use.
5. *Z*-buffer β -ME. Add β -ME (toxic!) to *Z*-buffer to final 0.27%. Prepare fresh.
6. 1 M Na_2CO_3. Prepare fresh with sterile ddH_2O.

7. 2× YPDA: 40 g L^{-1} D-glucose monohydrate, 0.2 g L^{-1} adenine hemisulfate, 40 g L^{-1} peptone, 20 g L^{-1} yeast extract. Autoclave at reduced pressure (10 PSI) for 10 min at 115 °C (*see* **Note 2**).

2.7 Preparation of Yeast Extracts for Immunoblotting

1. 1 M Tris–HCl pH 6.8.
2. Yeast cracking buffer: 8 M urea, 40 mM Tris–HCl pH 6.8, 1 mM EDTA, 0.4 mg mL^{-1} bromophenol blue, 5% sodium dodecyl sulfate (SDS). Heat at 37 °C overnight to fully dissolve SDS.
3. 100 mM phenylmethane sulfonyl fluoride (PMSF) (toxic!). Dissolve PMSF in isopropanol and store at room temperature.
4. 100× protease inhibitor cocktail. Dissolve one vial (Sigma I3911) in 10 mL sterile ddH_2O. Store at −20 °C.
5. 1 mm glass beads.

3 Methods

3.1 Purification of Phycocyanobilin (PCB) from Spirulina

Currently phycocyanobilin is commercially available (from Frontier Scientific, USA). This protocol, which is adapted from Kunkel et al. [6], is presented in case it becomes unavailable, as it has in the past. A typical yield of 50 μmol PCB can be obtained from 50 g starting material, enough for 2.5 L of yeast media. If upscaling, all volumes throughout the procedure must be increased. The procedure involves use of several toxic solvents; *the entire procedure should be performed within a fume hood*. Waste solvents should be collected separately, not mixed, and disposed of according to local regulations.

1. Measure 50 g dry *Spirulina* powder into a methanol-washed spherical flask.
2. Add 200 mL methanol, attach to a rotary evaporator and heat to 60 °C degrees using a water bath for 10 min while rotating to mix (*see* **Note 5**).
3. Pour the slurry into a filter funnel attached to a vacuum flask. Add a small amount of methanol to the spherical flask to wash remaining traces, and add to the filter funnel. Apply a vacuum until all methanol has been pulled through, leaving behind the solid material.
4. Return the *Spirulina* to the spherical flask using a spatula.
5. Repeat **steps 2–4** four times.
6. Repeat **steps 2–4** five times at an increased temperature of 65 °C and reduced incubation time of 5 min (*see* **Note 5**). The dry material can be stored for several days at 4 °C protected from moisture.

7. Transfer washed powder to a methanol-washed spherical flask.
8. Add 200 mL methanol, attach to a rotary evaporator and heat to 75 °C degrees using a water bath for 5 h while rotating to mix and using a condenser to return any methanol to the flask. Protect the fume hood from light (*see* **Note 6**).
9. Divert the condensed methanol to a collection flask, allowing the volume in the spherical flask to reduce to approximately 30 mL, remove from heat to cool to room temperature.
10. Cool the spherical flask on ice.
11. Collect the liquid fraction from the slurry using a methanol-cleaned filter funnel attached to a clean methanol-washed vacuum flask.
12. Place 70 mL butyl methyl-ether in a methanol cleaned separation flask.
13. Transfer the methanol fraction to the separation flask and swirl to mix.
14. Carefully add 300 mL ice-cold 1% (w/v) citric acid and 10 mL saturated NaCl onto the methanol/ether phase in the separation flask.
15. Very gently swirl.
16. Sit to allow phases to separate.
17. Discard the entire aqueous (lower) phase (*see* **Note 7**).
18. Collect the ether phase (70 mL) into a second methanol-cleaned separation flask.
19. Carefully add 100 mL ice-cold 0.5% (w/v) $NaHCO_3$ and swirl gently, proceed to the next step quickly (*see* **Note 8**).
20. Collect the aqueous (lower) phase into a beaker containing 2 g citric acid. Discard the chlorophyll-containing ether phase. Stir gently with a glass rod.
21. Add 50 mL chloroform to the neutralized aqueous phase. Transfer to a chloroform-cleaned separation flask and shake vigorously to transfer the PCB to the chloroform phase.
22. Allow the phases to separate. Collect the organic (lower) phase into a beaker containing 10 g ammonium sulfate to absorb remaining traces of water. Stir gently with a glass rod.
23. Transfer the chloroform phase (no solid material) to a clean spherical flask, attach to a rotary evaporator and heat to 80 °C with the condenser running. Allow chloroform to evaporate until only 10 mL remains. Allow to cool to room temperature.
24. Cool the spherical flask on ice.
25. Fill a glass beaker with 150 mL hexane. Add the chloroform–PCB slowly dropwise while stirring with a glass rod. Continue stirring until blue crystals have formed.

26. Transfer 15 mL to each of ten Corvex glass centrifuge tubes using a glass Pasteur pipette. Place inside rubber adapters and centrifuge for 10 min at 3600 × *g* at 4 °C (*see* **Note 3**).
27. Carefully remove the hexane using a glass pasture pipette, careful to not disturb the crystals.
28. Air-dry in a fume hood.
29. Dissolve crystals in 30 mL methanol then prepare 1 mL aliquots in 1.5 mL reaction tubes. Evaporate methanol by placing in a heat block, and heating to 80 °C until completely dry.
30. Store closed at −20 °C.
31. Resuspend a single aliquot in 100 μL DMSO. The volume for other aliquots can be calculated from **step 34**, for a final concentration of 10 mM.
32. Add 2 μL dissovled PCB to 998 μL 95% methanol–1.85% HCl in a disposable cuvette. Store the remainder of the PCB at −20 °C.
33. Measure the absorbance spectrum from 300 to 800 nm, blanking against 95% methanol–1.85% HCl–0.2% DMSO (*see* **Note 9**).
34. The PCB concentration = Ab_{680nm} × 13.2 mM (*see* **Note 9**).

3.2 Transformation of *Saccharomyces cerevisiae*

The following method is based on the method described in the Clontech yeast protocols handbook [10]. For larger numbers of transformations, the Frozen EZ-Yeast transformation II kit from Zymo Research is recommended. Either a single bait plasmid or both a bait and prey plasmid can be cotransformed. If only transforming a bait plasmid, then selective media should lack tryptophan in case of most *GAL4* based vectors. Suitable plasmids and yeast strains are detailed in Subheading 3.7.

1. Streak desired yeast strain onto a YPDA agar plate for single colonies (*see* **Note 10**). If using yeast already carrying a bait plasmid then selective media should be used instead.
2. Seal with Parafilm and incubate inverted at 30 °C for 3 days.
3. Scrape a single colony from the plate and suspend in a reaction tube containing 0.5 mL sterile ddH_2O.
4. Transfer 250 μL suspended yeast to each of two culture tubes containing 5 mL CSM complete media, or selective media in the case of already transformed yeast.
5. Incubate for 16–18 h at 30 °C 200 rpm.
6. Transfer both the overnight cultures to 100 mL prewarmed CSM complete in a 500 mL flask.
7. Incubate for 3–5 h until the OD_{600} is 0.5–0.6 (*see* **Note 1**).

8. Harvest yeast by centrifugation at 1000 × *g* for 5 min at room temperature using a swinging bucket centrifuge and 2 × 50 mL sterile plastic tubes.
9. Discard media and resuspend yeast by vortexing in 40 mL sterile ddH_2O, combining the two pellets from **step 8**.
10. Pellet yeast by centrifugation at 1000 × *g* for 5 min at room temperature.
11. Discard water, and resuspend yeast by vortexing in 40 mL fresh 1× TE–LiAc (5 mL 10× TE, 5 mL 10× LiAc, and 40 mL sterile ddH_2O).
12. Pellet yeast by centrifugation at 1000 × *g* for 5 min at room temperature.
13. Discard water, and resuspend yeast by vortexing in 2 mL fresh 1× TE–LiAc.
14. Incubate stationary at 30 °C for 15 min.
15. Denature 10 mg mL^{-1} single-stranded DNA by heating to 95 °C for 10 min and then placing on ice.
16. Prepare transformation master mix: 10 mL 50% (w/v) PEG 4000, 1.25 mL 10× TE, 1.25 mL 10× LiAc, 625 μL single-stranded DNA, and 2 mL yeast from **step 14**.
17. Add 100 μL transformation master mix with a maximum of 2.5 μL plasmid DNA (100–300 ng μL^{-1}). When cotransforming both bait and prey, use equal quantities of both vectors, totaling 2.5 μL.
18. Incubate stationary at 30 °C for 30 min. During this incubation prepare CSM –Leu –Trp plates by combining equal parts A and B and melting in a microwave, pouring into 10 cm petri dishes, and allowing to dry open in the laminar flow bench.
19. Heat-shock at 42 °C for 20 min.
20. Harvest yeast by centrifugation at maximum speed for 30 s at room temperature.
21. Discard the supernatant, and resuspend yeast in 50 μL ddH_2O by pipetting.
22. Spread yeast on CSM –Leu –Trp selective plates, allow to dry completely in a laminar flow bench.
23. Seal plates with Parafilm and incubate inverted at 30 °C for 3–5 days (*see* **Note 11**).
24. Replate transformants onto a fresh selective CSM agar plate to create a master plate for further experiments. Incubate inverted at 30 °C for 2 days, then store at 4 °C for up to 1 week. For long-term storage, create glycerol stocks by suspending yeast in sterile 25% glycerol and storing at –80 °C.

3.3 Filter Lift Assay

Steps 3 and **5–7** should be performed in weak green light.
Steps 8, 10, and **15–17** should be performed in a fume hood.

1. Working in a laminar flow bench, combine equal volumes CSM Agar part A and part B (−Leu −Trp). Loosen the lid and heat in a microwave until agar is completely molten and no foam is present.
2. Remove sufficient 10 mM PCB in DMSO from storage at −20 °C, wrap in foil to protect from light, and allow to thaw at 30 °C.
3. Returning to the laminar flow bench, prepare CSM + 20 μM PCB plates by pouring 15 mL molten CSM agar into a sterile 50 mL reaction tube, adding 30 μL 10 mM PCB, swirling to mix in the PCB, and pouring into a 10 cm petri dish. Allow plates to dry uncovered in the laminar flow bench in darkness, once set, close and place in a black box to protect from light (*see* **Note 12**).
4. Prepare yeast suspensions, taking a small amount of yeast from a transformant master plate with a pipette tip and suspending in sterile ddH_2O to a final OD_{600} of approximately 0.1.
5. Pipet 5 μL spots of yeast suspension onto the agar plates, only pressing to the first stop on the pipette to avoid creating droplets, and allow to completely dry before sealing plates with Parafilm, and inverting to prevent condensation from spreading yeast (*see* **Note 13**).
6. Place plates in a black box to protect from light and store in complete darkness at 26 °C for 24 h (*see* **Note 14**).
7. Transfer plates to desired light treatments. Recommended treatments include continued 24 h darkness, 24 h 1 μmol m^{-2} s^{-1} red light, and 24 h 10 μmol m^{-2} s^{-1} far-red light (*see* **Note 15**).
8. When incubations are near completion, working in a fume hood prepare 5 mL *Z*-buffer X-gal solution per plate to be assayed.
9. For each plate to be assayed, prepare a petri dish containing four sterile filter paper discs, and a second containing a single filter paper disc with a small edge cut off to allow orientation.
10. In the fume hood add 4.5 mL *Z*-buffer X-gal solution to each of the petri dishes containing four filter paper discs. Remove any air bubbles using a sterilized glass spreader. Incubate for 10 min.
11. Hold X-gal containing plates on an angle to allow excess solution to drain to one side, and remove it using a pipette.
12. Carefully remove Parafilm and lids from yeast plates, ensuring no condensation contacts the agar.

13. Pick up a single filter paper disc with sterile forceps by the top cut-off edge. Carefully place down onto the yeast plate, starting by placing the bottom edge of the filter, then allowing it to fall onto the plate. Immediately apply gentle pressure to the back of the filter with a sterile glass spreader until the yeast spots are visible, remove the filter and slowly submerge in liquid nitrogen, then place the filter with yeast upright in another petri dish (*see* **Note 16**).
14. Once all plates have been filter lifted, proceed to freeze in liquid nitrogen an additional four times to lyse yeast cells, allowing filters to thaw between each freezing.
15. Returning to the fume hood, hold a petri dish containing *Z*-buffer X-gal pre-soaked filters on an angle. Starting with the bottom edge of the yeast-bound filter, carefully touch to the upper side of the angled dish, and slowly place down while leveling the plate, being careful to avoid creating any air bubbles between the filters.
16. Hold plates on a slight angle in the light, the bound yeast should look slightly glossy, but not the filter paper. If yeast look matt, then a small amount of additional *Z*-buffer X-gal needs to be added at the edge of the filter, very slowly until the yeast are glossy (*see* **Note 17**). Close and place flat inside a plastic bag to prevent evaporation.
17. Check filters once every hour, noting any visible blue spots. The exact length of time for development will be dependent on the strength of your interaction. The assay should be stopped after a maximum of 8 h. To stop the reaction, transfer the upper filter to a new petri dish, and leave open in the fume hood to dry. Do not scan until all traces of β-mercaptoethanol have evaporated. Note that samples on the filter will be arranged as a mirror image of your agar plate.

3.4 Histidine Growth Selection Assay

Steps 3 and **5–6** should be performed in weak green light.

1. Working in a laminar flow bench, combine equal volumes CSM Agar part A and part B (−Leu −Trp −His), and for control plates CSM Agar part A and part B (−Leu −Trp). Loosen lids and heat in a microwave until agar is completely molten and no foam is present.
2. Remove sufficient 10 mM PCB in DMSO from storage at −20 °C, wrap in foil to protect from light, and allow to thaw at 30 °C. Also thaw sufficient 1 M 3-AT.
3. Returning to the laminar flow bench, prepare CSM + 20 μM PCB + 3-AT plates by pouring 15 mL molten CSM agar into a sterile 50 mL reaction tube, adding 30 μL 10 mM PCB, swirling to mix in the PCB, adding 0–750 μL 1 M 3-AT, swirling

again and pouring into a 10 cm petri dish. Allow plates to dry uncovered in the laminar flow bench in darkness, once set, close and place in a black box to protect from light (*see* **Note 12**). A range of 3-AT concentrations should be prepared 0 mM, 1 mM (15 μL 1 M 3-AT), 2 mM (30 μL), 5 mM (75 μL), 10 mM (150 μL), 20 mM (300 μL), and 50 mM (750 μL). 3-AT should not be added to the CSM –Leu –Trp control.

4. Prepare yeast suspensions, taking a small amount of yeast from a transformant master plate with a pipette tip and suspending in sterile ddH_2O to a final OD_{600} of 0.1 (1×10^6 yeast per mL) in a sterile microtiter plate.
5. Pipet 5 μL spots of yeast suspensions onto the agar plates using a multichannel pipette, only pressing to the first stop on the pipette to avoid creating droplets, and allow to completely dry before sealing plates with Parafilm, and inverting to prevent condensation from spreading yeast (*see* **Note 13**).
6. Incubate plates in either darkness, 1 μmol m^{-2} s^{-1} red light, or 10 μmol m^{-2} s^{-1} far-red light for 3 days at 26 °C (*see* **Note 15**).
7. Record results by scanning in positive film mode and inverting the image. Bait–prey combinations that are able to grow on higher 3-AT concentrations have higher *HIS3* reporter gene activation. Check that yeast transformants are able to grow normally on the CSM –Leu –Trp control.

3.5 ONPG Quantitative Assay

Steps 4–8 and **11–19** should be performed in weak green light. Safety glasses should be worn while working with liquid N_2.

1. Label three long culture tubes per bait–prey combination (A/B/C for independent transformants), and prepare three reaction tubes containing 50 μL sterile ddH_2O.
2. Prewarm CSM –Leu –Trp media (–Met for yeast three-hybrid) to 25 °C. You will need 6 mL per bait–prey pair.
3. Remove sufficient 10 mM PCB in DMSO from storage at –20 °C, wrap in foil to protect from light, and allow it to thaw at 30 °C.
4. Pour CSM media into a sterile 50 mL reaction tube and add 2 μL per mL 10 mM PCB for a final concentration of 20 μM, immediately screw on the lid and shake to fully suspend the PCB.
5. Using a plastic serological pipette, add 2 mL CSM PCB media to each of the labelled culture tubes.
6. Take a small amount of yeast (approximately 2 μL) from a master transformant plate, and suspend in the reaction tubes containing 50 μL sterile ddH_2O prepared during **step 1**, then transfer this suspension to a labelled culture tube. Proceed for all yeast samples.

7. Vortex each sample to fully suspend yeast.
8. Place on a roller drum at 40 rpm, and incubate in complete darkness for 16 h at 26 °C. Protect from light exposure during transport (*see* **Note 12**).
9. Label 3 new culture tubes per bait–prey combination for each desired light treatment.
10. Prewarm 6 mL 2×YPDA media per culture to 26 °C.
11. Add 6 mL 2× YPDA to a 2 mL CSM PCB culture, vortex to fully suspend yeast, and pipette 2 mL into new culture tubes (up to four different light treatments). This can be performed in the culture room, sterility does not need to be maintained, but samples cannot be exposed to light other than weak green light.
12. Return to the drum roller and perform desired light treatments (*see* **Note 15**). Recommended treatments include 5 min 20 μmol m^{-2} s^{-1} red light, and 20 μmol m^{-2} s^{-1} far-red light.
13. Following 4 h incubation, place culture tubes into a wire rack on ice.
14. Pipet 1 mL of each culture into a labelled 1.5 mL safe-lock reaction tube on ice.
15. Centrifuge samples at 5000 × *g* 4 °C for 1 min.
16. Carefully remove the media using a pipette. Do not disturb the pellet.
17. Add 1 mL ice-cold Z-buffer and suspend the pellet by pipetting.
18. Centrifuge samples at 5000 × *g* 4 °C for 1 min.
19. Remove *Z*-buffer using a pipette.
20. Add 200 μL *Z*-buffer and store at −80 °C.
21. Briefly vortex each original culture from **step 12** and mix 0.25 mL with 0.75 mL ddH_2O in a 1 mL cuvette, immediately measure the OD_{600}. Blank with 0.25 mL 2× YPDA mixed with 0.75 mL ddH_2O. Multiply the results by 4 and record the culture densities (*see* **Note 1**).
22. Thaw samples from **step 20** in a 30 °C heat block and vortex to suspend pellets.
23. Pour liquid N_2 into a cardboard cryobox placed within a Styrofoam container, up to the top of the insert (*see* **Note 18**).
24. Transfer the samples to the liquid N_2 for 2 min to freeze, then thaw in a 30 °C heat block.
25. Repeat **step 24** four additional times to lyse the cells. Add additional liquid N_2 as required.

26. Place lysed yeast samples on ice. Samples can be stored at −80 °C if not to be processed immediately.
27. Prepare sufficient *Z*-buffer β-ME, *Z*-buffer ONPG, and 1 M Na_2CO_3.
28. For each bait–prey combination and light treatment prepare 10 clean 1.5 reaction tubes, labeled 0–9, 10–19, 20–29, and so on.
29. To each tube with a number ending in zero 0, 10, 20, ..., add 50 μL *Z*-buffer as blanks.
30. If yeast lysates were stored at −80 °C, remove from storage and place on ice. If samples were not frozen, proceed to **step 32**.
31. One bait–prey–light treatment at a time, thaw the respective three biological samples (replicates A, B, C) in a heat block at 30 °C.
32. Mix well by pipetting, and pipette 50 μL into tubes ending in 1, 2, and 3 for transformant A, 4, 5, and 6 for B, and 7, 8, and 9 for C.
33. Record the number series (i.e., 20–29) and bait–prey–light treatment combination. In a fume hood add 250 μL *Z*-buffer β-ME to all 10 samples. Add 60 μL *Z*-buffer ONPG to all 10 samples, record the start time to the nearest minute, mix by inversion, and place in a 30 °C incubator.
34. Repeat **steps 31–33** for all samples, checking on samples in the 30 °C incubator every 30 min for development of a strong yellow color. Strong interactions take approximately 30 min, and weak interactions up to 24 h, negative reactions should not change color within 24 h.
35. Once samples have developed, or 24 h has passed, add 150 μL 1 M Na_2CO_3 to all ten samples in the set (i.e., 20–29 including the blank) record the stop time to the nearest minute, invert to mix, and store at 4 °C.
36. Centrifuge samples at 20,000 × *g* for 10 min at 4 °C.
37. Carefully transfer 250 μL of samples to clear 96 well plates, avoiding air-bubbles, and measure the absorbance at 420 nm.
38. Calculate the β-galactosidase activity in Miller units using the following formula [10]:

$$\text{Activity} = (1000 \times \text{Ab}_{420} \times D) / (V_1 \times C \times (V_2 / V_3) \times t \times \text{OD}_{600})$$

$D = 1.67$, this factor is due to the depth of 250 μL in the microtiter plate being only 0.6 cm.

V_1 = 1 mL, the volume of yeast culture harvested at **step 14**.

C = 5 fold as the yeast are suspended in only 0.2 mL at **step 20**.

V_2 = 0.05 mL, the volume of yeast lysate added to the reaction.

V_3 = 0.51 mL the total reaction volume.

t = elapsed time, in minutes.

OD_{600} = the culture density calculated in **step 21**.

Thus, activity = $(1670 \times Ab_{420}) / (0.49 \times t \times OD_{600})$.

3.6 Preparation of Yeast Extracts for Immunoblotting

Reduced or absent reporter gene activity may not only be a result of weak or the absence of bait–prey interactions but may also result if either the BD or AD fusions are inadequately expressed. To control for this, the abundance of BD and AD fusions can be compared by immunoblotting. This is vital when comparing truncated forms to map interaction domains.

Steps 1–6 should be performed in weak green light.

1. Grow yeast as described in Subheading 3.5, including light treatments (**steps 1–13**).
2. Transfer 1 mL of culture to a safelock 1.5 mL reaction tube.
3. Harvest yeast by centrifugation at 1000 × *g* 4 °C for 1 min, remove the supernatant using a pipette, careful not to disturb the pellet.
4. Repeat **steps 2** and **3** twice more to combine the yeast from 3 mL of culture into 1 pellet.
5. Resuspend the yeast in 1 mL sterile ddH_2O using a pipette.
6. Centrifuge again at 1000 × *g* 4 °C for 1 min, remove the supernatant using a pipette, and freeze the tube containing the pellet in liquid N_2. Store at −80 °C.
7. Mix 0.25 mL of the original culture with 0.75 mL ddH_2O in a 1 mL cuvette, immediately measure the OD_{600}. Blank with 0.25 mL 2× YPDA mixed with 0.75 mL ddH_2O.
8. Multiply the measured result by 12 (3/0.25) to obtain the OD units present in the frozen pellet.
9. Working in a fume hood, add β-mercaptoethanol to 1% in sufficient cracking buffer. Add protease inhibitor cocktail to a final 5× concentration.
10. Heat cracking buffer to 60 °C.
11. For each sample, prepare a 1.5 mL reaction tube containing 50 μL 1 mm glass beads and place on ice.
12. Add PMSF to prewarmed cracking buffer to a 5 mM final concentration. PMSF is only stable for 15 min.
13. To a frozen yeast pellet, add 100 μL prewarmed cracking buffer per 7.5 OD units, suspend and transfer to a tube containing glass beads on ice.

Proceed only once all samples have passed stage 13, and only with four samples at a time.

14. Heat samples to 60 °C for 60 s. Vortex immediately for 60 s then place on ice.

    Proceed only once all samples have passed stage 14, and only with four samples at a time.

15. Heat samples to 60 °C for 60 s. Centrifuge at 20,000 × *g* 4 °C for 30 s, and immediately transfer the supernatant to a new tube on ice.

    Proceed only once all samples have passed stage 15.

16. Heat samples to 80 °C for 3 min, place on ice.
17. Proceed with separation by SDS-PAGE and immunoblotting following standard procedures, loading 20 μL of each sample.

3.7 Experimental Design

When using the phytochromes in yeast two-hybrid experiments several factors need to be taken into account. Phytochromes cannot be used as *N*-terminal fusions due to the requirement of the *N*-terminus to be free during folding of a figure-of-eight knot structure, thus commercial yeast two-hybrid vectors must be modified during cloning [4]. Commonly used are the vectors pD153 (R. Brazas) [11] and pCGADT7 [12], however, it should be noted that the empty pD153 vector should not be used as a negative control and an alternate vector such as pGBKT7 (Clontech) should be used instead. This is due to the inclusion of an HA tag, which when fused to the GAL4 DNA binding domain in the absence of an additional fusion, is weakly autoactivating [1]. Full length *Arabidopsis* phytochromes are toxic to *E. coli* when fused to the *N*-terminus of the LEXA DNA binding domain; however, they could be used as an activation domain fusion in this system. Compatible yeast strains for the *GAL4* systems include Y190, AH109, Y187 (*LacZ* reporter only), and Y2HGold (*HIS3* reporter only).

Full-length *Arabidopsis* phyA does possess weak autoactivation if exposed to strong red light, or if yeast is allowed to reach stationary phase, so care must be taken to avoid these situations. Full-length phyB, C, and E are strongly autoactivating. This autoactivation is caused by the C-terminal PAS- and histidine kinase-like domains. Therefore to use phyB, C, or E as a bait proteins, truncations should be used, ending after the PHY domain (aa 1-624 for phyB), phytochrome dimerization is preserved as the GAL4 DNA binding domain forms dimers [13, 14]. Full-length phytochromes can be used as prey when the activation domain is fused to their C-terminus [15]. However, some interactions cannot be detected in this orientation, particularly for phyA, for unknown reasons.

The yeast three-hybrid system is a modification of the yeast two-hybrid system that allows for testing the effects of a third protein on the interaction of two others. This includes (1) bridging interactions; where the two other proteins independently bind to the third, but not each other, (2) competition;

where the third and one of the other two proteins compete for the same binding site on the last, (3) modification; where either binding or covalent modification, such as phosphorylation, is mediated by the third that alters the binding activity of the pair of proteins. Yeast three-hybrid experiments use the same reporter strains as for two-hybrid, however, for the commonly used pBridge vector (Clontech), strains lacking a functional *MET25* gene cannot be used (Y187). pBridge encodes for both the bait protein, and an additional protein under control of the *MET25* promoter, which is repressed by methionine, though basal expression still occurs on media containing methionine. A number of control experiments must first be performed before beginning a yeast three-hybrid experiment. Firstly, all possible pairwise interactions must be tested (i.e., 1–2, 2–1, 1–3, 3–1, 2–3, 3–2). Secondly immunoblots should be performed to establish how well each of the three proteins express in yeast. Once this information has been established the following situations must be considered; only a nonautoactivating protein can be used as the bait, if a protein is autoactivating it also cannot be used as the tertiary protein if it can independently bind to the selected bait protein, and finally the ideal expression levels from lowest to highest would be bait–prey–tertiary. A total of four pBridge vectors will need to be used as bait plasmids; the empty control vector, one expressing both the bait and tertiary proteins, and one each expressing only either the bait or tertiary proteins. When interpreting results, be aware that expression of any tertiary protein will likely result in a reduction of reporter activity compared to the control, due to the high expression from the *MET25* promoter. In this respect, light-regulated interactions have an advantage; the phytochromes can be inactivated while still expressed. Alternatively a nonfunctional tertiary protein could be used as a control.

4 Notes

1. Optical density measurements will vary between different makes of spectrophotometers. It is important to check if an OD_{600} of 1.0 equates to 1×10^7 yeast cells per mL using a cell counter, and if not, adjust OD measurements accordingly.
2. Glucose will burn if not fully dissolved or autoclaved for too long, which will strongly affect results. If it is not possible to autoclave at a lower pressure, omit the glucose and 10% of the water. Glucose can be filter-sterilized separately as a 200 g L^{-1} solution (10×). CSM agar is prepared as two parts as agar will lose its strength if autoclaved in an acidic solution, particularly if not used immediately. Yeast minimal base and amino acid

mixtures can be obtained from either MP Bio (YNB and CSM supplements) or Takara (SD base—contains glucose and DO supplements).

3. Inspect glass centrifuge tubes for any scratches, chips, or imperfections before use. If any are found, discard the tube. If a tube breaks during a run, the adapter must also be discarded. Glass centrifuge tubes are used without lids (the rotor lid must always be used), and do not exceed 15 mL per tube.
4. *Spirulina* powder can be purchased from health food stores. Store at 4 °C protected from moisture.
5. Phycocyanobilin is covalently attached to protein in *Spirulina*, making it insoluble in methanol. The chromophore is freed by the process of methanolysis which begins to occur at 67 °C; therefore, while removing chlorophyll it is important to stay below this temperature. Once the methanol begins to turn slightly blue you must stop washing and proceed to the next step. You will not be able to remove all chlorophyll during washing.
6. Dependent on the make of your rotary evaporator, you may not be able to divert condensed methanol to return to the reaction flask. If this is the case, then the collected methanol can be returned through the glass tube intended for checking the vapor temperature within the reaction flask.
7. Phycocyanobilin will appear green while in butyl methyl-ether.
8. The chromophore is highly unstable at alkaline pH in the presence of water. Hydrolysis must be avoided; if it becomes purple in color then the phycocyanobilin has been degraded.
9. Phycocyanobilin has a molar absorbance coefficient of 37,900 M^{-1} cm^{-1} at its absorbance peak of 680 nm [16]. The presence of contaminating chlorophyll can be seen through the presence of an additional peak at 660 nm.
10. Yeast are stored at −80 °C, using a pipette or sterile loop, scrape a small amount of material (approximately 1–2 μL) from the stock and spread on a small area of an agar plate. Do not allow frozen stocks to thaw. Using a sterile loop, streak to obtain single colonies. To create a stock, suspend a large yeast colony in sterile 25% glycerol, and place directly into the −80 °C freezer.
11. The number of days the yeast should be grown for will depend on if your constructs affect the growth speed of the yeast, transformants should normally grow within 3 days.
12. The chromophore is highly sensitive to all wavelengths of light, especially when plates have been freshly poured. Keep exposure to even green light to a minimum. The chromophore is

also less stable in the presence of high concentrations of 3-amino triazole. When using the strain Y190, omit 3-AT concentrations below 5 mM. If working on other photoreceptors, such as cryptochromes, PCB should be omitted.

13. Practice in preventing aerosols is essential. If after growing yeast you notice small colonies outside the spots you plated then you have created tiny droplets during pipetting. Only submerge your tip far enough into the liquid as required, ensure no air bubbles are present in the tip, and pipet within close proximity to the agar (but not touching it), and only to the first stop on the pipette. Avoid placing strong positives near samples of interest. Spacing can be controlled through the use of a multichannel pipette.

14. The chromophore is taken up by the yeast and incorporated into the phytochromes during the 24 h dark incubation, greatly increasing light stability. 26 °C is used as higher temperatures result in faster break down of the chromophore. Phytochromes have also been reported as inhibited by higher temperatures [17, 18]. If performing a yeast three-hybrid experiment, extend this incubation to 48 h to accommodate slower growth on –Met media.

15. For chromophore containing agar plates a maximum light intensity of 1 μmol m^{-2} s^{-1} red light is recommended to prevent photobleaching. For the filter lift assay or liquid cultures, an alternate light treatment of 6 h 20 μmol m^{-2} s^{-1} red/far-red light can be used after 42 h growth in darkness (plates), or after YPDA addition (liquid cultures). If performing a yeast three-hybrid experiment, extend this incubation on –His media to 5 days to accommodate slower growth on –met media. If working on other photoreceptors, use appropriate light treatments, such as 20 μmol m^{-2} s^{-1} blue light for cryptochromes.

16. It is possible to prepare more than one filter lift from a single plate if desired. Do not allow filters to become completely soaked.

17. If you add too much *Z*-buffer X-gal, yeast will begin to spread, too little and weak interactions may be missed.

18. This setup is to avoid the tops of tubes being submerged in liquid N_2, which often results in exploding tubes.

References

1. Criekinge WV, Beyaert R (1999) Yeast two-hybrid: state of the art. Biol Proced Online 2:1–38
2. Harper JW, Adami GR, Wei N et al (1993) The p21 Cdk-interacting protein Cip1 is a potent inhibitor of G1 cyclin-dependent kinases. Cell 75:805–816
3. Johnston M, Flick JS, Pexton T (1994) Multiple mechanisms provide rapid and stringent glucose repression of *GAL* gene expression in *Saccharomyces cerevisiae*. Mol Cell Biol 14:3834–3841
4. Burgie ES, Bussell AN, Walker JM et al (2014) Crystal structure of the photosensing module from a red/far-red light-absorbing plant phytochrome. Proc Natl Acad Sci U S A 111:10179–10184
5. Hanzawa H, Shinomura T, Inomata K et al (2002) Structural requirement of bilin chromophore for the photosensory specificity of phytochromes A and B. Proc Natl Acad Sci U S A 99:4725–4729
6. Kunkel T, Tomizawa K-I, Kern R et al (1993) *In vitro* formation of a photoreversible adduct of phycocyanobilin and tobacco apophytochrome B. Eur J Biochem 215:587–594
7. Kami C, Mukougawa K, Muramoto T et al (2004) Complementation of phytochrome chromophore-deficient *Arabidopsis* by expression of phycocyanobilin:ferredoxin oxidoreductase. Proc Natl Acad Sci U S A 101:1099–1104
8. Rockwell NC, Su YS, Lagarias JC (2006) Phytochrome structure and signaling mechanisms. Annu Rev Plant Biol 57:837–858
9. Kim D, Yukl ET, Moënne-Loccoz P et al (2006) Fungal heme oxygenases: functional expression and characterization of Hmx1 from *Saccharomyces cerevisiae* and CaHmx1 from *Candida albicans*. Biochemistry 45:14772–12780
10. Clontech Laboratories Inc (2009) Yeast protocols handbook. Available via Clontech. www.clontech.com/xxclt_ibcGetAttachment.jsp?cItemId=17602. Accessed July 2009
11. Shimizu-Sato S, Huq E, Tepperman JM et al (2002) A light-switchable gene promoter system. Nat Biotechnol 20:1041–1044
12. Rausenberger J, Tscheuscler A, Nordmeier W et al (2011) Photoconversion and nuclear trafficking cycles determine phytochrome A's response profile to far-red light. Cell 146:813–825
13. Hughes RM, Vrana JD, Song J et al (2012) Light-dependent, dark-promoted interaction between *Arabidopsis* cryptochrome 1 and phytochrome B proteins. J Biol Chem 287:22165–22172
14. Gadhavi P, Morgan PJ, Alefounder P et al (1996) A physico-chemical investigation of the self-association of the DNA binding domain of the yeast transcriptional activator GAL4. Eur Biohys J 24:405–412
15. Sheerin DJ, Menon C, zur Oven-Krockhaus S et al (2015) Light-activated phytochrome A and B interact with members of the SPA family to promote photomorphogenesis in *Arabidopsis* by reorganizing the COP1/SPA complex. Plant Cell 27:189–201
16. Cole WJ, Chapman DJ, Siegelman HW (1967) Structure of phycocyanobilin. J Am Chem Soc 89:3643–3645
17. Legris M, Klose C, Burgie SE et al (2016) Phytochrome B integrates light and temperature signals in Arabidopsis. Science 354:897–900
18. Jung JH, Domijan M, Klose C et al (2016) Phytochromes function as thermosensors in Arabidopsis. Science 354:886–889

Chapter 2

Rapid Examination of Phytochrome–Phytochrome Interacting Factor (PIF) Interaction by In Vitro Coimmunoprecipitation Assay

Inyup Paik and Enamul Huq

Abstract

Phytochromes are red/far-red light receptors that perceive and transduce light signals to regulate various physiological responses. A large number of phytochrome interacting proteins have been identified using genetic and biochemical approaches. A direct interaction between phytochrome and its interacting proteins, therefore, defines one of the critical steps to initiate light signaling cascades in plants. Thus it is important to thoroughly examine the light-dependent interaction between phytochromes and putative phytochrome-interacting proteins. In this chapter, a protocol for rapid and simple light dependent in vitro coimmunoprecipitation between phytochromes and phytochrome interacting factors is described. In principle, this protocol can be adapted for other putative phytochrome interacting proteins.

Key words Protein–protein interaction, Phytochrome interacting factors, In vitro coimmunoprecipitation assay, TnT system

1 Introduction

Phytochromes (phys) are red and far-red light receptors with pivotal roles in plant growth and development. They utilize a chromophore to perceive light and switch between two interconvertible conformers (Pr and Pfr) in response to far-red and red light, respectively. Phytochromes are synthesized in the cytosol and red light perception by phytochromes leads to a translocation of phytochromes into the nucleus [1, 2] and triggers light signaling cascade primarily by direct interaction with multiple phytochrome interacting proteins [3]. A variety of phytochrome interacting proteins were identified including multiple transcription factors, translational regulators, protein kinases, phosphatases, E3 ligases, other photoreceptors, and hormone signaling regulators [4–14].

Among them, the basic helix–loop–helix transcription factors called phytochrome interacting factors (PIFs) are well conserved in

Andreas Hiltbrunner (ed.), *Phytochromes: Methods and Protocols*, Methods in Molecular Biology, vol. 2026,
https://doi.org/10.1007/978-1-4939-9612-4_2,

plants and have been shown to function as master regulators of transcription in response to light and environmental cues [15].

Myriad of evidence have indicated that physical interaction between phytochromes and their binding proteins triggers multiple biochemical responses including but not limited to conformational changes, sequestration, phosphorylation, and degradation of the phytochrome interacting proteins [8, 16–19]. Many of these phytochrome interacting proteins exert conformer specific bindings indicating the complex light regulation of the binding between phytochrome and its interacting proteins.

The biological significance of physical interaction between phytochrome and its binding proteins is critical; therefore, it is essential to examine direct phytochrome-protein interaction in vitro. However, the high molecular weight of the phytochromes (~125 kDa) and the requirement of the chromophore make it laborious to purify functional phytochromes for simple in vitro interaction assay. In this chapter, we provide a detailed protocol for simple and rapid light dependent co-immunoprecipitation assay between phytochrome and PIFs, a bHLH class of transcription factors from Arabidopsis, by utilizing a T&T quick-coupled transcription/translation system (Promega, Madison, WI).

The T&T quick-coupled transcription/translation system provides a rapid method for synthesis of phytochrome and its binding proteins simply by mixing and incubating template DNA in the T&T reaction. This system also allows assembly of holoproteins by expressing the apoprotein in the TnT reaction and incubation with a chromophore in case of phytochromes. Eukaryotic nature of the T&T mixture enables putative post-translational modifications of the protein of interest and helps proper folding of the proteins. On the other hand, the system only allows for relatively small amount of translation product. Therefore, it may not be adequate for large-scale biochemical analysis.

2 Materials

2.1 Expression of Protein Using the TnT System and ^{35}S-Methionine Labeling

1. TnT® Quick Coupled Transcription/Translation Systems (Promega, Madison, WI; Catalog #L1170).
2. ^{35}S-Methionine (PerkinElmer; Catalog #NEG709A).
3. Plasmid DNA (e.g., pET17b) or PCR-generated DNA template that contains coding sequence for the proteins of interest cloned under the control of the T7 or SP6 RNA polymerase promoter. See the TnT® manual for details.
4. 6× SDS gel loading buffer: 300 mM Tris–HCl, pH 6.8, 12% SDS, 0.6% Bromophenol blue, 60% Glycerol, 600 mM DTT.
5. 2× SDS gel loading buffer: Add 1 mL 6× SDS gel loading buffer and 2 mL H_2O.

6. 4× lower gel buffer: 1.5 M Tris–HCl, pH 8.8, 0.4% SDS.
7. 4× upper gel buffer: 0.5 M Tris–HCl, pH 6.8, 0.4% SDS.
8. 10% sodium dodecyl sulfate–polyacrylamide gel electrophoresis (SDS-PAGE) gel:
   (a) Lower gel (total 10 mL): Add 3.4 mL 30% acrylamide–bis solution, 2.5 mL 4× lower gel buffer, 5 μL tetramethylethylenediamine (TEMED), 50 μL 10% ammonium persulfate (APS), 4.05 mL H_2O.
   (b) Upper gel (total 5 mL): Add 0.83 mL 30% acrylamide–bis solution, 1.25 mL 4× lower gel buffer, 5 μL TEMED, 50 μL 10% APS, 2.87 mL H_2O.
9. Fixing solution (total 100 mL): Add 10 mL of glacial acetic acid to a mixture of 50 mL of methanol and 40 mL H_2O.
10. 10% glycerol.
11. Phycocyanobilin (PCB): Dissolve PCB (Frontier Scientific, Catalog# P14137) in DMSO to 600 μM final concentration. Protect from light and store at −20 °C.
12. Whatman 3 MM filter paper.
13. Standard equipment to run and dry SDS-PAGE gels.
14. Typhoon imaging system (GE Healthcare Life Sciences).

2.2 Coimmunoprecipitation

1. Antibody for the immunoprecipitation of the bait protein; we used GAL4-TA antibody (C-10; Santa Cruz; Catalog #sc-1663).
2. Dynabeads™ Protein A for Immunoprecipitation (Thermo Fisher Scientific; Catalog #10002D).
3. Magnetic rack.
4. 10× PBS buffer: To make 1 L add 25.6 g $Na_2HPO_4 \cdot 7\,H_2O$, 80 g NaCl, 2 g KCl, 2 g KH_2PO_4.
5. 50 mg/mL BSA (bovine serum albumin).
6. 10% NP-40 (Tergitol, Sigma; Catalog #127087-87-0).
7. Water.
8. 2× PBS binding buffer (2× PBS, 1 μg/mL BSA, 0.1% NP-40): To make 50 mL, add 10 mL 10× PBS, 1 mL 50 mg/mL BSA, 0.5 mL 10% NP-40, 38.5 mL water.
9. 1× PBS binding buffer (PBS, 1 μg/mL BSA, 0.1% NP-40): To make 50 mL, add 5 mL 10× PBS, 1 mL 50 mg/mL BSA, 0.5 mL 10% NP-40, 43.5 mL water; in the following, 1× PBS binding buffer always refers to PBS binding buffer containing BSA.
10. 1× PBS binding buffer without BSA (PBS, 0.1% NP-40): To make 50 mL, add 5 mL 10× PBS, 0.5 mL 10% NP-40, 44.5 mL water; in the following, it is explicitly mentioned if 1× PBS binding buffer without BSA is to be used.

3 Methods

3.1 Expression of Proteins with T&T Reaction and ^{35}S-Methionine Labeling

3.1.1 T&T Reaction

1. Quickly thaw TnT® Quick Coupled Transcription/Translation mix on tap water and place it on ice. Store rest of mix in aliquots at −80 °C. Avoid freeze–thaw cycle more than once.
2. Prepare DNA that contains T7/SP6 promoter (*see* **Note 1**).
3. Prepare T&T reactions for prey and bait proteins as following in 1.5 mL test tubes: 40 μL T&T quick master mix, 2 μL ^{35}S-methionine, 1 μg template plasmid DNA, nuclease-free water to 50 μL total volume.
4. Incubate tubes in 30 °C for 60–90 min.

3.1.2 Preparing Prey (Phytochrome) Protein

1. Add 49 μL 2× PBS buffer to T&T prey mix, resulting in 98 μL T&T prey mix.
2. Add 2 μL of the 600 μM phycocyanobilin (PCB) to the 98 μL T&T prey mix to a final concentration of 12 μM PCB (in the green safe light).
3. Incubate for 1 h on ice in the dark.
4. Save 10 μL of the translation product for input control.

3.1.3 Preparing Bait (GAD-PIFs) Protein

1. Dilute the bait T&T mix with 50 μL 2× PBS buffer, resulting in 100 μL T&T bait mix.
2. Attach the GAD antibody to the Dynabeads: add 5 μL (1 μg) antibody to 20 μL beads; incubate for 30 min at 4 °C.
3. Collect Dynabeads on magnetic rack. Wash twice with 500 μL 1× PBS binding buffer and resuspend into 40 μL 1× PBS binding buffer.
4. Add 40 μL beads to the bait T&T mix, incubate for 2 h at 4 °C, wash twice with 500 μL 1× PBS binding buffer and resuspend the pellet into 40 μL 1× PBS binding buffer.
5. Save 10 μL of the translation product for input control.

3.2 In Vitro Coimmuno-precipitation

3.2.1 Light Treatment and Immunoprecipitation

1. Combine 12.5 μL of prepared T&T bait and 12.5 μL of original T&T prey in a total volume of 50–75 μL (make up rest of the volume with 1× PBS buffer) (*see* **Note 2**).
2. Place test tubes on ice in a way that the light can reach to the solutions in the tubes (*see* **Note 3**).
3. Treat tubes with far-red (10 μmol m^{-2} s^{-1}) and red (7 μmol m^{-2} s^{-1}) light for 5 min to convert phyB to the inactive Pr and the active Pfr state, respectively.
4. Incubate for 3 h at 4 °C with gentle rotation in the dark (*see* **Note 4**).
5. Wash twice with 1 mL 1× PBS binding buffer in green safe light using the magnetic rack, then once with 1× PBS binding

buffer without BSA. Wash by inverting tubes upside down for 10–20 times and make sure the Dynabead pellets are resuspended completely.

6. Resuspend the pellet in 25 µL 1× PBS binding buffer without BSA and 5 µL 6× SDS gel loading buffer.

3.2.2 Analysis of Coimmunoprecipitation with SDS-PAGE

1. Heat pellets from coimmunoprecipitation at 65 °C for 10 min. Heat previously prepared input controls with 10 µL of 2× SDS gel loading buffer.
2. Load samples onto a 10% SDS-PAGE gel. Percentage of the PAGE gel should be adjusted according to the size of the protein product used in the co-immunoprecipitation assay. Use 100 V/20 mA to run the gel until the blue dye runs off the gel. This typically takes up to 2 h.
3. Gel fixation: shake the gel in fixing solution for 30 min followed by shaking in 10% glycerol for 10 min (*see* **Note 5**).
4. Place the gel on two sheets of Whatman 3 MM filter paper. Cover with plastic wrap and dry on a gel dryer. Dry gel under vacuum at 80 °C for 2 h.
5. Expose the dried gel on a PhosphorImaging screen overnight (*see* **Note 6**). Scan the image the next morning using a PhosphorImager (*see* **Note 7**). An example of the results produced is shown in Fig. 1.
6. Use Image J (https://imagej.nih.gov/ij/) software to quantify protein bands detected from the assay.

4 Notes

1. For phytochrome cloning, we used the pET3B vector (Novagen, Catalog# 69419-3), which contains the T7 promoter. For baits cloning, PIFs in this case, we used modified pET17b/GAD vector [11]. pET17b/GAD vector harbors N-terminal GAL4 activation domain (GAD) that can be used for immunoprecipitation. Any type of tag (e.g., FLAG, MYC, HA) that is compatible with the protein of interest can be used for the experiment.
2. The total volume of the coimmunoprecipitation sample is small such that only the bottom of the 1.5 mL tube is occupied by the solution. It is important to briefly spin down tubes and keep the solutions at the tip of the tubes during the 3 h incubation.
3. Use small ice buckets and put tubes on the ice parallel so that the light can reach into the tubes from the clear side of the 1.5 mL tubes.

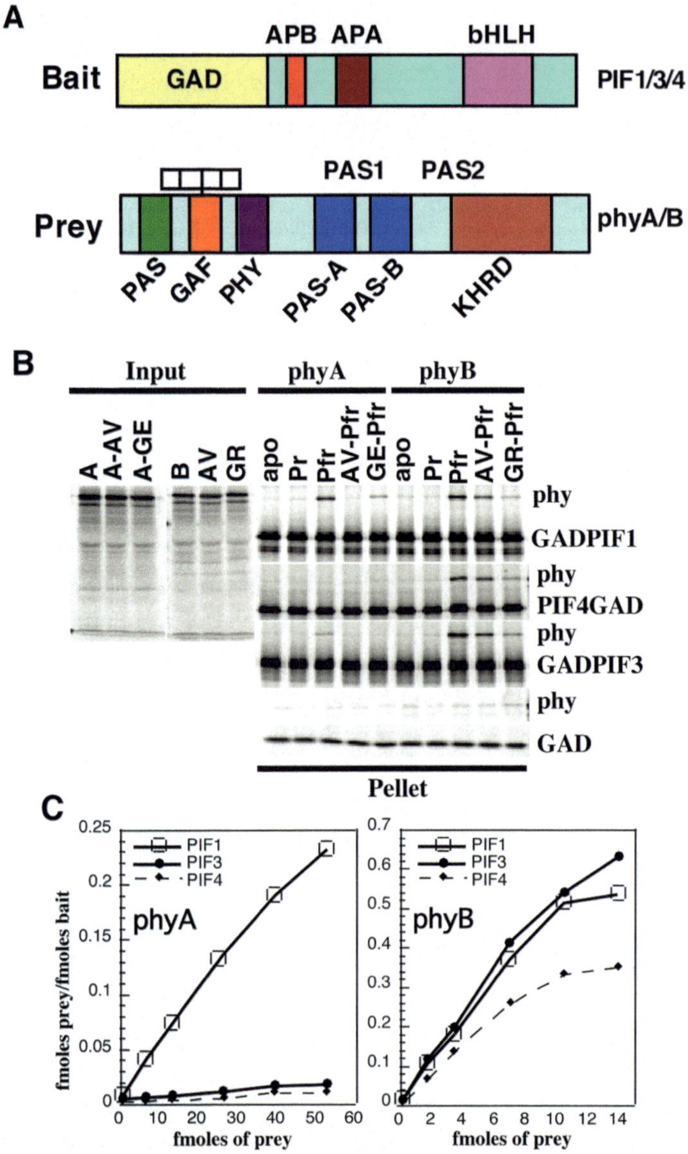

Fig. 1 In vitro coimmunoprecipitation assay between phytochromes and PIFs. (**a**) Schematic diagram showing the bait and prey constructs used for the experiment. PIFs were cloned into pET17b/GAD vector as described in the method section. Empty vector containing GAD only was used as a negative control. Phytochromes are transcribed/translated followed by incubation with phycocyanobilin to reconstitute the holoprotein. (**b**) PIF1, PIF3, and PIF4 show red light-dependent (Pfr conformer-specific) interactions with phytochromes. *apo* Apoprotein, *Pr* Pr conformer of phytochrome, *Pfr* Pfr conformer of phytochrome, *AV/GE* missense mutations in phyA, *AV/GR* missense mutations in phyB. (**c**) Quantification of the coimmunoprecipitation results. Coimmunoprecipitated proteins were quantified using Image J software. From Huq E, Al-Sady B, Hudson M, Kim C, Apel K, Quail PH (2004) PHYTOCHROME-INTERACTING FACTOR 1 is a critical bHLH regulator of chlorophyll biosynthesis. Science 305:1937–1941. Reprinted with permission from AAAS

4. Gentle, non-upside-down rotation is preferred, since total immunoprecipitation reaction is only about 50 μL.
5. Shaking in 10% glycerol prevents gels from cracking during the procedure. Gel drying makes radioisotope signals sharper; however, you need to carefully adjust drying condition to prevent cracking and masking (or damaging) signals from co-IP results.
6. Dried gels still could be sticky on the surface and damage the phosphorscreen. This can be prevented by wrapping the gel with a plastic wrap.
7. We used Typhoon FLA phosphorimager for the imaging. We typically used scanning parameters as follows: Laser: 650 nm, Filter: IP, PMT: 950 V, Pixel size: 25 nm.

Acknowledgments

We thank members of the Huq laboratory for critical reading of the manuscript. This work was supported by grants from the National Institute of Health (NIH) (GM-114297) and National Science Foundation (MCB-1543813) to E.H.

References

1. Kircher S, Kozma-Bognar L, Kim L, Adam E, Harter K, Schäfer E, Nagy F (1999) Light quality-dependent nuclear import of the plant photoreceptors phytochrome a and B. Plant Cell 11:1445–1456
2. Yamaguchi R, Nakamura M, Mochizuki N, Kay SA, Nagatani A (1999) Light-dependent translocation of a phytochrome B-GFP fusion protein to the nucleus in transgenic *Arabidopsis*. J Cell Biol 145:437–445
3. Bae G, Choi G (2008) Decoding of light signals by plant phytochromes and their interacting proteins. Annu Rev Plant Biol 59:281–311
4. Paik I, Yang S, Choi G (2012) Phytochrome regulates translation of mRNA in the cytosol. Proc Natl Acad Sci U S A 109:1335 1340
5. Ni W, Xu S-L, González-Grandío E, Chalkley RJ, Huhmer AFR, Burlingame AL, Wang Z-Y, Quail PH (2017) PPKs mediate direct signal transfer from phytochrome photoreceptors to transcription factor PIF3. Nat Commun 8:15236
6. Ni W, Xu S-L, Tepperman JM, Stanley DJ, Maltby DA, Gross JD, Burlingame AL, Wang Z-Y, Quail PH (2014) A mutually assured destruction mechanism attenuates light signaling in *Arabidopsis*. Science 344:1160–1164
7. Shi H, Shen X, Liu R, Xue C, Wei N, Deng XW, Zhong S (2016) The red light receptor phytochrome B directly enhances substrate-E3 ligase interactions to attenuate ethylene responses. Dev Cell 39:597–610
8. Sheerin DJ, Menon C, zur Oven-Krockhaus S, Enderle B, Zhu L, Johnen P, Schleifenbaum F, Stierhof Y-D, Huq E, Hiltbrunner A (2015) Light-activated phytochrome A and B interact with members of the SPA family to promote photomorphogenesis in Arabidopsis by reorganizing the COP1/SPA complex. Plant Cell 27:189–201
9. Lu X-D, Zhou C-M, Xu P-B, Luo Q, Lian H-L, Yang H-Q (2015) Red-light-dependent interaction of phyB with SPA1 promotes COP1–SPA1 dissociation and photomorphogenic development in *Arabidopsis*. Mol Plant 8:467–478
10. Hughes RM, Vrana JD, Song J, Tucker CL (2012) Light-dependent, dark-promoted interaction between *Arabidopsis* cryptochrome 1 and phytochrome B proteins. J Biol Chem 287:22165–22172
11. Huq E, Al-Sady B, Hudson M, Kim C, Apel K, Quail PH (2004) Phytochrome-interacting factor 1 is a critical bHLH regulator of chlorophyll biosynthesis. Science 305:1937–1941

12. Oh E, Kim J, Park E, Kim J-I, Kang C, Choi G (2004) PIL5, a phytochrome-interacting basic helix-loop-helix protein, is a key negative regulator of seed germination in *Arabidopsis thaliana*. Plant Cell 16:3045–3058
13. Ni M, Tepperman JM, Quail PH (1999) Binding of phytochrome B to its nuclear signalling partner PIF3 is reversibly induced by light. Nature 400:781–784
14. Huq E, Quail PH (2002) PIF4, a phytochrome-interacting bHLH factor, functions as a negative regulator of phytochrome B signaling in *Arabidopsis*. EMBO J 21:2441–2450
15. Paik I, Kathare PK, Kim J-I, Huq E (2017) Expanding roles of PIFs in signal integration from multiple processes. Mol Plant 10:1035–1046
16. Luo Q, Lian H-L, He S-B, Li L, Jia K-P, Yang H-Q (2014) COP1 and phyB physically interact with PIL1 to regulate its stability and photomorphogenic development in *Arabidopsis*. Plant Cell 26:2441–2456
17. Park E, Park J, Kim J, Nagatani A, Lagarias JC, Choi G (2012) Phytochrome B inhibits binding of PHYTOCHROME INTERACTING FACTORs to their target promoters. Plant J 72:537–546
18. Shen H, Moon J, Huq E (2005) PIF1 is regulated by light-mediated degradation through the ubiquitin-26S proteasome pathway to optimize photomorphogenesis of seedlings in Arabidopsis. Plant J 44:1023–1035
19. Al-Sady B, Ni W, Kircher S, Schäfer E, Quail PH (2006) Photoactivated phytochrome induces rapid PIF3 phosphorylation prior to proteasome-mediated degradation. Mol Cell 23:439–446

Chapter 3

Characterization of Light-Regulated Protein–Protein Interactions by In Vivo Coimmunoprecipitation (Co-IP) Assays in Plants

Ling Zhu and Enamul Huq

Abstract

In light signaling pathways, protein complexes play essential roles in light perception and signal transduction. The phytochrome family of photoreceptors perceives red/far-red region of the light spectrum and then translocates into the nucleus to form protein complexes. Many phytochrome interacting proteins have been identified based on yeast two-hybrid screening and other protein–protein interaction methods. However, it is essential to demonstrate that these proteins interact with phytochromes in vivo to be functionally relevant. In this chapter, a protocol for demonstrating light dependent in vivo interactions between phytochromes and phytochrome interacting proteins is described. This protocol can be adapted for any putative phytochrome interacting protein for validation of their interaction in vivo.

Key words Arabidopsis, Protein–protein interaction, Phytochrome interacting factors, In vivo coimmunoprecipitation

1 Introduction

Light provides energy source for plant growth and development. However, light also acts as a pivotal informational signal that regulates plant adaptation to prevailing environmental conditions. Plants perceive light signal through multiple classes of photoreceptors responding to different spectrum of light. These include UVR8 (for sensing UV-B light), phototropins (phot1/2), cryptochromes (cry1/2/3) and Zeitlupe family of receptors (ZTL/LKP1/FKF) (for sensing UV-A/Blue lights), and phytochromes (for sensing blue/red/far-red lights) [1]. Among these photoreceptors, phytochromes (phys) are well-characterized light perception system in Arabidopsis and other plants [2]. Phytochromes are chromoproteins with an attached bilin chromophore that perceives light. There are two light-switchable forms of phytochromes: Pr, an inactive form that perceives light in the red region of the spectrum and Pfr, a biologically active form that perceives far-red

Andreas Hiltbrunner (ed.), *Phytochromes: Methods and Protocols*, Methods in Molecular Biology, vol. 2026, https://doi.org/10.1007/978-1-4939-9612-4_3,

region of the spectrum. These two forms can interconvert and exist as an equilibrium in plant cells. In darkness, the Pr form of phytochromes is present in the cytosol and is photoinactive. Upon light exposure, phytochromes undergo conformational changes to the Pfr form and translocate into the nucleus [3]. The photoactivated phytochromes in the nucleus induce a large change in protein–protein complex formation and transcriptional reprogramming that drive photomorphogenesis.

Within the nucleus, phytochromes interact with bHLH transcription factors called the phytochrome interacting factors (PIFs) [4, 5], circadian clock factors [6, 7], splicing factors [8], E3 ubiquitin ligases [9], and others. Among the best-characterized phytochrome interacting proteins are the PIFs. PIFs are negative regulators in the light signaling pathways. The interaction between phytochromes and PIFs results in rapid phosphorylation, ubiquitination, and degradation of the PIFs through the 26S proteasome system [2]. PIFs are phosphorylated by various kinases in response to light. The phosphorylated forms are recognized by various E3 ubiquitin ligases and then rapidly degraded to promote photomorphogenesis [2]. Thus, multiple dynamic and rapid changes in protein–protein interactions and protein complex formation are necessary for rapid responses to light signal. Therefore, identification and in vivo demonstration of protein–protein interaction under both dark and light conditions are necessary to understand the mechanisms by which plants respond to light signal. In this chapter, we provide a detailed protocol for demonstrating in vivo light-dependent interaction between phytochrome and PIF1 as an example. The in vivo coimmunoprecipitation (co-IP) method is suitable for detecting protein interacting partners under both dark and light conditions.

Although, the in vivo coimmunoprecipitation (co-IP) method demonstrates interaction between phytochromes and putative interaction partners, there are several advantages and disadvantages for this method. The advantages include the following: (1) this method provides robust verification of interactions in vivo where the interaction is functionally relevant, (2) proteins can be posttranslationally modified in vivo if the posttranslational modification is necessary for interaction with phytochromes, and (3) the interaction between phytochromes and the putative phytochrome interacting partner might still be detected even if the interaction is indirect. The disadvantages include the following: (1) the interaction between phytochromes and the putative phytochrome interacting partner might be indirect, as a bridging protein can facilitate the interaction, (2) this method cannot be used to distinguish interactions in a cell- and/or tissue-specific manner in vivo; although, a modified version of this method can be used for this purpose.

2 Materials

2.1 Plant Material Preparation

1. 10 cm ø petri dishes.
2. 7.5 cm ø filter paper.
3. Standard dark room and light chambers (Model: E30LED, Percival Scientific, Inc.).
4. MS medium: dissolve 4.33 g Murashige and Skoog basal medium in 1 L H_2O, adjust pH to 5.7 by NaOH, add 9 g agar and autoclave.
5. Liquid MS medium: dissolve 4.33 g Murashige and Skoog basal medium in 1 L H_2O, adjust pH to 5.7 by NaOH and autoclave.
6. Bortezomib stock: make 40 mM bortezomib (Fisher Scientific, #B1408100MG) stock in DMSO; aliquot to 20 μL volume and store at −20 °C.
7. 40 μM bortezomib in liquid MS media: mix 5 μL 40 mM bortezomib stock with 5 mL liquid MS in 10 cm ø petri dish.
8. Liquid nitrogen.
9. Mortar and pestle.
10. Photometer (e.g., Beckman DU530 UV-VIS Life Science Spectrophotometer).

2.2 Coimmunoprecipitation

1. 100 mM phosphate buffer, pH 7.8: Mix 90.8 mL 1 M K_2HPO_4 and 9.2 mL 1 M KH_2PO_4, and dilute the combined 1 M stock solutions to 1 L with distilled H_2O.
2. Native extraction buffer/IP buffer: 100 mM phosphate buffer, pH 7.8, 150 mM NaCl, 0.1% NP40, 1× protease inhibitor cocktail (Sigma-Aldrich Co., St Louis, MO, Cat. No: P9599), 1 mM PMSF, 10 mM iodoacetamide, 40 μM bortezomib, 25 mM β-glycerophosphate, 10 mM sodium fluoride (NaF), and 2 mM sodium orthovanadate (Na_3VO_4).
3. IP wash buffer: 100 mM phosphate buffer, pH 7.8, 150 mM NaCl, 0.5% NP40, 1× protease inhibitor cocktail, 1 mM PMSF, 40 μM bortezomib, 25 mM β-glycerophosphate, 10 mM sodium fluoride (NaF), and 2 mM sodium orthovanadate (Na_3VO_4).
4. Dynabeads Protein A (Life Technologies Co., Carlsbad, CA, Cat. No: 10002D).
5. Magnetic rack.
6. Primary antibodies used for immunoprecipitation.
7. Bio-Rad Protein Assay Dye Reagent Concentrate (Bio-Rad Laboratories, Inc., Hercules, CA, Cat. No: 500-0006).

2.3 Western Blot

1. 6× SDS gel loading buffer: 300 mM Tris–HCl, pH 6.8, 12% SDS, 0.6% bromophenol blue, 60% glycerol, 600 mM DTT.
2. 1.5× SDS gel loading buffer: 75 mM Tris–HCl, pH 6.8, 3% SDS, 0.15% bromophenol blue, 15% glycerol, 150 mM DTT.
3. 4× lower gel buffer: 1.5 M Tris–HCl, pH 8.8, 0.4% SDS.
4. 4× upper gel buffer: 0.5 M Tris–HCl, pH 6.8, 0.4% SDS.
5. 6.5% sodium dodecyl sulfate–polyacrylamide gel electrophoresis (SDS-PAGE) gel:
   (a) Lower gel (total 10 mL): Add 2.2 mL 30% acrylamide–bis solution, 2.5 mL 4× lower gel buffer, 5 μL tetramethylethylenediamine (TEMED), 50 μL 10% ammonium persulfate (APS), 5.25 mL H_2O.
   (b) Upper gel (total 5 mL): Add 0.83 mL 30% acrylamide–bis solution, 1.25 mL 4× lower gel buffer, 5 μL TEMED, 50 μL 10% APS, 2.87 mL H_2O.
6. PVDF membrane.
7. Transfer buffer for semidry method: 48 mM Tris, 39 mM glycine, 20% methanol.
8. Whatman 3 MM filter paper.
9. Standard equipment to run SDS-PAGE gels.
10. Semidry electroblotter.
11. Primary and secondary antibodies.
12. Dry milk.
13. Bovine serum albumin (BSA).
14. 1× TBST: 20 mM Tris, 150 mM NaCl, 0.05% Tween 20; adjust pH to 7.6 by HCl.
15. LumiGLO Reserve® Chemiluminescent Substrate Kit (SeraCare, Milford, MA, Cat. No: 54-71-00).
16. X-ray film or imaging system for detection of chemiluminescent signals (e.g., FluorChem Q, ProteinSimple, San Jose, CA, Cat. No: 92-14095-00).

3 Methods

3.1 Plant Material Preparation

3.1.1 Seed Plating and Germination

1. Sterilize 0.04 g Arabidopsis seeds for each sample, and plate onto 7.5 cm ø circular filter in a 10 cm ø MS/agar plate. Try to align the seeds on the filter paper in a line and keep ~1.5 cm space between the lines.
2. Keep the plates at 4 °C for 4 days for stratification.
3. Expose the seeds for 3 h to white light at room temperature to stimulate germination. Keep the plates open in the hood for the first half an hour to reduce water accumulation in the plates.

4. Keep the plates at 21 °C for 21 h and give 3000 μmol m^{-2} far-red light total afterward to photoinactivate phytochromes. Align all the plates in the dark and wrap with aluminum foil. Place plates vertically at 21 °C for 3–4 days, so seedlings will grow vertically.
5. The seedlings will grow flat on the filter paper.

3.1.2 Bortezomib Pretreatment and Sample Collection

1. Prepare final concentration of 40 μM bortezomib with liquid MS media in 10 cm ø plates.
2. In the dark room, with very dim safe green light, carefully transfer the filter paper with seedlings grown on MS agar plates into MS liquid media with 40 μM bortezomib. Make sure the seedlings are fully soaked in the liquid but not floating around (*see* **Notes 1** and **2**).
3. Keep all the samples in MS liquid media supplemented with 40 μM bortezomib for 4 h.
4. Either keep seedlings in darkness or treat seedlings with light pulse (*see* **Note 3**). Collect individual samples and dry seedlings very well using paper towels. Samples can be wrapped with aluminum foil and stored at −80 °C or directly ground for protein extraction.

3.2 Coimmunoprecipitation

3.2.1 Binding Primary Antibodies to Dynabeads

1. Resuspend the Dynabeads by vortexing for 30 s.
2. Transfer 20 μL Dynabeads per sample to an eppendorf tube.
3. Place the tubes in a magnetic rack for 1 min and discard the supernatant.
4. Wash the Dynabeads with 1 mL IP buffer per tube by inverting the tubes multiple times to fully resuspend beads in IP buffer.
5. Place the tubes in a magnetic rack for 1 min and discard the supernatant.
6. Resuspend beads in 20 μL IP buffer per sample and add 1 μg of the desired antibody into individual tube.
7. Couple Dynabeads and antibodies at 4 °C by end-to-end rotation for 30–60 min.
8. Place the tube in a magnetic rack for 1 min and discard the supernatant.
9. Wash Dynabeads with 1 mL IP buffer by fully resuspending Dynabeads; place tubes back in a magnetic rack for 1 min and discard the supernatant.
10. Repeat washing step as described in **step 9** for another two times.
11. After three times wash, add 50 μL IP buffer per tube and store on ice. The antibody precoupled Dynabeads are ready to use.

3.2.2 Preparation of Crude Extracts

1. Grind samples into powder in liquid nitrogen using a mortar and pestle; add three times liquid nitrogen for grinding to make sure samples turning into fine powder and maximize the total protein yield.
2. Add 800 μL of native extraction buffer per sample directly into the mortar and keep grinding the sample till the sample melts into liquid form (*see* **Notes 4** and **5**).
3. Collect individual samples into 1.5 mL eppendorf tubes and keep on ice.
4. Centrifuge samples at 16,000 × *g* in a microcentrifuge at 4 °C in the dark for 15 min.
5. Transfer 1 mL of the supernatant for each sample into new eppendorf tubes as cell lysate (*see* **Note 6**).

3.2.3 Measurement of Protein Concentration in Cell Lysate

1. Pipette samples for calibration curve for measurement of protein concentration as shown in Table 1; mix the samples well and incubate for 5 min at room temperature before measuring the absorption at 595 nm using a photometer (e.g., Beckman DU530 UV-VIS Life Science Spectrophotometer) (*see* **Note 7**); plot the protein concentration of the samples against the absorption at 595 nm.
2. Measure protein concentration in cell lysate; add 795 μL H_2O, 200 μL Bio-Rad Protein Assay Dye Reagent Concentrate and 5 μL of protein crude extracts for each sample. Mix well by pipetting up and down a few times and incubate for 5 min at room temperature before measuring the absorption at 595 nm using a photometer; calculate the protein concentration in the protein crude extracts using the calibration curve from the previous step.

3.2.4 Coimmunoprecipitation Using Antibody Precoupled Dynabeads

1. For each sample add 20 μL antibody precoupled Dynabeads to equal amount (usually 1 mL) of crude protein extract (total protein); compensate for different volumes of crude protein

Table 1
Calibration curve for measurement of protein concentration

Protein concentration (μg/mL)	0.5 mg/mL BSA (μL)	H_2O (μL)	IP buffer (μL)	Protein assay dye reagent concentrate (μL)
0	0	795	5	200
2.5	5	790	5	200
5	10	785	5	200
10	20	775	5	200
20	40	755	5	200

extracts by adding IP buffer; work in the darkroom using safety green light.

2. Incubate the samples by end-to-end rotation at 4 °C in the dark for 3 h.
3. After 3 h IP in the dark, all the wash steps need to be done in darkness.
4. Place the tubes in a magnetic rack for 1 min.
5. Save 50 μL supernatant from each sample as input controls and add 10 μL 6× SDS gel loading buffer into each tube.
6. Discard the rest of supernatant and wash the Dynabeads with 1 mL wash buffer per tube by inverting the tubes multiple times to fully resuspend beads in wash buffer.
7. Place the tubes in a magnetic rack for 1 min and discard the supernatant.
8. Repeat washing step as described in **steps 6** and 7 for another two times.
9. After three times wash, centrifuge tubes briefly with 500 rpm for 1 min and place tubes back in a magnetic rack for 1 min.
10. Discard any wash buffer left over and add 20 μL 1.5× SDS gel loading buffer.
11. Wrap IP sample and input control samples with aluminum foil. Freeze them together at −80 °C.

3.3 Detection of Protein–Protein Interactions

3.3.1 Electrophoresis of Protein Complex

1. Thaw input and IP samples from −80 °C quickly.
2. Mix and incubate samples at 65 °C for 10 min.
3. Load 10 μL input and 20 μL IP samples onto 6.5% SDS-PAGE gel. Using a constant voltage of 120 V, run gel until the bromophenol blue dye runs off the bottom of the gel (*see* **Note 8**).

3.3.2 Transfer Protein Complex Onto PVDF Membrane Using Semidry Method

1. Dissemble the gel cassette and cut off the upper gel.
2. Cut a piece of PVDF membrane and eight pieces of Whatman 3 MM filter paper with slightly larger size than the gel.
3. Rinse the PVDF membrane thoroughly with methanol and keep it in transfer buffer.
4. Prewet eight pieces of Whatman 3 MM filter paper with transfer buffer.
5. Clean the semidry electroblotter using 70% ethanol; dry it very well with Kimwipes.
6. Put four pieces of prewet Whatman 3 MM filter paper and get rid of any bubbles using a glass Pasteur pipette.

7. Gently lay the gel, PVDF membrane, and the remaining four pieces of prewet Whatman 3 MM filter papers sequentially. Make sure to remove any bubbles within all layers.
8. Wipe off any excess transfer buffer and carefully attach the cover of the semidry electroblotter; do not move the cover afterward.
9. Tight screws at all three positions evenly; start transfer using a constant current of 50 mA per gel; transfer at room temperature for 90 min.

3.3.3 Western Blot

1. After transfer, discard the SDS-PAGE gel and directly put the PVDF membrane in 1× TBST/5% dry milk; incubate under constant shaking at room temperature for 1 h.
2. Wash the membrane with 1× TBST multiple times to wash off residual milk.
3. Rotate the membrane with primary antibody in 1× TBST/1% BSA overnight at 4 °C.
4. Wash the membrane with 1× TBST three times for 10 min at room temperature.
5. Rotate the membrane with appropriate secondary antibody in 1× TBST/0.5% dry milk for 1 h at room temperature.
6. Wash the membrane with 1× TBST three times for 10 min at room temperature.
7. Mix 300 μL substrate A and 300 μL substrate B of the LumiGLO Reserve® Chemiluminescent Substrate Kit in an 1.5 mL eppendorf tube.
8. Slightly dry the membrane on a paper towel and put in a transparent plastic folder.
9. Pipet mixed chemiluminescent substrate equally onto the member; the whole membrane should be covered with substrate.
10. Let the substrate react with membrane for 90 s at room temperature and image the luminescent signal using an X-ray film or an imaging system such as FluorChem Q.

4 Notes

1. During the bortezomib treatment, the liquid should be minimized to just cover the surface of vertically grown seedlings on the filter paper. Too little media will cause inefficient treatment which results in not blocking the protein degradation process. Too much media will cause seedlings floating around and make

sample collection difficult. Usually 5 mL liquid MS is used for 10 cm ø plate. If other sizes of plates are used, the exact amount of liquid MS needed should be tested first.

2. MG132 is a common inhibitor for 26S proteasome mediated degradation. Due to its instability, MG132 efficiency relies on the storage time and condition strictly. Bortezomib is a therapeutic broad proteasome inhibitor used in human and works more efficiently than MG132 for PIF1 studies.
3. This protocol is designed to assess the protein–protein interacting complexes PIF1 mainly is involved in. As mentioned in the introduction, PIF1 is rapidly degraded upon light exposure. Thus the light treatment is with a very short period time of saturating light (3000 μmol m^{-2} red light) and samples are collected after 5 min dark incubation. However, this Co-IP assay is suitable for all other factors in the light signaling pathway including other PIFs. The amount of light and the time of dark incubation should be modified related to the protein focused on. An example of in vivo coimmunoprecipitation between PIF1 and phyA/B is shown in Fig. 1.
4. For the native extraction buffer/IP buffer, adding 10 mM iodoacetamide is essential to keep intact phytochromes during preparation of the extract and co-IP. Failure to add 10 mM iodoacetamide might result in truncated version of phyA.
5. When studying the protein complexes containing COP1, adding 0.125–0.25 mM EDTA in the native extraction buffer/IP buffer helps to stabilize the COP1 mediated protein–protein interactions. Lack of EDTA dramatically reduced the Co-IP efficiency.
6. It is very important to keep samples and extracts in darkness all the time before adding 1.5× SDS gel loading buffer into the samples. Any excessive light exposure will affect the proteins interacting complex especially in the study of PIF1 interacting partners.
7. The protein concentration can be measured using different kits. This protocol used the Bio-Rad Protein Assay Dye Reagent Concentrate. Other protein concentration assays such as BCA protein assay could also be used.
8. The concentration of acrylamide in the SDS-PAGE gel should be adjusted according to the size of proteins of interest. Generally with the increase of the acrylamide–bis concentration in the lower gel, the separation is improved for proteins with small molecular weight. Commercially available gradient gel can be used for optimal separation.

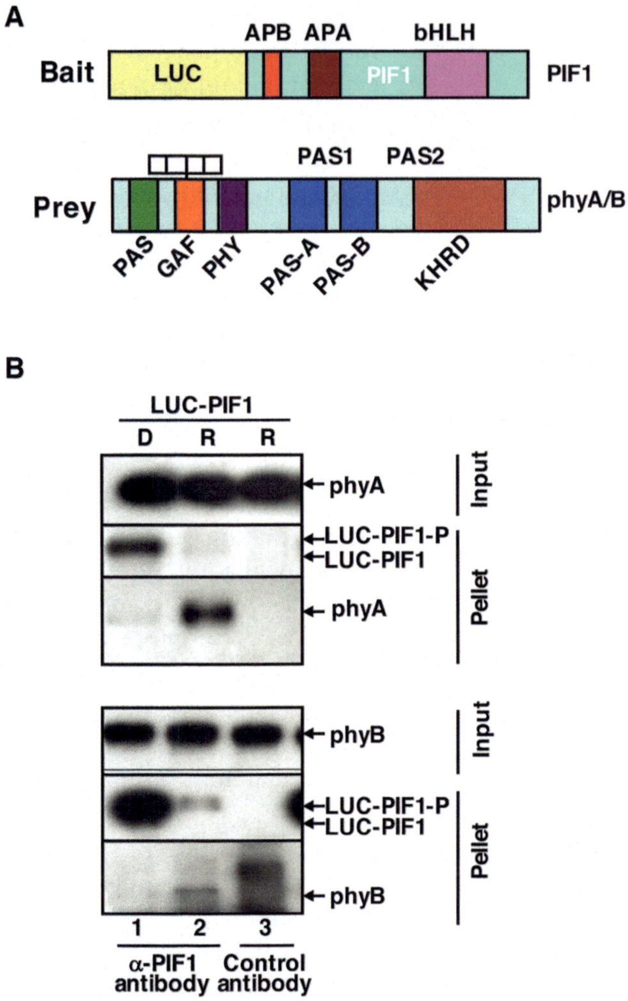

Fig. 1 In vivo coimmunoprecipitation of PIF1 with phyA/B. (**a**) Schematic diagram showing the bait (LUC-PIF1) and the preys (native phyA and phyB proteins). Firefly luciferase is fused with PIF1 at the N-terminus. APB, phyB binding domain; APA, phyA binding domain; bHLH, basic helix–loop–helix domain. (**b**) LUC-PIF1 interacts with the Pfr forms of phyA and phyB in in vivo coimmunoprecipitation assays. The input and pellet fractions from in vivo co-immunoprecipitation assays are indicated. Total protein was extracted from four-day old dark-grown seedlings either exposed to a short red light pulse (R; 3000 μmol m^{-2}) or kept in the dark. Coimmunoprecipitations were carried out using the anti-PIF1 antibody (lanes 1 and 2) or with an unrelated IgG as a control (lane 3). The immunoprecipitated samples were then probed with anti-phyA, anti-phyB, or anti-LUC antibodies. Reproduced in part from "Shen H, Zhu L, Castillon A, Majee M, Downie B, Huq E (2008) Light-induced phosphorylation and degradation of the negative regulator PHYTOCHROME-INTERACTING FACTOR1 from *Arabidopsis* depend upon its direct physical interactions with photoactivated phytochromes. Plant Cell 20:1586–1602; www.plantcell.org; Copyright American Society of Plant Biologists"

Acknowledgments

We thank the Huq laboratory for critical reading of the manuscript. This work was supported by grants from the National Institute of Health (NIH) (GM-114297) and National Science Foundation (MCB-1543813) to E.H.

References

1. Galvão VC, Fankhauser C (2015) Sensing the light environment in plants: photoreceptors and early signaling steps. Curr Opin Neurobiol 34:46–53
2. Pham VN, Kathare PK, Huq E (2018) Phytochromes and phytochrome interacting factors. Plant Physiol 176:1025–1038
3. Klose C, Viczián A, Kircher S, Schäfer E, Nagy F (2015) Molecular mechanisms for mediating light-dependent nucleo/cytoplasmic partitioning of phytochrome photoreceptors. New Phytol 206:965–971
4. Castillon A, Shen H, Huq E (2007) Phytochrome interacting factors : central players in phytochrome- mediated light signaling networks. Trends Plant Sci 12:514–521
5. Leivar P, Quail PH (2011) PIFs: pivotal components in a cellular signaling hub. Trends Plant Sci 16:19–28
6. Leivar P, Monte E (2014) PIFs: systems integrators in plant development. Plant Cell 26:56–78
7. Paik I, Kathare PK, Kim J-I, Huq E (2017) Expanding roles of PIFs in signal integration from multiple processes. Mol Plant 10:1035–1046
8. Xin R, Zhu L, Salomé PA, Mancini E, Marshall CM, Harmon FG, Yanovsky MJ, Weigel D, Huq E (2017) SPF45-related splicing factor for phytochrome signaling promotes photomorphogenesis by regulating pre-mRNA splicing in *Arabidopsis*. Proc Natl Acad Sci U S A 114:E7018–E7027
9. Xu X, Paik I, Zhu L, Huq E (2015) Illuminating progress in phytochrome-mediated light signaling pathways. Trends Plant Sci 20:641–650

Chapter 4

Detection of Phytochrome Phosphorylation in Plants

Eva Klement, Péter Gyula, and András Viczián

Abstract

Posttranslational modification (PTM) of proteins occurs during or after translation and in most cases means covalent binding of a functional group to certain amino acid side chains. Among PTMs, phosphorylation is extensively studied for decades. During phosphorylation, a phosphate group is added to the target residue that is dominantly serine, threonine, and tyrosine in eukaryotes. The phosphate group attachment is catalyzed by kinases, whereas the removal of phosphate (dephosphorylation) is performed by phosphatases. Phosphorylation of phytochrome photoreceptors alters light signaling in multiple ways, thus the examination of this PTM is an expanding aspect of light signaling research. Although this chapter presents methods for detecting phosphorylated phytochrome B molecules, it can be applied on other phytochrome species. The first presented protocol of this chapter shows how the phosphorylation state of phytochrome photoreceptors can be monitored in a modified polyacrylamide gel electrophoresis system. The second protocol describes in detail how phosphorylated amino acids of a target molecule can be identified using mass spectrometry analysis.

Key words Posttranslational modification, Phosphate, Immunoprecipitation, Immunoblot, Phos-tag, Mass spectrometry

1 Introduction

Phosphorylation is a highly dynamic posttranslational protein modification, which can alter the function of the modified protein in many ways. The negative charge of the phosphate group changes the electrostatic surface and the conformation of the protein, which may affect enzyme activity and/or protein–protein interactions either positively or negatively. The addition or removal of a phosphate to a specific sequence position frequently acts as a signal, inducing a chain of events, biological processes [1, 2]. Phosphorylation and dephosphorylation reactions are catalyzed by kinases and phosphatases, respectively. The affinity of these enzymes for protein side chains and the surrounding sequences also differs, and the same target protein can be simultaneously modified at many sites by various kinases and phosphatases.

Andreas Hiltbrunner (ed.), *Phytochromes: Methods and Protocols*, Methods in Molecular Biology, vol. 2026, https://doi.org/10.1007/978-1-4939-9612-4_4, © Springer Science+Business Media, LLC, part of Springer Nature 2019

The first demonstration of phytochrome phosphorylation came from studies performed on phytochrome A (phyA) [3], and it turned out that phosphorylation of a set of serine residues in phyA attenuates signaling [4, 5] and destabilizes the photoreceptor [6]. It was also revealed that the C-terminal part of phytochromes shows close homology to histidine kinases [7], and phyA is a functional protein kinase [8, 9]. Recent studies demonstrated that not only phyA but also phyB is phosphorylated on a set of serines, threonines, and tyrosines. The phosphorylation of the examined residues results in attenuated phyB signaling via complex responses involving (1) disturbed binding to signaling partners, (2) improper nuclear localization and protein complex formation, and (3) accelerated inactivation by dark reversion [10, 11]. The fine details of phytochrome phosphorylation/dephosphorylation have not been elucidated yet, although phytochrome autophosphorylation has already been reported [6, 8] and several phosphatases which dephosphorylate phytochrome have been identified [12–14]. In the past few years, numerous studies demonstrated that phytochromes are not only phosphoproteins but can also phosphorylate signaling partners, and that the proper phosphorylation state of these proteins is required for appropriate light signal transduction [9, 15–21].

This solid set of data indicates that monitoring the phosphorylation state of phytochromes is essential for understanding the fine details of signaling driven by these receptors. This chapter provides two protocols for the detection of phytochrome phosphorylation: the first one shows how the general phosphorylation status of the molecule can be monitored using a modified version of sodium dodecyl sulfate polyacrylamide gel electrophoresis (SDS-PAGE), whereas the second one describes how to identify the phosphorylated amino acid residues by mass spectrometry.

The combinatorial action of kinases and phosphatases creates different pools of the target protein (i.e., hypophosphorylated and hyperphosphorylated forms) having diverse functions. To measure the ratio of these forms under certain conditions, a good separation technique is necessary. In the case of small or medium-sized proteins, the phosphorylated isoforms usually can be separated by normal SDS-PAGE. However, phytochromes are relatively large proteins (~125 kDa), and conventional SDS-PAGE does not separate clearly the unmodified and the phosphorylated protein populations. Phos-tag (1,3-bis[bis(pyridine-2-ylmethyl)amino] propane-2-olato dizinc(II) complex) is a synthetic moiety that selectively binds phosphomonoester dianions ($R\text{–}OPO_3^{2-}$) in aqueous solutions at neutral pH [22]. This moiety can be used in various ways: it can be linked to biotin for immunoblotting purposes, can be coupled to agarose beads for the purification and enrichment of phosphopeptides to be analyzed by mass-spectrometry, or can be incorporated into an acrylamide gel matrix enabling separation of the differently phosphorylated pools of proteins during gel electrophoresis by slowing down the migration speed of the phosphorylated forms (Fig. 1). Since its introduction [23], the Phos-tag chemistry has been improved [24, 25]. Thus, here we present how a Phos-tag-incorporating SDS-PAGE can be used in order to confirm/visualize the presence of phosphorylated phyB.

To decipher how phosphorylation regulates distinct processes, we need to map which amino acid residues in the investigated protein are phosphorylated under certain conditions. Phosphoproteomics provides a solution to this. Continuous technological advances in protein mass spectrometry enabled large-scale phosphorylation studies and resulted in an explosion of phosphorylation data: the Plant Protein Phosphorylation DataBase (P3DB) contains nearly 50,000 modification sites [26]. However, large-scale phosphorylation analyses are somewhat stochastic, and the detailed analysis of the role of phosphorylation in distinct processes requires the isolation of the subset of proteins involved in that process, at least in the discovery phase. This approach was used during the examination of the light-dependent phosphorylation of cryptochrome 2, photoreceptor of blue light [27] or the phytochrome-interacting transcription factor, PHYTOCHROME INTERACTING FACTOR 3 (PIF3, [20]).

The analysis of protein phosphorylation by mass spectrometry (MS) is generally performed at the peptide level. The proteins are digested with a specific protease, usually trypsin; thus, knowing the

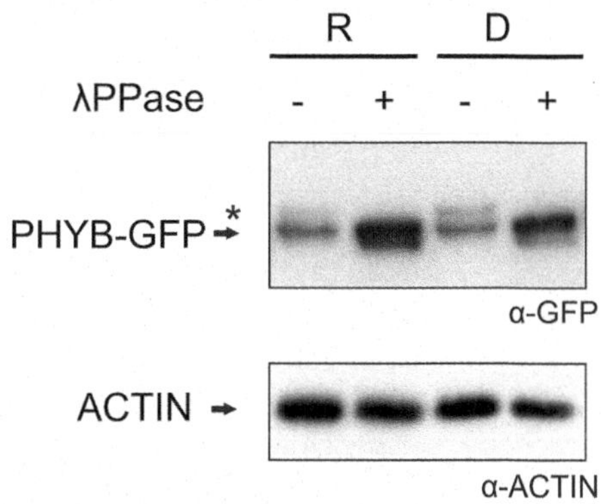

Fig. 1 Phosphorylated form of PHYB-GFP is detectable both in etiolated and in red light-illuminated seedlings. Arabidopsis plants expressing PHYB-GFP were grown on half-strength Murashige–Skoog agar plates in the dark for 4 days and were irradiated for 24 h with red light (R, 2 μmol m^{-2} s^{-1}) or kept further in the dark (D). The samples were processed in the dark, under safety green light. The cleared lysates were either treated (+) or not treated (–) with λ phosphatase and then run on Phos-tag acrylamide gel. The upper part of the gel was probed with anti-GFP-HRP (α-GFP) antibody to detect PHYB-GFP, whereas the lower part was tested for actin with anti-ACTIN (α-ACTIN) antibody as a loading control. The asterisk symbol denotes the phosphorylated PHYB-GFP isoforms

amino acid sequence of the protein, the resulting peptide masses are predictable. Then the digest mixture is subjected to reversed phase fractionation and the eluent is monitored on-line by a mass spectrometer as the detector. The masses of the eluting components are determined nowadays usually with high resolution and high mass accuracy and computer-selected components are subjected to fragmentation analyses (MS/MS data are acquired) in order to obtain amino acid sequence information.

Even in the analysis of highly purified proteins, the enrichment of the modified peptides is highly recommended. A tryptic digestion of *Arabidopsis thaliana* phytochrome B may result in over 200 tryptic peptides considering one missed cleavage. Since phosphorylation at individual sites is substoichiometric, the modified sequences often display low-intensity signals, which disfavors their selection for tandem mass spectrometry (MS/MS), that is, peptide fragmentation analysis in the presence of coeluting, unmodified peptides of high abundance. The lower ionization efficiency of phosphopeptides also renders their detection more difficult. The most commonly applied enrichment techniques are based on the affinity of the phosphate group toward certain metal ions (e.g., iron, gallium, or titanium). These techniques include immobilized metal affinity chromatography (IMAC) and enrichment on titanium dioxide [28, 29]. The presented workflow applies IMAC on Fe(III)-nitrilotriacetic acid (NTA)-functionalized magnetic beads. The acidic environment and the high organic content in the loading solvent ensure selective binding of the phosphopeptides, whereas the recovery of the phosphopeptides is accomplished by elution under basic conditions. Previous analyses on the phosphorylation of Arabidopsis phyB applied titanium dioxide enrichment [10, 11].

The MS characteristics of the phosphopeptides facilitate their identification and the localization of the modification site. The phosphate group increases the mass of the precursor and the respective fragment ions by 79.966 Da. Additionally, upon collisional activation, phosphoserine and phosphothreonine containing peptides usually undergo beta-elimination of the phosphate group from the precursor and fragment ions. This results in neutral loss products of 97.977 Da in collision-induced dissociation (ion trap CID) and higher energy C-trap dissociation (HCD, beam-type CID) spectra. The phosphotyrosine-containing peptides cannot lose the phosphate via beta-elimination, but may feature a characteristic immonium ion at m/z of 216.042. In electron transfer dissociation (ETD), the "collision" of multiply charged peptide ions with fluoranthene anions induces electron transfer and radical ion formation that leads to fragmentation along the peptide backbone, but the side chain modifications remain intact. This alternative fragmentation might provide additional information and assist site assignment. Phosphorylation site assignment is especially crucial if the sequence contains multiple potential modification sites in close proximity (Fig. 2).

Despite of efficient enrichment, high sensitivity detection and improved data analysis, there may be phosphorylation sites that remain hidden. For phyB we observed 75–95% sequence coverage by tryptic digestion. The use of alternative enzymes or chemical cleavage may increase sequence coverage, and thus, lead to the complete phosphosite mapping of the protein.

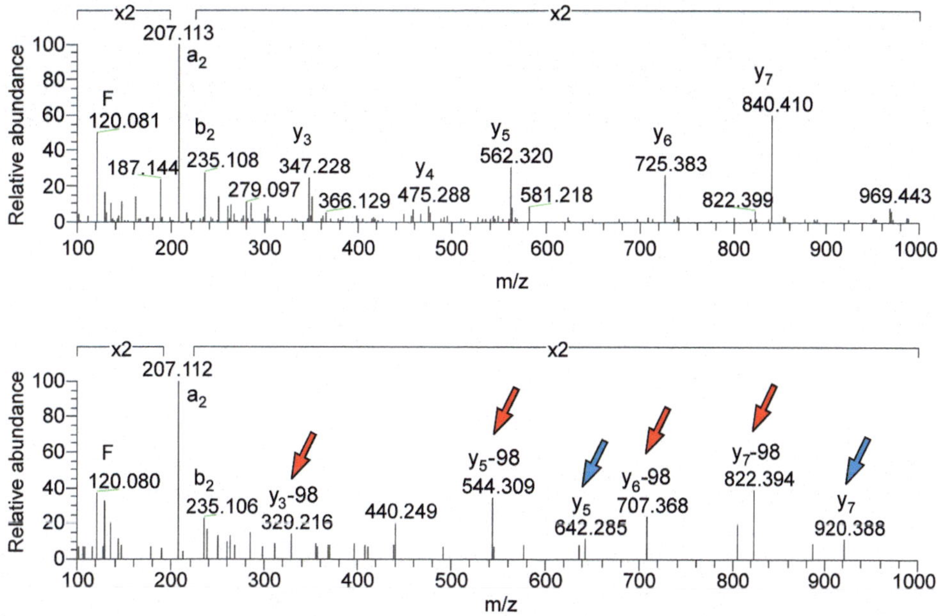

Fig. 2 The *Arabidopsis thaliana* phytochrome B is phosphorylated at serine 86. Transgenic Arabidopsis seedlings expressing the *35S:PHYB-GFP* transgene were grown for 10 days under 12 h white light (100 μmol m^{-2} s^{-1})/12 h dark cycles on Murashige–Skoog medium containing 3% sucrose. Total protein extract was prepared from seedlings collected at the end of the day, and the PHYB-GFP fusion protein was immunoprecipitated using GFP-TRAP_A agarose beads. The protein band representing PHYB-GFP was cut out from the SDS-containing polyacrylamide gel, digested with trypsin and subjected to MS analysis after phosphopeptide enrichment. Upper panel: MS/MS spectrum of the doubly charged precursor ion detected at *m/z* 537.758 representing the unmodified tryptic peptide ^{80}SFDYSQSLK^{88}. Lower panel: MS/MS spectrum of the doubly charged precursor ion detected at *m/z* 577.742 representing the monophosphorylated ^{80}SFDYSQpSLK^{88} peptide. Assignment of the fragment ions is according to the nomenclature in reference [35]. The one-letter-code (F) stands for the immonium ion of a particular amino acid; b and a ions indicate fragments from the peptide *N*-terminus, whereas y ions are fragments from the *C*-terminus. Fragment ions that underwent beta-elimination of phosphoric acid (−98 Da) are labeled with red arrows, the y_5 and y_7 fragments retaining the modification and thus shifted by 80 Da are indicated by blue arrows. x2 stands for a twofold magnification of the relative abundance in the corresponding *m/z* range

2 Materials

2.1 Materials and Equipment for Zn^{2+}-Phos-tag SDS-PAGE

2.1.1 Sample Preparation

1. Water: ultrapure deionized water with 18.2 MΩ·cm resistance (Millipore). Whenever we refer to water, we refer to water of this purity throughout the entire protocol.
2. Protein extraction buffer (50 mM Tris–HCl pH 7.5, 100 mM NaCl, 1 mM DTT, 1× protease inhibitor cocktail) (*see* **Note 1**): mix 8 mL water, 500 μL 1 M Tris–HCl pH 7.5, 200 μL 5 M NaCl, 10 μL 1 M DTT and 100 μL protease inhibitor cocktail (Sigma-Aldrich). Fill up to 10 mL with water and chill on ice before use. Always prepare fresh.
   (a) 1 M Tris–HCl pH 7.5: dissolve 6.05 g of Tris base in 30 mL water. Adjust pH to 7.5 with HCl. Adjust final volume to 50 mL with water and filter through 0.2 μm membrane into a sterile plastic bottle. Store at room temperature.
   (b) 5 M NaCl: dissolve 14.61 g NaCl in 45 mL of water. Adjust the volume to 50 mL and filter through 0.2 μm membrane into a sterile plastic bottle. Store at room temperature.
   (c) 1 M dithiothreitol (DTT): dissolve 0.1542 g DTT in ~800 μL water in a 1.5 mL microcentrifuge tube. Vortex until the white powder dissolves, spin briefly and then fill up to 1 mL with water. Make 200 μL aliquots and store them at −20 °C.
3. 1.5 mL microcentrifuge tubes.
4. Disposable plastic pestles that fit into the bottom of the 1.5 mL microcentrifuge tube.
5. Microcentrifuge.
6. Liquid nitrogen.
7. Thermoblock set at 95 °C.

2.1.2 Phosphatase Treatment

1. Lambda protein phosphatase supplied with 10× phosphatase buffer, storage buffer and 10 mM $MnCl_2$ (New England Biolabs).
2. 10 mM adenosine triphosphate (ATP).
3. Water bath set at 30 °C.
4. Microcentrifuge.
5. 2× LDS loading buffer: to obtain 1 mL 2× LDS loading buffer, combine 500 μL NuPAGE™ 4× LDS buffer (Invitrogen), 200 μL 1 M DTT, and 300 μL water. Make 200 μL aliquots and store them at −20 °C.

2.1.3 Gel Casting and Running of Zn^{2+}-Phos-tag Polyacrylamide Gels

1. Gel casting and running apparatus. There are several different brands available. Here we use Mini-Protean II Cell Gel System (Bio-Rad).
2. 1.25 M bis–Tris–HCl pH 6.8: dissolve 13.08 g bis–Tris in 40 mL water in a beaker, set the pH with HCl to 6.8, then fill it up to 50 mL. Pass through a 0.2 μm filter into a sterile 50 mL plastic bottle.
3. 30% [w/v] acrylamide–bis-acrylamide 37.5:1 solution (Sigma-Aldrich). Store at 4 °C in the dark.
4. 10 mM $ZnCl_2$: dissolve 0.01363 g $ZnCl_2$ in water, fill up to 10 mL, and pass through a 0.2 μm filter into a sterile plastic bottle.
5. 5 mM Phos-tag acrylamide solution: add 100 μL methanol to 10 mg Phos-tag acrylamide AAL-107 residue (Wako Chemicals), vortex thoroughly. Transfer the solution to a 15 mL conical centrifuge tube. Rinse the tube used for shipment with 1 mL ultrapure water to take up the residual material, vortex and transfer the solution to the 15 mL conical centrifuge tube. Add 2.2 mL water to the solution and vortex until the white, cloudy material dissolves.
6. 10% [v/w] ammonium persulfate (APS): dissolve 0.1 g APS in 1 mL water. Always prepare fresh.
7. *N*,*N*,*N*′,*N*′-tetramethylethylenediamine (TEMED).
8. 2-propanol
9. 4% [w/v] bromophenol blue: dissolve 0.4 g bromophenol blue in water and fill up to 10 mL.
10. 1 M sodium bisulfite: dissolve 5.2 g sodium bisulfite (Acros) in water and fill up to 50 mL (*see* **Note 2**).
11. 10% sodium dodecyl sulfate (SDS) (*see* **Note 3**).
12. 20× 3-(*N*-morpholino)propanesulfonic acid (MOPS) Running Buffer (1 M MOPS, 1 M Tris base, 2% SDS): for 1 L solution dissolve 209.2 g MOPS, 121.2 g Tris base, and 20 g SDS in water (*see* **Note 4**).
13. Precision Plus Protein™ Dual Color Standards (Bio-Rad) (*see* **Note 5**).

2.1.4 Electroblotting of the Zn^{2+}-Phos-tag Polyacrylamide Gel

1. Wet electroblotting system. In this protocol, we use Mini-Trans Blot system (Bio-Rad).
2. 20× NuPAGE™ Transfer Buffer (Invitrogen) (*see* **Note 6**).
3. Methanol.
4. PVDF membrane (Immobilon-P^{SQ}, 0.2 μm pore size, Millipore).
5. 3MM Whatman filter paper (Sigma-Aldrich).
6. Powder-free disposable gloves (*see* **Note 7**).

7. Glass rod.
8. Magnetic stirrer.
9. Cold room.

2.1.5 Staining of the PVDF Membrane

1. 0.05% Reactive Brown 10 solution: dissolve 0.05 g Reactive Brown 10 (Sigma-Aldrich) in 100 mL water. Alternatively use PROCION RED BROWN MX-5BR (Kemtex).
2. 0.1% [w/v] NaOH solution: dissolve 0.1 g NaOH pellet in 100 mL water.

2.1.6 Western Blot Analysis

1. 10× Phosphate-buffered saline (PBS; 100 mM Na_2HPO_4, 18 mM KH_2PO_4, 1.37 M NaCl, 27 mM KCl): dissolve 14.2 g Na_2HPO_4, 2.4 g KH_2PO_4, 80.0 g NaCl, 2 g KCl in 800 mL water and fill up to 1 L (*see* **Note 8**).
2. 1× PBST (PBS + 0.1% Tween 20) buffer: for 1 L buffer, dilute 100 mL 10× PBS with water and add 1 mL Tween 20.
3. Low-fat milk powder (obtained at a local store).
4. Blocking solution (5% low-fat milk in PBST): dissolve 2.5 g low-fat milk powder in 40 mL 1× PBST, then fill up to 50 mL with 1× PBST. Prepare it fresh, before use (*see* **Note 9**).
5. Anti-GFP antibody linked to horseradish peroxidase (HRP) (Anti-GFP-HRP, Miltenyi Biotech).
6. ECL Western Blotting Substrate (Pierce). Other brands are also available.
7. Hyperfilm ECL (Amersham). Other brands are also available.
8. Dark photo room equipped either with developing solutions, developing trays etc., or with a developing machine.
9. Film developer and fixing solutions.
10. Film cassettes.
11. Timer.
12. Powder-free disposable gloves (*see* **Note 7**).
13. Paper towel.
14. Saran wrap.
15. Adhesive tape.

2.2 Isolation and Immunoprecipitation of PHYB-GFP from Transgenic Arabidopsis Plants for MS Analysis

2.2.1 Protein Isolation and Immunoprecipitation

1. Transgenic Arabidopsis plants expressing PHYB-GFP protein at high levels by the constitutive viral *35S* promoter. Having high protein levels is essential, because here we have to isolate the protein, not only detect its presence by western blot analysis.
2. All the necessary materials and equipment to grow and handle Arabidopsis plants (not listed here).
3. Protein Extraction Buffer (50 mM Tris–HCl, pH 8.8; 1% SDS; 1% Triton X-100; 50 mM sodium bisulfite; 1 mM PMSF;

100× phosphatase inhibitor mixture; 1 protease inhibitor cocktail tablet in every 10 mL buffer): Mix: 333 μL 1.5 M Tris–HCl, pH 8.8; 2 mL 10% SDS; 200 μL 1% Triton X-100; 1 mL 1 M sodium bisulfite; 100 μL 0.2 M PMSF; 200 μL HALT phosphatase inhibitor mixture; 2 protease inhibitor cocktail tablets (Roche mini PI tablets) and fill up to 20 mL. Always make a fresh buffer and cool it on ice before use.

(a) 1.5 M Tris–HCl, pH 8.8: dissolve 181.7 g Tris in 900 mL water. Set the pH to 8.8 using concentrated HCl and fill up to 1000 mL.

(b) 10% SDS, *see* Subheading 2.1.3.

(c) 1 M sodium bisulfite (*see* **Note 2** and Subheading 2.1.3).

(d) 0.2 M phenylmethylsulfonyl fluoride: dissolve 0.035 g PMSF in 1 mL methanol.

(e) Phosphatase inhibitor mix (e.g., HALT from Thermo Scientific) (*see* **Note 10**).

4. Anti-GFP agarose beads: Anti-GFP VHH coupled to agarose for immunoprecipitation of GFP-tagged proteins (GFP-TRAP_A, Chromotek) (*see* **Note 11**).
5. 2× SDS PAGE Sample Buffer (Bio-Rad) (*see* **Note 12**).

2.2.2 SDS PAGE and Protein Staining Reagents

1. Resolving gel solution (*see* **Note 13**): To obtain 5 mL 8% solution, mix 2.3 mL water; 1.3 mL 30% [w/v] acrylamide–bis-acrylamide 37.5:1 solution (*see* **Note 14**); 1.3 mL 1.5 M Tris–HCl pH 8.8 solution; 50 μL 10% SDS; 50 μL 10% APS, and 3 μL TEMED.

   (a) 1.5 M Tris–HCl pH 8.8 solution, *see* Subheading 2.2.1.

   (b) 10% SDS, 10% [w/v] APS, and TEMED, *see* Subheading 2.1.3.

2. Stacking gel solution: To obtain 2 mL stacking gel solution, mix thoroughly 1.4 mL water; 330 μL 30% [w/v] acrylamide–bis-acrylamide; 250 μL 1 M Tris–HCl pH 6.8; 20 μL 10% SDS; 20 μL 10% APS, and 2 μL TEMED.

   (a) 1 M Tris–HCl pH 6.8: dissolve 121.2 g Tris in 900 mL water. Set the pH to 6.8 using concentrated HCl and fill up to 1 L.

   (b) 10% SDS, 10% [w/v] APS and TEMED, *see* Subheading 2.1.3.

3. Electrophoresis running buffer: 25 mM Tris base, 192 mM glycine, 0.1% SDS. Dissolve 3.03 g Tris base, 14.4 g glycine, and 1 g SDS in water and fill up to 1 L.

2.2.3 Further Materials and Laboratory Equipment

1. Prestained protein molecular weight marker.
2. 2-propanol.
3. Fixation solution: mix 25 mL 2-propanol and 10 mL acetic acid with 65 mL water.
4. Protein staining solution (PageBlue™, Thermo Fisher).
5. Mortar, pestle.
6. Rotator.
7. Refrigerated microcentrifuge.
8. Rocking platform.
9. pH meter.
10. Block heater, heated up to 95 °C.
11. Glass plates, plate holders, gel running apparatus, power supply for SDS-PAGE. We use Mini Protean II electrophoresis system (Bio-Rad), but there are also other suitable systems from different manufacturers available.
12. Adjustable micropipettes, tips.
13. 1.5 mL microcentrifuge tubes and 2 mL safe-lock microcentrifuge tubes (Eppendorf).
14. Ice.
15. Liquid N_2.

2.3 Materials and Equipment for Mass Spectrometry Analysis

For the digestion and phosphopeptide enrichment prepare solutions with HPLC grade solvents and analytical grade reagents. For the LC-MS/MS analysis use ULC-MS grade solvents. Wear protective gloves and sleeves during digestion and while preparing digestion solutions to avoid contamination of the samples with keratin. Working under a laminar flow hood also reduces the contamination of the samples.

2.3.1 In-Gel Digestion

1. Microcentrifuge tubes (*see* **Note 15**).
2. 25 mM ammonium bicarbonate in water: dissolve 200 mg ammonium bicarbonate in 100 mL water.
3. 25 mM ammonium bicarbonate in 50% acetonitrile: dissolve 200 mg ammonium bicarbonate in 50 mL water, then add 50 mL acetonitrile.
4. 10 mM dithiothreitol (DTT) in 25 mM ammonium bicarbonate: dissolve 1.5 mg DTT in 1 mL 25 mM ammonium bicarbonate. Prepare fresh.
5. 55 mM iodoacetamide (IAM) in 25 mM ammonium bicarbonate: dissolve 10 mg IAM in 1 mL 25 mM ammonium bicarbonate. Prepare fresh.
6. Trypsin: porcine trypsin, side-chain protected by reductive methylation (*see* **Note 16**).

7. 1 mM trifluoroacetic acid (TFA): add 7.6 μL TFA to 100 mL water (*see* **Note 17**).
8. 2% formic acid in 50% acetonitrile: using glass pipette with rubber bulb, add 2 mL formic acid to 48 mL water, then mix with 50 mL acetonitrile (*see* **Note 17**).
9. Laminar flow hood.
10. Fume hood.
11. Block heater.
12. Vacuum centrifuge.

2.3.2 Preparation of Fe-NTA Beads for Phosphopeptide Enrichment

1. Ni-NTA Magnetic Agarose Beads, 5% suspension (Qiagen).
2. Magnetic stand for microcentrifuge tubes.
3. 100 mM ethylenediaminetetraacetic acid disodium salt (EDTA): dissolve 37.2 mg EDTA dihydrate in 1 mL water (*see* **Note 18**).
4. 100 mM $FeCl_3$: dissolve 27 mg $FeCl_3$ in 1 mL water.
5. Storage solution: 100 mM acetic acid in water–methanol–acetonitrile (1:1:1) mixture. Add 17 μL acetic acid to 1 mL water, then mix with 1 mL methanol and 1 mL acetonitrile (*see* **Note 17**).

2.3.3 Phosphopeptide Enrichment by IMAC on Fe-NTA

1. Loading buffer: 0.1% trifluoroacetic acid (TFA) in 80% acetonitrile. Add 10 μL TFA to 2 mL water and 8 mL acetonitrile (*see* **Note 17**).
2. Elution buffer: 1% NH_4OH in 50% acetonitrile. Dilute 4 μL 25% NH_4OH in 50 μL water, then add 50 μL acetonitrile.
3. 10% formic acid in water: add 1 mL formic acid to 9 mL water (*see* **Note 17**).
4. Vacuum centrifuge.

2.3.4 RP-HPLC with Online Mass Spectrometry

An LC-MS/MS setup using Waters Acquity U-HPLC on-line coupled to a Thermo Orbitrap Elite mass spectrometer is described here. Other instrumentation can also be used with appropriate modifications.

1. HPLC system: Waters Acquity U-HPLC equipped with a trap and an analytical column (nanoAcquity UPLC Symmetry C18, 100 Å, 5 μm, 180 μm × 20 mm, 10 K-2D-V/Mtrap and nanoAcquity UPLC Peptide BEH C18, 130 Å, 1.7 μm, 75 μm × 250 mm, 10 K psi, respectively).
2. Solvent A: 0.1% formic acid in water. Add 100 μL formic acid to 100 mL water (*see* **Note 17**).
3. Solvent B: 0.1% formic acid in acetonitrile. Add 100 μL formic acid to 100 mL acetonitrile (*see* **Note 17**).

4. Mass spectrometer: Thermo Orbitrap Elite equipped with Nanospray Flex ion source and ETD.
5. Spray capillary: uncoated silica emitter, OD 360 μm, ID 20 μm, tip ID 10 μm, 12 cm (New Objective).

2.3.5 Data Analysis

1. Peak-picking software: Proteome Discoverer (or equivalent).
2. Database search engine: Protein Prospector (or equivalent).

3 Methods

3.1 Methods for Zn^{2+}-Phos-tag SDS-PAGE

3.1.1 Sample Preparation

During sample preparation avoid using phosphate-containing buffers or components (Tris-phosphate buffer, phosphatase inhibitor cocktail with sodium pyrophosphate, β-glycerophosphate, etc.) because phosphate ions and monoacyl (but not diacyl) phosphate groups bind Phos-tag. Phosphatase inhibitors can be used but expect less clear gel separation. Also, avoid EDTA throughout sample preparation (phosphatase or protease inhibitor cocktails may contain EDTA), because it chelates the Zn^{2+} in the gel and prevents separation of phosphorylated species. However, if the sample contains EDTA, add an equivalent amount of $ZnCl_2$ to the sample before loading to get rid of the free EDTA. Do not add too much $ZnCl_2$, because zinc precipitates SDS.

1. During sample collection, use paper tissues to remove the water from the plant surface. Place 0.1–0.2 g portions of plant tissue into 1.5 mL microcentrifuge tubes and snap-freeze them in liquid N_2. Work in darkness (under safety green light), if phytochrome should be in the inactive state.
2. Grind frozen plants in liquid N_2 using the tissue grinder pestles in 1.5 mL microcentrifuge tubes for about 1 min. Add 500 μL ice-cold Protein extraction buffer (Subheading 2.1.1) for every 0.1 g plant material. Grind on ice until a homogenous extract is obtained (roughly 3 min). Incubate the samples at 95 °C for 10 min.
3. Centrifuge samples for 10 min at 20,000 × *g*, transfer the clear supernatant to a new tube. Samples can be stored at −20 °C.

3.1.2 Phosphatase Treatment

1. For each sample, prepare a phosphatase-treated and a non-phosphatase-treated reaction. Make 25 μL reactions in the following way: mix 20 μL cleared extract, 1.25 μL water, 0.5 μL 10× phosphatase buffer, 2.5 μL 10 mM $MnCl_2$, 0.25 μL 10 mM ATP. Add 0.5 μL phosphatase to the treated samples and 0.5 μL enzyme storage buffer to the nontreated samples.
2. Incubate at 30 °C for 30 min.
3. Add 25 μL 2× LDS loading buffer, vortex briefly, and incubate the samples at 70 °C for 10 min. Let them cool to room temperature, spin briefly. The samples can be stored at −20 °C.

3.1.3 Gel Casting for Zn^{2+}-Phos-tag SDS-PAGE

1. Wash the glass plates and the combs thoroughly with detergent, rinse with distilled water and then with ethanol. Let them dry.
2. Assemble the gel casting frame according to the manufacturer's instructions.
3. Prepare a 7%, 50 μM Phos-tag separating gel by pipetting the gel components (except the APS and TEMED) into a 15 mL conical centrifuge tube: for a Protean minigel (83 × 60 × 1 mm separating gel) add 2288 μL water, 1430 μL 1.25 M bis-Tris–HCl pH 6.8, 1167 μL 30% acrylamide–bis-acrylamide 37.5:1 premixed solution, 50 μL 10 mM $ZnCl_2$, 50 μL 5 mM Phos-tag acrylamide. Vortex the mixture well and let it warm up to room temperature for optimal acrylamide polymerization.
4. Add 12.5 μL 10% APS, vortex, then add 2.5 μL TEMED, vortex again until getting a homogenous solution (*see* **Note 19**). Pipet the separating gel between the glass plates leaving enough space for the stacking gel.
5. Gently pipet 2-propanol onto the top of the gel to seal it from the air and to get a flat, bubble-free surface. Let the gel stand for about half an hour until a discernible sharp line can be seen between the gel and the 2-propanol layer.
6. Prepare the stacking gel solution: in a 15 mL conical centrifuge tube. Mix 1412 μL water, 715 μL 1.25 M bis–Tris–HCl pH 6.8, 333 μL 30% acrylamide–bis-acrylamide 37.5:1, 25 μL 4% [w/v] bromophenol blue (*see* **Note 20**). Vortex well.
7. Pour the 2-propanol off and rinse the top of the gel with ultrapure water, drain the remaining water. Do not allow the surface to dry.
8. Add 12.5 μL 10% APS to the stacking gel solution, vortex, add 2.5 μL TEMED, vortex again until getting a homogenous solution (*see* **Note 19**). Pour onto the top of the resolving gel. Place combs into the stacking gel by inserting it at a convenient angle to prevent bubbles trapping under the teeth. Let it stand for at least half an hour.

3.1.4 Running of the Zn^{2+}-Phos-tag Gels

This gel system is quite sensitive to the salt content, and even small disturbances can lead to the distortion of the protein bands. It is recommended to load the same volume with the same ionic composition to each well of the gel. Instead of having empty wells on the gel, fill them with mock samples mixed from the Protein extraction buffer (Subheading 2.1.1), phosphorylation buffer and 2× LDS loading buffer. Be sure that the protein molecular weight marker does not contain Tris-phosphate buffer. If the sample lacks divalent cations, then either add an equivalent amount of $ZnCl_2$ or prepare EDTA-free loading buffer. Avoid high salt buffers, because high salt concentration can distort bands.

1. Freshly prepare 0.5 L 1× Running Buffer by mixing 25 mL 20× MOPS Running Buffer and 2.5 mL 1 M sodium bisulfite, then fill it up to 0.5 L with water.
2. Remove the gels from the casting apparatus and place them into the running unit. Put the unit into the tank.
3. Fill the upper buffer container with running buffer and check if it leaks. If it does not, remove the combs carefully. If the walls between the wells are distorted, adjust them with a small spatula. Rinse the wells with running buffer using a syringe.
4. Load 10–20 μL samples onto the gel. Load the same volume of mock samples into the unused wells. Mock samples must have the same buffer composition as the real ones (save some extraction buffer for this purpose) to prevent distortion of bands.
5. Run the gel at 100 V (at constant voltage) for 1.5 h for a minigel (*see* **Note 21**). Check the position of the bands of the prestained protein molecular weight marker (*see* **Note 22**).

3.1.5 Blotting of the Zn^{2+}-Phos-tag Gels

1. Prepare 1 L 1× NuPAGE™ transfer buffer by mixing 50 mL 20× NuPAGE™ transfer buffer, 100 mL methanol and 5 mL 1 M sodium bisulfite with water.
2. Cut the PVDF membrane and six pieces of 3 MM filter papers for each gel (*see* **Note 23**).
3. Rinse the PVDF membrane in methanol for 1 min, then in water until it is evenly saturated (it sinks). Wash the membrane in 1× NuPAGE Transfer Buffer for at least 15 min before use.
4. After the run is complete, remove the gel, and cut the stacking gel off. Soak the separating gel in 2× 100 mL 1× NuPAGE Transfer Buffer for 30 min each (*see* **Note 24**).
5. Assemble the blotting sandwich. Fill a suitable container with 1× NuPAGE Transfer Buffer and place the plastic pad in it. Submerge one piece of foam pad into the buffer, drive out the bubbles. Place three layers of filter paper onto the foam pad, gently driving the bubbles out with hand or a glass rod. Place the gel gently onto the filter papers, avoiding wrinkles or bubbles. Put the PVDF membrane carefully onto the gel and remove the bubbles. Place three layers of filter paper onto the gel, and remove the bubbles again. Apply the second pad and secure the sandwich with the clamps. Put a magnetic stirrer and then the sandwich into the blotting tank and fill it with ~800 mL 1× NuPAGE Transfer Buffer. Remove the cooling block from the freezer and insert into the tank. Carry the whole unit into the cold room and place it onto a magnetic stirring platform.

6. Blot the proteins at constant 150 mA (24–26 V) for 12–20 h with slow stirring using the magnetic stirrer.
7. After blotting, disassemble the blotting unit, discard the buffer, remove the PVDF membrane and rinse it with water a few times. Perform the washing and rinsing steps on a tilting rocking platform with constant but gentle agitation.

3.1.6 Staining the Membrane for Total Protein Visualization

This part of the protocol is an optional step. It demonstrates how the total protein amount loaded on the gel and the blotting efficiency can be checked with a few simple steps. The procedure can be omitted proceeding with **step 1** of Subheading 3.1.7.

1. Stain the membrane with Reactive Brown 10 solution for about 30 s (*see* **Note 25**).
2. Destain with ultrapure water until no background is visible.
3. The membrane can be dried at this point for documentation (place the membrane between two layers of transparent film and scan). Wet the membrane again by rinsing in methanol and then in water.
4. Rinse the membrane in 0.1% NaOH to remove the stain completely and then in water for 5 min.

3.1.7 Specific Detection of GFP-Tagged Phytochrome

1. Rinse the membrane in PBST for 5 min and proceed with blocking.
2. Block the membrane in 50 mL 5% low-fat milk in PBST for 1 h at room temperature or for 12–16 h at 4 °C.
3. Add primary antibody at the dilution recommended by the manufacturer (usually 1:500–10,000; in this case, we use 1:5000 of anti-GFP-HRP) to 10 mL PBST and incubate the membrane for 1 h. Be sure that the entire membrane is covered with the hybridization solution.
4. Wash the membrane in 50 mL PBST for 10 min. Repeat it at least three times. If the primary antibody is linked to HRP, proceed with **step 7**.
5. Incubate the membrane in 10 mL secondary HRP-linked antibody diluted to 1:3000–30,000 (according to the manufacturer's recommendation) in PBST for 1 h.
6. Wash the membrane in 50 mL PBST for 10 min at least three times.
7. Wash the membrane twice for 10 min with 50 mL PBS without Tween 20 (Tween 20 may contain peroxides that can interfere with detection).
8. Mix 2 mL each of solution 1 and 2 of the ECL Western Blotting Substrate in a clean tray and mix them well.

9. Remove membrane from PBS and place it on a paper towel for a few seconds to remove the PBS, but do not allow the membrane to dry out.
10. Wet both sides of the membrane in ECL solution. Incubate for 1 min with gentle agitation and be sure that the solution covers the membrane completely.
11. Remove the membrane and remove the ECL substrate solution briefly with a paper tissue. Lay the membrane on a sheet of Saran wrap and cover it with another piece of wrap. Remove wrinkles. Put the wrapped membrane with the protein side up into the cassette and fix it with a piece of adhesive tape.
12. In a dark photo room, switch off the white light and switch on film-safe red lamp. Cut pieces of Hyperfilm ECL and lay one piece on the membrane. Close the cassette quickly and start the timer. Expose it for 30 s, remove the film, put in another for 5 min and then another for 1 h (or whatever time you find suitable).
13. Develop the films either manually or using a film developer equipment (*see* **Note 26**).

3.2 Isolation of PHYB-GFP from Transgenic Arabidopsis Plants for MS Analysis

3.2.1 Protein Isolation and Immunoprecipitation

1. Grow Arabidopsis plants expressing phytochrome fused to GFP. The growth conditions and applied treatments (light, hormone, etc.) depend on the aim of the experiment.
2. Harvest 1 g of seedlings, blot the water from the plant surface with a paper tissue. Wrap the seedlings into a piece of aluminum foil and snap-freeze them in liquid N_2 (*see* **Note 27**).
3. Mix the anti-GFP agarose beads and pipet 30–50 μL bead slurry per each sample to a 1.5 mL microcentrifuge tube. Use low-retention tips to avoid beads sticking to the tip surface. Wash the beads with 1 mL ice-cold Protein Extraction Buffer (Subheading 2.2.1), suspend them by turning the tube up and down a few times. Spin the tubes in a microcentrifuge for 1 min at 1300 × *g* at 4 °C. Remove the supernatant carefully using a micropipette. Wash two more times and suspend the beads in 30 μL Protein Extraction Buffer.
4. Chill a mortar and a pestle in liquid N_2. Place 1 g frozen plant material into the mortar and grind until it becomes a fine powder. Add 2 mL ice-cold freshly made Protein Extraction Buffer to the powder. Let the buffer freeze in the cold mortar and crush the frozen buffer with the pestle, grind it to a fine powder and mix thoroughly with the plant powder.
5. Let the mixture thaw slowly on the bench at room temperature and occasionally stir with the pestle. Avoid making foam.
6. When the mixture is thawed, pipet immediately into an ice-cold 2 mL microcentrifuge tube and place it on ice.

7. Centrifuge for 20 min at 4 °C at 20,000 × *g*. Transfer the supernatant into the microcentrifuge tube which contains the washed anti-GFP agarose beads.
8. Incubate the supernatant with the agarose beads for 1 h at 4 °C on a tube rotator.
9. Collect the beads by spinning the tubes in a microcentrifuge at 4 °C at 1300 × *g* for 1 min.
10. Wash the beads with 1 mL ice-cold Protein Extraction Buffer. Gently turn the tubes up and down to suspend the beads and collect them again in a microcentrifuge at 4 °C at 1300 × *g*. Perform four washes altogether and leave about 50 μL Protein Extraction Buffer on the beads.
11. Elute PHYB-GFP from the agarose beads. Add 50 μL 2× SDS PAGE Sample Buffer into the tube, mix gently by knuckling the tube. Avoid smudging the beads all over the tube. Incubate the sample for 5 min at 95 °C.
12. Spin the tubes at room temperature at 20,000 × *g* for 1 min and transfer the supernatant to a new tube.
13. Proceed immediately to the SDS-PAGE or freeze the samples in liquid N_2 and store them at −80 °C.

3.2.2 SDS-PAGE and Protein Staining in Acrylamide Gel

1. Clean and assemble the glass plates in the provided holders (*see* Subheading 3.1.3).
2. Prepare the resolving gel solution (*see* **Note 13**). Pipet or pour the gel solution between the glass plates; leave enough space for the stacking gel and the comb. Pipet 2 mL 100% 2-propanol on top of the resolving gel. Wait at least 30 min until the resolving gel polymerizes.
3. Pour off the 2-propanol and rinse the gel surface carefully with a few mL of water using a micropipette.
4. For each 1-mm-thick gel, 2 mL stacking gel solution is needed in our gel running system. Pipet or pour the solution on the resolving gel. Avoid bubble formation and insert combs immediately (*see* **Note 28**). Wait for at least 30 min for complete gel polymerization (*see* **Note 13**).
5. Remove combs slowly, keeping the borders between the wells intact. Put the gel into the companion running module and place it into the electrophoresis tank. Fill the tank with fresh 1× Running buffer. Be sure that the buffer level is a few mm higher than the highest point of the gel. Rinse the wells by pipetting the Running buffer up and down to wash away the acrylamide pieces.
6. Thaw the protein samples and incubate them at 95 °C for 2 min.

7. Pipet the protein molecular weight marker to an unmodified well and load the protein extracts into the wide (modified) wells (*see* **Note 28**).
8. Start the electrophoresis using 40–60 V. When the dye front reaches the resolving gel, increase the voltage to 80–120 V. Continue the electrophoresis for about 2–2.5 h.
9. Lift the companion running module from the tank and separate the glass plates gently using a spatula.
10. Discard the stacking gel and immerse the resolving gel into a tray filled with water (*see* **Note 29**). Use plenty of water. Wash three times for 10 min. Perform these steps together with the following ones applying gentle agitation on a rocking platform.
11. Remove all the water after the last wash and pour Fixation solution on the gel to cover it completely. Incubate with gentle agitation for 15 min.
12. Pour off the Fixation solution and wash with water for 5 min three times.
13. Cover the gel completely with Protein staining solution (PageBlue) and stain for 1 h. In most cases this time is sufficient, but can be extended even to overnight staining in case of low-level expression or limited amount of processed plant material.
14. Rinse the gel 2–3 times with water for 1 min to remove the excess of the Protein staining solution.
15. Wash the gel with water for 10 min. This step can be repeated until the stained protein band appears and the background becomes clear (*see* **Note 30**).
16. Cut out the band containing PHYB-GFP using a scalpel (*see* **Note 31**).
17. Place the gel slice into a 1.5 mL microcentrifuge tube and cover it with water. It can be stored for a few days at 4 °C before the MS analysis.

3.3 Detection of Phosphorylated Amino Acids of phyB Using Mass Spectrometry

3.3.1 In-Gel Digestion

1. Cut the gel band to 1 mm × 1 mm pieces.
2. Add 50 μL (or enough volume to cover the gel pieces) 25 mM ammonium bicarbonate in 50% acetonitrile to the gel pieces, vortex for 10 min and discard supernatant to remove detergent and salts. Repeat this step (*see* **Note 32**).
3. Add 25 μL (or enough volume to cover the gel pieces) 10 mM DTT in 25 mM ammonium bicarbonate to the gel pieces and incubate at 56 °C for 30 min.
4. Let the gel pieces settle down and discard the supernatant.
5. Add 25 μL 55 mM IAM in 25 mM ammonium bicarbonate. Incubate at room temperature in the dark for 30 min.

6. Discard supernatant and add 50 μL (or enough to cover) 25 mM ammonium bicarbonate in 50% acetonitrile to the gel, vortex for 10 min, let the gel pieces settle down and discard supernatant. Repeat this step.
7. Dry the gel pieces completely in a vacuum centrifuge.
8. Dissolve trypsin in 25 mM ammonium bicarbonate to a concentration of 5 ng/μL, add 20 μL solution, that is, 100 ng trypsin to the dry gel pieces and keep at 4 °C for 15 min (*see* **Note 33**).
9. Cover the gel with 25 mM ammonium bicarbonate if needed and proceed with the digestion at 37 °C for 4 h.
10. Add 20 μL 2% formic acid in 50% acetonitrile and sonicate for 5 min.
11. Spin the tubes in a microcentrifuge at 2000 × *g* for 1 min and transfer supernatant containing the digested peptides into a clean microcentrifuge tube.
12. Add 50 μL 2% formic acid in 50% acetonitrile to the gel pieces, vortex for 20 min, then sonicate for 5 min. Collect supernatant again and add to the previously collected supernatant. Repeat this step once more.
13. Dry the supernatant in a vacuum centrifuge.

3.3.2 Preparation of Fe-NTA Beads

Fe-NTA beads can be prepared in batch in advance and stored at 4 °C for at least 1 month.

1. Mix Ni-NTA bead suspension thoroughly before taking out the necessary amount. Calculate 2–5 μL bead slurry for each sample (*see* **Note 34**).
2. Remove supernatant using a magnetic stand (*see* **Note 35**).
3. Add water to the beads and vortex for 1 min. Collect the beads with the magnetic stand and remove supernatant. Repeat the wash two more times.
4. Add 100 mM EDTA to the beads and vortex for 30 min to remove Ni^{2+} ions, collect the beads with the magnetic stand and discard the supernatant.
5. Wash the beads with water three times as in **step 3** to remove EDTA.
6. Add 100 mM $FeCl_3$ solution to the beads and vortex for 30 min to load NTA functional groups with Fe^{3+} ions. Collect the beads with the magnetic stand and discard the supernatant.
7. Wash the beads with water three times (vortex for 1 min) to remove residual $FeCl_3$.
8. Suspend the beads in the storage solution. Use the initial volume to restore the original slurry concentration.

3.3.3 Phosphopeptide Enrichment

1. Remove storage solution using the magnetic stand and wash Fe-NTA beads twice with 10 μL loading buffer.
2. Dissolve the sample in 10 μL loading buffer (*see* **Note 36**).
3. Add the sample to the Fe-NTA beads and vortex for 30 min.
4. Collect the supernatant and transfer to a clean microcentrifuge tube (*see* **Note 37**).
5. Wash the beads three times with 10 μL loading buffer.
6. Add 10 μL elution buffer to the Fe-NTA beads and vortex for 30 min.
7. Collect the beads with the magnetic stand and pipet the supernatant into a clean microcentrifuge tube.
8. Acidify the eluate with 1 μL 10% formic acid (*see* **Note 38**).
9. Dry flow-through and eluate in a vacuum centrifuge.

3.3.4 RP-HPLC with Online Mass Spectrometry

1. Equilibrate the columns with 100% solvent A at a flow rate of 300 nL/min for 20 min. Apply the same flow rate during the entire chromatography.
2. Dissolve the sample in 10 μL solvent A and load 5 μL to the column (*see* **Note 39**).
3. Separate the peptides with the following gradient: increase solvent B from 0 to 5% in 1 min, then up to 40% in 50 min and to 90% in 2 min. Keep at 90% for 2 min, then decrease to 3% in 1 min. Increase solvent B again to 90% in 1 min, keep at 90% for 2 min, and finally decrease to 0% in 1 min and keep at 0% to equilibrate the column (*see* **Note 40**).
4. In the MS/MS analysis use data-dependent acquisition mode with higher energy collisional activation (HCD) and electron transfer dissociation (ETD, optional) on the 3–5 most abundant multiply charged precursor ions. Measure survey scans and HCD spectra in the Orbitrap, ETD spectra in the linear ion trap. Additional parameters: transfer capillary temperature: 300 °C, AGC target: 10^6, 5×10^4 and 10^6 for survey scans, HCD/ETD scans and reagent ions, respectively; Orbitrap resolution: 60000 and 15,000 for survey and HCD scans, respectively; dynamic exclusion: 15 s; signal threshold to trigger MS/MS: 10^4; isolation width: 3 Da; normalized collision energy for HCD: 30; charge state dependent ETD time enabled, that is, 100 ms, 66.6 ms, and 50 ms for charge state 2, 3, and 4, respectively; supplemental activation for ETD: 15 (*see* **Note 41**).

3.3.5 Data Analysis

1. Convert raw data to peak list(s) using Proteome Discoverer or other equivalent software. Separate HCD and ETD spectra to different nodes.
2. Perform a general search with the HCD data of the flow-through fraction. In Protein Prospector set the following

parameters: database: SwissProt (concatenated with the randomized sequences); taxonomy: all; digest: trypsin; max. missed cleavages: 1; constant modification: carbamidomethyl of Cys; variable modifications: Met-loss, acetyl and the combination of these on the protein *N*-terminus, pyroGlu formation of *N*-terminal Gln, oxidation of Met; masses: monoisotopic; parent tolerance: 5 ppm, fragment tolerance: 20 ppm; instrument: ESI-Q-high-res. The result of this search gives information on sample purity and may reveal copurified contaminants (*see* **Note 42**).

3. Perform a species-specific search with the HCD (and ETD) data of the enriched fraction. Select Arabidopsis (or your species of interest) from the list provided under "Taxonomy" or type in the appropriate box found under "Pre-Search Parameters." Include contaminants identified in the general search of the flow-through as "Additional Accession Numbers." To the variable modifications listed above add phosphorylation of serine, threonine, and tyrosine residues. If ETD was also performed on the selected precursor ions, use the above settings for ETD data except that set 0.6 Da as fragment tolerance and ESI-ETD-low-res as instrument type (*see* **Note 43**).
4. The assessment of the site of the modification within the identified phosphopeptide is performed simultaneously with the database search using Protein Prospector. A Site Localization in Peptide (SLIP) score above 6 indicates reliable (> 95%) site assignment [30]. It also displays if there is ambiguity concerning modification site localization within the phosphopeptide and lists potential modification sites (*see* **Notes 44** and **45**).
5. Perform manual validation of the site assignments. Open up the spectrum in 'Search Compare' and compare the fragmentation of different isoforms entered under "Additional Sequences." Protein Prospector uses only the 40 most abundant ions for database searches (20 from each halves of the spectrum), and only this information is used also for the site assignment. However, the less abundant fragments may provide additional information that may confirm or override the original assessment. Introducing more masses from the peaklist or even the unprocessed spectrum and altering the mass accuracy considered are available options.

4 Notes

1. This buffer composition is recommended for assays including phosphatase treatment of crude protein extract. Detergents (Triton X-100, Na-deoxycholate etc.) were omitted, because we found that at least Triton X-100 can induce artificial dimerization of GFP-tagged phytochromes that is manifested in high

molecular weight bands stuck at the top of the gel, especially in light-grown plants.

2. Always prepare fresh solution prior to use.
3. Alternatively, dissolve 10 g SDS in water to make 100 mL 10% [w/v] solution. We found it more convenient to use the commercially available solution.
4. No need to set the pH. Do not autoclave, because MOPS degrades at high temperature (becomes yellow). The solution can be sterilized by filtering, but we found it unnecessary.
5. This Protein molecular weight marker contains EDTA. Thus it is essential to add an equivalent amount of $ZnCl_2$ before loading it onto the gel. Add 2.5 μL 10 mM $ZnCl_2$ to 5 μL marker, which contains 5 mM EDTA.
6. This buffer contains 500 mM Bicine, 500 mM bis-Tris free base, 20 mM EDTA, pH 7.2. It is more convenient to use the commercially available solution.
7. Do not use regular gloves, because the powder can result in high background signal.
8. No need to set the pH. The 10× stock solution has a pH ~ 6.8, but when diluted to 1× the pH should be 7.4. There is no need to autoclave either.
9. In case of using phospho-specific antibody in the Western blot assay, consider applying an alternative blocking agent (e.g., 3% [w/v] bovine serum albumin) and buffer (TBST), because milk contains casein, which is a phosphoprotein that interferes with the detection.
10. Other brands with different composition are also available. These mixes are the combinations of different phosphatase inhibitors and substrates. It is worth trying different products. The presence of phosphatase inhibitors is essential for successful phosphopeptide isolation.
11. The magnetic bead version of this product (GFP-TRAP_M) is also available from Chromotek. Using a magnetic stand is faster than the collection of the beads with centrifugation.
12. The 2× SDS PAGE Sample Buffer can also be prepared (100 mM Tris pH 6.8, 40 mM DTT, 2% SDS, 20% glycerol, 0.2% bromophenol blue). Store at −20 °C.
13. It is advised to prepare more than one gel at a time when the number of the samples justifies this. Calculate the necessary volume accordingly. It is also a good idea to leave a few mL of gel solution in the mixing beaker to monitor the solidification of the gel.
14. Store at 4 °C in the dark.

15. Use low-retention microcentrifuge tubes to reduce the loss due to adhesion of the peptides to the plastic tube. Rinse the tubes with methanol and let them dry.
16. In general, tryptic digestion is performed at an enzyme to protein ratio of 1/100 [w/w]. As the total protein amount in a gel band is difficult to calculate, side chain protected trypsin is added to the samples. Should trypsin be in excess to the protein, reductive methylation of the lysine residues of trypsin prevents its autolysis.
17. Work with concentrated acids under the fume hood and wear protective gloves and goggles.
18. The disodium salt of EDTA dissolves slowly in water. Heating the solution or adjusting the pH to 8 by adding NaOH accelerates the process.
19. This step is crucial. Vortex until a clear solution is formed free of density gradients. Usually, it takes about 10 s of vigorous vortexing.
20. Adding bromophenol blue to the stacking gel is not essential, but makes the sample loading much easier and does not interfere with separation.
21. Consider using lower voltage and longer time for nicer separation. The excess heat produced during the run can make the bands distorted, especially when the samples contain a high concentration of salt.
22. Some bands may run at unusual positions on a Phos-tag gel because they are phosphorylated (e.g., casein), or may contain specific glutamine residues that can bind to the Phos-tag [31]. In general, proteins run slower on a Phos-tag gel than on a normal gel, because Zn^{2+} binds protein-associated SDS [23].
23. The size should be somewhat larger than the gel itself, because the gel swells a bit in the transfer buffer.
24. In case of using a home-made buffer, be sure that it contains 1 mM EDTA, whose presence is essential to chelate the Zn^{2+} in the gel (it facilitates the transfer of phosphorylated proteins).
25. Reactive Brown 10 staining is much faster and simpler than any other common protein staining method; its sensitivity is also superior compared to them, it is completely reversible, and it does not interfere with the immunodetection [32].
26. Alternatively, instead of using conventional films, sensitive digital cameras or gel detection systems can also be used.
27. Here we use Arabidopsis seedlings which harbor the *35S:PHYB-GFP* transgene and express the PHYB-GFP protein at a high level, but this protocol can be applied on other GFP-tagged phytochromes. Most of the cases, collecting 1 g

of transgenic plant material is sufficient for the MS analysis. During experimental design, consider collecting technical repetitions and perform biological replicates. Thus it is advised to bulk up the seed stocks before starting the experiments.

28. Keep in mind that the whole protein sample of about 100 μL should fit into one well. Using combs with wide wells (5-well, 8-well combs) has the disadvantage of limited sample number per gel, and using these wells would require using a large volume of expensive protein molecular weight marker in order to load a visible amount. We recommend to use a 10-well or a 15-well comb and join 2 or 3 wells together with thin adhesive tape. Leave one well intact for the protein molecular weight marker and leave one well empty between the samples to prevent cross-contamination.
29. Rinse the tray with methanol before the washing procedure. Use powder-free gloves during the entire procedure, and do not touch the gel with bare hands. These precautions help to minimize the keratin contamination of the sample.
30. It is important to know the position of the target protein band. Compare the position of the stained bands to the positions of the known protein molecular weight markers. Use a transparent tray for the detection (e.g., a rectangular petri dish) and place a sheet of white paper underneath. Illuminate the gel from the side for better visibility.
31. Cut the band as tightly as possible to reduce contamination with other proteins. Additionally, nonspecific binding to excess polyacrylamide during in-gel digestion reduces detection sensitivity of the peptides.
32. Washing the gel pieces with a buffer containing 50% acetonitrile causes their dehydration and the gel pieces shrink.
33. A 1 μg/μL stock solution of trypsin can be prepared in 1 mM TFA and stored at −20 °C. Dilute this stock solution to 5 ng/μL with 25 mM ammonium bicarbonate just before digestion. Rehydrating the gels with the trypsin solution at 4 °C reduces the autolysis of the enzyme.
34. The beads settle fast. Pipetting up and down several times helps to keep the suspension mixed. In most cases, 2–5 μL of the 5% bead suspension should have the capacity for a tryptic digest of a 1D SDS-polyacrylamide gel band. However, the sample–bead ratio might need to be adjusted.
35. The magnetic stand collects the beads to one side of the microcentrifuge tube, which helps to remove the supernatant completely, without loss of the beads.
36. If it is difficult to dissolve the sample in the loading buffer, dissolve it in 0.5% TFA and then dilute it with acetonitrile to a final concentration of 0.1% TFA and 80% acetonitrile.

37. Use the flow-through of the Fe-NTA phosphopeptide enrichment to identify the proteins present in your sample, that is, your protein of interest and also the copurified, if any, contaminants.
38. Extended storage in a highly basic solution might cause degradation of the peptides; therefore, the supernatant should be acidified.
39. Very hydrophilic peptides—for example, those on the *N*-terminus of *Arabidopsis thaliana* phyB—might not be retained by the trap column; therefore, direct injection is preferred in the phosphorylation analysis of phytochromes.
40. The gradient depends on sample complexity. For more complex samples use a shallow gradient. The extra wash at 90% solvent B at the end of the chromatography helps to reduce carryover. Alternatively, blank injections can be included between the samples.
41. ETD spectra can also be acquired in the Orbitrap with high resolution and high mass accuracy. This increases confidence in phosphopeptide identification at the expense of detection sensitivity.
42. If your protein of interest is not present in the SwissProt database, append its sequence to the SwissProt entries as a user-defined sequence. Alternatively, perform a species-specific search against the UniProt (Tremble) database, and consider contaminants identified in the first search against the SwissProt entries and specify these contaminants as additional accession numbers.
43. If ETD was acquired in the Orbitrap with high resolution use the same settings as for HCD spectra except that set ESI-ETD-high-res as instrument type.
44. Ambiguity in the localization of the phosphorylation site can arise from incomplete fragmentation of a phosphopeptide with multiple potential modification sites, or from coelution of the different phospho-isoforms during chromatography.
45. There are other software tools available to localize the modification site within a phosphopeptide (e.g., A-score or phosphoRS) [33, 34]. These tools use the output from the database search and look for fragments in the MS/MS spectra that discriminate between potential modification sites.

Acknowledgments

We thank Katalin F. Medzihradszky for critical reading of the manuscript. This work was supported by grants from the Economic Development and Innovation Operative Program (GINOP-2.3.2-15-2016-00015 and GINOP-2.3.2-15-2016-00032).

References

1. Cohen P (2002) The origins of protein phosphorylation. Nat Cell Biol 4:E127–E130
2. Humphrey SJ, James DE, Mann M (2015) Protein phosphorylation: a major switch mechanism for metabolic regulation. Trends Endocrinol Metab 26:676–687
3. Wong YS, Cheng HC, Walsh DA, Lagarias JC (1986) Phosphorylation of *Avena* phytochrome *in vitro* as a probe of light-induced conformational changes. J Biol Chem 261:12089–12097
4. Stockhaus J, Nagatani A, Halfter U, Kay S, Furuya M, Chua NH (1992) Serine-to-alanine substitutions at the amino-terminal region of phytochrome A result in an increase in biological activity. Genes Dev 6:2364–2372
5. Kim JI, Shen Y, Han YJ, Park JE, Kirchenbauer D, Soh MS, Nagy F, Schafer E, Song PS (2004) Phytochrome phosphorylation modulates light signaling by influencing the protein-protein interaction. Plant Cell 16:2629–2640
6. Han YJ, Kim HS, Kim YM, Shin AY, Lee SS, Bhoo SH, Song PS, Kim JI (2010) Functional characterization of phytochrome autophosphorylation in plant light signaling. Plant Cell Physiol 51:596–609
7. Schneider-Poetsch HA, Braun B, Marx S, Schaumburg A (1991) Phytochromes and bacterial sensor proteins are related by structural and functional homologies. Hypothesis on phytochrome-mediated signal-transduction. FEBS Lett 281:245–249
8. Yeh KC, Lagarias JC (1998) Eukaryotic phytochromes: light-regulated serine/threonine protein kinases with histidine kinase ancestry. Proc Natl Acad Sci U S A 95:13976–13981
9. Shin AY, Han YJ, Baek A, Ahn T, Kim SY, Nguyen TS, Son M, Lee KW, Shen Y, Song PS, Kim JI (2016) Evidence that phytochrome functions as a protein kinase in plant light signalling. Nat Commun 7:11545
10. Nito K, Wong CC, Yates JR 3rd, Chory J (2013) Tyrosine phosphorylation regulates the activity of phytochrome photoreceptors. Cell Rep 3:1970–1979
11. Medzihradszky M, Bindics J, Adam E, Viczian A, Klement E, Lorrain S, Gyula P, Merai Z, Fankhauser C, Medzihradszky KF, Kunkel T, Schafer E, Nagy F (2013) Phosphorylation of phytochrome B inhibits light-induced signaling via accelerated dark reversion in Arabidopsis. Plant Cell 25:535–544
12. Kim DH, Kang JG, Yang SS, Chung KS, Song PS, Park CM (2002) A phytochrome-associated protein phosphatase 2A modulates light signals in flowering time control in Arabidopsis. Plant Cell 14:3043–3056
13. Ryu JS, Kim JI, Kunkel T, Kim BC, Cho DS, Hong SH, Kim SH, Fernandez AP, Kim Y, Alonso JM, Ecker JR, Nagy F, Lim PO, Song PS, Schafer E, Nam HG (2005) Phytochrome-specific type 5 phosphatase controls light signal flux by enhancing phytochrome stability and affinity for a signal transducer. Cell 120:395–406
14. Phee BK, Kim JI, Shin DH, Yoo J, Park KJ, Han YJ, Kwon YK, Cho MH, Jeon JS, Bhoo SH, Hahn TR (2008) A novel protein phosphatase indirectly regulates phytochrome-interacting factor 3 via phytochrome. Biochem J 415:247–255
15. Fankhauser C, Yeh KC, Lagarias JC, Zhang H, Elich TD, Chory J (1999) PKS1, a substrate phosphorylated by phytochrome that modulates light signaling in Arabidopsis. Science 284:1539–1541
16. Colon-Carmona A, Chen DL, Yeh KC, Abel S (2000) Aux/IAA proteins are phosphorylated by phytochrome *in vitro*. Plant Physiol 124:1728–1738
17. Shen H, Zhu L, Castillon A, Majee M, Downie B, Huq E (2008) Light-induced phosphorylation and degradation of the negative regulator PHYTOCHROME-INTERACTING FACTOR1 from Arabidopsis depend upon its direct physical interactions with photoactivated phytochromes. Plant Cell 20:1586–1602
18. Shen Y, Zhou Z, Feng S, Li J, Tan-Wilson A, Qu LJ, Wang H, Deng XW (2009) Phytochrome A mediates rapid red light-induced phosphorylation of Arabidopsis FAR-RED ELONGATED HYPOCOTYL1 in a low fluence response. Plant Cell 21:494–506
19. Chen F, Shi X, Chen L, Dai M, Zhou Z, Shen Y, Li J, Li G, Wei N, Deng XW (2012) Phosphorylation of FAR-RED ELONGATED HYPOCOTYL1 is a key mechanism defining signaling dynamics of phytochrome A under red and far-red light in Arabidopsis. Plant Cell 24:1907–1920
20. Ni W, Xu SL, Chalkley RJ, Pham TN, Guan S, Maltby DA, Burlingame AL, Wang ZY, Quail PH (2013) Multisite light-induced phosphorylation of the transcription factor PIF3 is necessary for both its rapid degradation and concomitant negative feedback modulation of photoreceptor phyB levels in Arabidopsis. Plant Cell 25:2679–2698
21. Yue J, Qin Q, Meng S, Jing H, Gou X, Li J, Hou S (2016) TOPP4 Regulates the Stability of PHYTOCHROME INTERACTING

FACTOR5 during Photomorphogenesis in Arabidopsis. Plant Physiol 170:1381–1397
22. Kinoshita E, Takahashi M, Takeda H, Shiro M, Koike T (2004) Recognition of phosphate monoester dianion by an alkoxide-bridged dinuclear zinc(II) complex. Dalton Trans 8:1189–1193
23. Kinoshita E, Kinoshita-Kikuta E, Takiyama K, Koike T (2006) Phosphate-binding tag, a new tool to visualize phosphorylated proteins. Mol Cell Proteomics 5:749–757
24. Kinoshita E, Kinoshita-Kikuta E (2011) Improved Phos-tag SDS-PAGE under neutral pH conditions for advanced protein phosphorylation profiling. Proteomics 11:319–323
25. Kinoshita-Kikuta E, Kinoshita E, Koike T (2015) Neutral Phosphate-Affinity SDS-PAGE system for profiling of protein phosphorylation. Methods Mol Biol 1295:323–354
26. Yao Q, Ge H, Wu S, Zhang N, Chen W, Xu C, Gao J, Thelen JJ, Xu D (2014) P(3)DB 3.0: From plant phosphorylation sites to protein networks. Nucleic Acids Res 42(Database issue):D1206–D1213
27. Liu Q, Wang Q, Deng W, Wang X, Piao M, Cai D, Li Y, Barshop WD, Yu X, Zhou T, Liu B, Oka Y, Wohlschlegel J, Zuo Z, Lin C (2017) Molecular basis for blue light-dependent phosphorylation of Arabidopsis cryptochrome 2. Nat Commun 8:15234
28. Posewitz MC, Tempst P (1999) Immobilized gallium(III) affinity chromatography of phosphopeptides. Anal Chem 71:2883–2892
29. Larsen MR, Thingholm TE, Jensen ON, Roepstorff P, Jorgensen TJ (2005) Highly selective enrichment of phosphorylated peptides from peptide mixtures using titanium dioxide microcolumns. Mol Cell Proteomics 4:873–886
30. Baker PR, Trinidad JC, Chalkley RJ (2011) Modification site localization scoring integrated into a search engine. Mol Cell Proteomics 10:M111.008078
31. Kinoshita E, Kinoshita-Kikuta E, Karata K, Kawano T, Nishiyama A, Yamato M, Koike T (2017) Specific glutamic acid residues in targeted proteins induce exaggerated retardations in Phos-tag SDS-PAGE migration. Electrophoresis 38:1139–1146
32. Yonan CR, Duong PT, Chang FN (2005) High-efficiency staining of proteins on different blot membranes. Anal Biochem 338:159–161
33. Beausoleil SA, Villen J, Gerber SA, Rush J, Gygi SP (2006) A probability-based approach for high-throughput protein phosphorylation analysis and site localization. Nat Biotechnol 24:1285–1292
34. Taus T, Kocher T, Pichler P, Paschke C, Schmidt A, Henrich C, Mechtler K (2011) Universal and confident phosphorylation site localization using phosphoRS. J Proteome Res 10:5354–5362
35. Biemann K (1990) Appendix 5. Nomenclature for peptide fragment ions (positive ions). Methods Enzymol 193:886–887

Chapter 5

Detection of SUMOylated Phytochromes in Plants

Beatriz Orosa and András Viczián

Abstract

Posttranslational modifications (PTMs) happen after or during protein translation. Small Ubiquitin-like Modifier (SUMO) proteins are covalently attached to certain lysine residues of the target proteins to modify their activity, stability, or localization. This process is called SUMOylation, which is a reversible PTM: SUMO protease enzymes can cleave SUMOs off the target protein backbone. Although many ubiquitinated proteins are targeted for degradation, SUMOylation does not necessary lead to the degradation of the modified protein but lead to the regulation of various physiological responses. SUMOylation of the examined protein cannot simply be monitored by immunoblotting techniques performed on total protein extracts, due to the SUMO-specific signals derived from other modified molecules. Furthermore, the fact that only a limited fraction of the target protein pool is SUMOylated makes the detection of SUMOylated proteins challenging. This protocol shows how SUMOylated phytochrome B (phyB) molecules can be detected using homologous and heterologous experimental systems *in planta*.

Key words SUMO, SUMOylation, Posttranslational modification, Immunoprecipitation, Immunoblot

1 Introduction

A wide range of posttranslational modifications (PTMs) can alter the functionality of the synthesized plant proteins [1]. SUMOylation is one among them and was identified about 2 decades ago in mammals [2]. Since then, it has also been found that there are hundreds of SUMO targets in the plant proteome [3–5], and their SUMOylation increases when the plants are subjected to various biotic and abiotic stresses, during certain developmental phases or together with definite metabolic and hormonal responses (for reviews see [6–8]). In the widely used model plant *Arabidopsis thaliana*, out of the identified eight SUMO paralog genes, the expression of four SUMO proteins (SUMO1, 2, 3, and 5) was detected [9]. The actual SUMOylation of the examined protein pool is finely tuned by the balance of SUMO conjugation and SUMO protease (deconjugation) enzymes [10, 11].

Andreas Hiltbrunner (ed.), *Phytochromes: Methods and Protocols*, Methods in Molecular Biology, vol. 2026, https://doi.org/10.1007/978-1-4939-9612-4_5,

A recent study demonstrated that phytochrome B (phyB) is also a subject of SUMOylation [12]. These authors show that (1) SUMOylation of the phyB pool preferably increases when the receptor is in the active state, that is, the Pfr form, (2) the SUMO is conjugated to the *C*-terminal part of phyB, (3) the phyB[Lys996Arg] mutant shows drastically reduced SUMOylation compared with the wild-type counterpart, (4) phyB SUMOylation attenuates light signaling most probably by altering the stability/ formation of complexes consisting of phyB and PHYTOCHROME INTERACTING FACTORs (PIFs), which are bHLH transcription factors involved in phytochrome downstream signalling, and (5) the SUMOylation of phyB is reversible: the OVERLY TOLERANT TO SALT (OTS) 1 and 2 SUMO proteases can remove SUMO from phyB [12]. This work not only demonstrates that SUMOylation and de-SUMOylation of the phytochrome pool is an important regulatory mechanism of light signaling but together with other findings [13] also suggests that besides phyB, other phytochromes and further components of light signaling cascades are potential SUMOylation targets waiting to be identified.

SUMOylation of a chosen plant protein can be examined using several different approaches:

1. Computer software tools can predict which conserved lysine residues [14] are SUMO targets with high probability [15–17]. The identified positions can be tested in the following assays using mutant proteins.
2. SUMOylation of a target protein can be reconstituted in vitro using purified components of the SUMOylation pathway [18–20].
3. *Escherichia coli* does not possess enzymes of SUMOylation pathways, thus it is a suitable artificial system to test the SUMOylation of a chosen protein expressed together with SUMO and the necessary components of the SUMOylation cascade [21–23].
4. Novel techniques like mass spectrometry (MS) proved to be helpful not only in the identification of SUMOylated proteins at the proteome level [24, 25] but also in the identification of SUMO target lysines of the examined protein [22].
5. SUMOylation of the chosen protein can be examined in transgenic plants. Different ways of this approach are discussed in this work.

We present here how SUMOylation of the phyB pool can be detected and monitored *in planta*, involving the combination of plant protein extraction, immunoprecipitation and immunodetection of phyB-SUMO conjugates. An efficient immunoprecipitation method is required to separate the phyB-SUMO from the thousands of other SUMOylated proteins. Unfortunately, anti-PHYB antibodies commercially available at this moment are not suitable to precipitate PHYB from solution, thus we need to express chimeric phyB with an easily detectable protein tag. We chose GREEN FLUORESCENT PROTEIN (GFP), due to the highly specific antibodies available from different suppliers, and knowing that the functionality of the phyB-GFP chimeras has been investigated and proved in several studies [26–28]. Conclusively, we can assume that because phyB-GFP is a functional photoreceptor, its SUMOylation pattern resembles that of endogenous phyB. The first part of this protocol describes the detection of transiently expressed SUMOylated phyB in *Nicotiana benthamiana* (tobacco). The real advantage of this system is its speed: the SUMOylation of many different proteins can be tested within a few weeks without the need of generating homozygous stable transgenic lines [29]. Expressing hemagglutinin-tagged SUMO (HA-SUMO) together with phyB-GFP gives the opportunity to detect only those phyB-GFP molecules which are tagged with transgenic SUMO protein (but not the tobacco counterpart) with the application of a tag-specific antibody (Fig. 1). The second part of this protocol demonstrates how phyB-GFP SUMOylated with endogenous SUMO can be detected in stable transgenic *Arabidopsis thaliana* plants. The obvious advantages of this method are that (1) the homologous Arabidopsis system is used with the endogenous SUMO molecules, (2) high level of phyB-GFP expression can be achieved, and (3) the transgenic plants can be examined at various developmental stages and under different experimental conditions (Fig. 2).

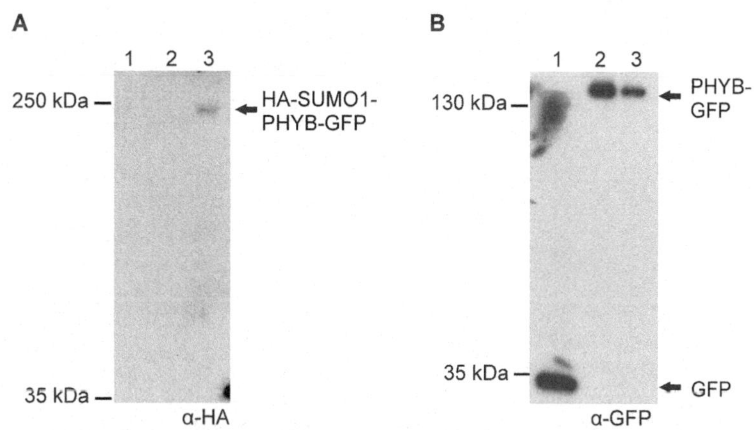

Fig. 1 phyB is SUMOylated in *Nicotiana benthamiana*. Transiently transformed (coinfiltration) tobacco leaves, expressing GFP + HA-SUMO1 + P19 (lane 1) or phyB-GFP + P19 (lane 2) or phyB-GFP + HA-SUMO1 + P19 (lane 3) proteins were subjected to total protein extraction and immunoprecipitation using GFP-TRAP-M beads. Identical amounts of samples were subjected to SDS-PAGE and subsequent western blot analysis using anti-HA (α-HA, panel A) and anti-GFP (α-GFP, panel B) to detect the presence of SUMOylated phytochromes and GFP-containing proteins, respectively

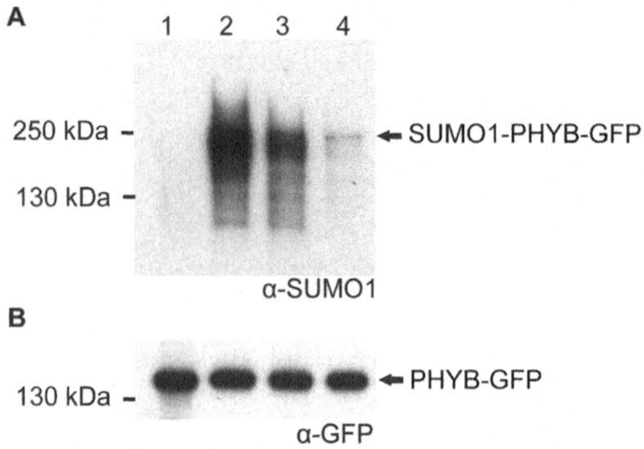

Fig. 2 phyB is SUMOylated in Arabidopsis plants grown in light/dark cycles. Arabidopsis *phyB-9* seedlings, expressing the PHYB-GFP chimeric protein under the control of the *35S* promoter were grown for 5 days on Murashige and Skoog medium under 12 h white light (100 μmol m^{-2} s^{-1})–12 h dark cycles at 22 °C. Samples were harvested on day 6 at the middle of the night (lane 1); at the middle of the day (lane 2); at the end of the day (lane 3) and at the end of the night (lane 4). PHYB-GFP was immunoprecipitated from the total protein extracts using GFP-TRAP-M beads and samples containing identical amounts of PHYB-GFP were separated by SDS-PAGE. Western blot analysis was performed using anti-SUMO1 (α-SUMO1, panel A) and anti-GFP (α-GFP, panel B) to detect SUMO-PHYB-GFP and PHYB-GFP, respectively

2 Materials

2.1 Transient Expression of PHYB in Tobacco

1. *Agrobacterium tumefaciens* strains hosting plant expression constructs. Successful expression of the proteins of interest requires the use of a strong promoter like the *35S* promoter from cauliflower mosaic virus. For the presented assay we have to express in the same tobacco plant (1) PHYB fused to GFP (PHYB-GFP); (2) SUMO1 with the HA tag fused to its N-terminus (HA-SUMO1); and (3) the P19 viral suppressor protein, which ensures high foreign (nontobacco) protein levels in the treated tissues [30].
2. Four-week-old tobacco plants (*see* **Note 1**).
3. 10 mM magnesium chloride ($MgCl_2$): Dissolve 2.033 g $MgCl_2 \times 6\ H_2O$ in 1 L water (*see* **Note 2**).
4. LB (Luria-Bertani) or YEB medium: To obtain 1 L LB medium, dissolve 10 g tryptone, 5 g yeast extract and 10 g NaCl in 950 mL water. Adjust the pH of the medium to 7.0 using 1 N NaOH, and bring the volume up to 1 L. Autoclave on liquid cycle for 20 min at 15 psi. Add the necessary antibiotics after cooling to room temperature. For LB agar-plates prepare LB medium and add 15 g/L agar before autoclaving. Cool to 50 °C in a temperature-controlled water bath before adding the necessary antibiotics. Store at 4 °C. Alternatively,

instead of LB, YEB medium can also be used for Agrobacterium propagation. Dissolve 5 g beef extract, 1 g yeast extract, 5 g peptone, 5 g sucrose, 1 mL 1 M NaOH, and 0.75 mL 1 M $MgCl_2$ in 900 mL water, fill up to 1 L and autoclave for 20 min at 15 psi.

5. 0.2 M acetosyringone (3′, 5′-dimethoxy-4′-hydroxyacetophenone): dissolve 0.039 g acetosyringone in 1 mL dimethyl sulfoxide (DMSO). Always prepare fresh stock.

2.2 Propagation of Transgenic Arabidopsis Plants Expressing SUMOylated phyB

1. Transgenic plants expressing recombinant phyB-GFP protein under the control of the *35S* promoter (*see* **Note 3**).
2. Materials and equipment necessary to grow and handle Arabidopsis plants. As they are changing from laboratory to laboratory, we do not list here all necessary equipment and chemicals.

2.3 Immuno precipitation of SUMOylated PHYB

1. Immunoprecipitation buffer: 100 mM Tris–HCl, pH 8.0, 0.1% [w/v] SDS, 0.5% [w/v] sodium deoxycholate, 1% [v/v] glycerol, 50 mM sodium bisulfite, 20 mM NEM, 1 protease inhibitor cocktail tablet (Roche mini PI tablets) in every 10 mL buffer.
   (a) 1 M Tris–HCl, pH 8.0: dissolve 121.2 g Tris in 900 mL water. Set the pH to 8.0 using concentrated HCl and fill up to 1000 mL.
   (b) 10% [w/v] sodium dodecyl sulfate (SDS): dissolve 10 g SDS in 80 mL water and make up to 100 mL.
   (c) 10% [w/v] sodium deoxycholate: dissolve 10 g sodium deoxycholate in 90 mL water and make up to 100 mL. Protect the solution from light.
   (d) 1 M sodium bisulfite stock solution (*see* **Note 4**): Make 10 mL stock solution by dissolving 1.90 g Na-bisulfite in water.
   (e) 1 M *N*-ethylmaleimide (NEM) stock solution (*see* **Note 5**): To obtain 1 M stock solution, dissolve 0.125 g *N*-ethylmaleimide (NEM) in 1 mL ethanol.
2. Magnetic anti-GFP beads. Anti-GFP VHH coupled to beads for immunoprecipitation of GFP-fusion proteins (GFP-TRAP-M, Chromotek).
3. Polyvinylpolypyrrolidone (PVPP).
4. 2× sodium dodecyl sulfate polyacrylamide gel electrophoresis (SDS PAGE) Sample Buffer: 100 mM Tris pH 6.8, 40 mM DTT, 2% SDS, 20% glycerol, 0.2% bromophenol blue (*see* **Note 6**).

2.4 SDS PAGE and Western Blotting Materials and Reagents

1. Resolving gel solution: To obtain 5 mL 8% solution, mix 2.3 mL water; 1.3 mL 30% [w/v] acrylamide–bis-acrylamide (Sigma-Aldrich) (*see* **Notes 7** and **8**); 1.3 mL 1.5 M Tris–HCl pH 8.8 solution; 0.05 mL 10% SDS; 0.05 mL 10% APS; and 0.003 mL *N, N, N, N′*-tetramethyl-ethylenediamine (TEMED) (Sigma-Aldrich) (*see* **Note 9**).
   (a) 1.5 M Tris–HCl pH 8.8: dissolve 181.7 g Tris in 900 mL water. Adjust the pH to 8.8 with concentrated HCl and fill up to 1000 mL.
   (b) 10% [w/v] SDS: *see* Subheading 2.3.
   (c) 10% [w/v] ammonium persulfate (APS): 10% solution in water, always prepare fresh.
2. Stacking gel solution: To obtain 2 mL stacking gel solution, mix 1.4 mL water; 0.33 mL 30% [w/v] acrylamide–bis-acrylamide; 0.25 mL 1 M Tris–HCl pH 6.8; 0.02 mL 10% SDS; 0.02 mL 10% APS, and 0.002 mL TEMED.
   (a) 1 M Tris–HCl pH 6.8: dissolve 121.2 g Tris in 900 mL water. Set the pH to 6.8 using concentrated HCl and fill up to 1000 mL.
   (b) 10% [w/v] SDS: *see* Subheading 2.3.
   (c) 10% APS: *see* Subheading 2.4, Resolving gel solution.
3. Electrophoresis running buffer: 25 mM TRIS base, 192 mM glycine, 0.1% SDS. Dissolve 3.03 g TRIS base, 14.4 g glycine, and 1 g SDS in 1 L water. No need to set the pH (*see* **Note 10**).
4. Transfer buffer: 25 mM Tris–HCl, 192 mM glycine, 20% [v/v] methanol. Dissolve 3.03 g TRIS base and 14.4 g glycine in 800 mL water, and add 200 mL methanol (*see* **Note 11**).
5. Tris buffered saline supplied with Tween 20 (TBST): 20 mM Tris–HCl pH 7.6, 137 mM NaCl, 0.1% Tween 20. To obtain 1 L TBST buffer, dissolve 2.423 g Tris, 8.006 g NaCl in 800 mL water. Set the pH to 7.6 using HCl. Add 1 mL Tween 20 and fill up to 1 L (*see* **Note 12**).
6. Prestained protein molecular weight marker (available from different suppliers).
7. Polyvinylidene difluoride (PVDF) membrane with 0.2 μm pore size (Immobilon-PSQ, Sigma-Aldrich).
8. 2-propanol.
9. Methanol.
10. Chemiluminescent substrate for peroxidase signal detection (Immobilon Western Chemiluminescent HRP substrate, Millipore).
11. 3MM Whatman filter paper (Sigma-Aldrich).

12. Fat-free milk powder.
13. Antibodies:
    (a) Anti-GFP antibody (mouse, Clontech).
    (b) Anti-HA antibody (mouse, Abcam).
    (c) Anti-SUMO1 antibody (rabbit, Agrisera).
    (d) Anti-rabbit IgG peroxidase conjugated antibody (Thermo Scientific).
    (e) Anti-mouse IgG peroxidase conjugated antibody (Thermo Scientific).

2.5 Laboratory Equipment

1. Refrigerated microcentrifuge.
2. pH meter.
3. Spectrophotometer.
4. Rocking platform.
5. Rotator.
6. Syringes, 1 mL.
7. Magnetic stand for the recovery of magnetic anti-GFP beads.
8. Heat block set at 95 °C.
9. Glass plates, combs, gel holders, gel running tanks, blotting devices, and power supply. There are many different brands available on the market; here we use Mini Protean system (Bio-Rad).
10. Sensitive cooled charge-coupled device (CCD) camera (e.g., Micromax from Roper Scientific or ORCA from Hamamatsu). Alternatively, any gel documentation system capable of handling low luminescent signals or chemiluminescent film (Kodak) with film developer can also be used.
11. Adjustable micropipettes, tips.
12. 1.5 mL microcentrifuge tubes and 2 mL safety lock microcentrifuge tubes (Eppendorf).
13. Ice.
14. Liquid N_2.

3 Methods

3.1 Expression and Extraction of SUMOylated phyB-GFP in Tobacco

1. Incubate the Agrobacterium strains harboring the appropriate binary vectors at 28 °C for 2 days on LB plates containing the necessary antibiotics.
2. Using a sterilized pipette tip or inoculation loop, take a single *Agrobacterium* colony into a sterile Erlenmeyer flask containing 10 mL LB medium with the necessary antibiotics. Incubate at

28 °C with constant shaking at 200 rpm for 12–16 h, allowing the *Agrobacterium* cells to reach the exponential growth phase.

3. Centrifuge the culture for 10 min at 3000 × *g* at 25 °C to pellet the cells. Discard the supernatant, wash the cells with 10 mM $MgCl_2$ (turning the centrifuge tube up and down). Centrifuge again and suspend the cells in 10 mL 10 mM $MgCl_2$ solution. Measure the optical density (OD) of the suspension at 610 nm (*see* **Note 13**) using a spectrophotometer.
4. Prepare the Infiltration solution. Dilute the Agrobacterium suspensions using 10 mM $MgCl_2$ and mix them to have the final OD_{610} of each bacterial culture between 0.2 and 0.3. Combine the following Agrobacteria containing the binary vectors necessary for the expression of the following proteins: PHYB-GFP + HA-SUMO1 + P19 (mixture 1); PHYB[K996R]-GFP + HA-SUMO1 + P19 (mixture 2); GFP + HA-SUMO1 + P19 (mixture 3). Calculate at least 1 mL Infiltration solution per tobacco leaf. Add 0.2 M acetosyringone stock solution to a final concentration of 0.2 mM. Incubate the Infiltration solution for 2–6 h at room temperature without agitation.
5. Mix the Infiltration solution gently (turning the tube up and down), before sucking it up into a 1 mL syringe. Push the end of the syringe to the abaxial side (lower surface) of the leaf and slowly inject the solution into the mesophyll tissue. Try to avoid damaging the leaf. The infiltrated area can easily be seen as a dark, buffer-soaked region. Repeat this procedure until the whole leaf is "filled" with Infiltration solution.
6. Keep the plants under the same conditions after the infiltration as before for 3 days.
7. Collect the injected leaves in the middle of the day. At this time of the day a large fraction of the PHYB pool is SUMOylated (Fig. 2 and [12]). Wrap 1.5–2 g of leaf material into aluminum foil and freeze immediately in liquid N_2. All samples must have the same weight. Store them at −80 °C (*see* **Note 14**).
8. Prepare the Immunoprecipitation buffer; always make fresh buffer every time and cool it on ice prior to use.
9. For immunoprecipitation, vortex the magnetic anti-GFP beads (GFP-TRAP-M) and pipet the necessary amount for the whole assay (all the samples) to a 1.5 mL microcentrifuge tube. Calculate 30–50 μL beads for 1000 μL of plant extract supernatant (which will be equivalent to one plant sample). Wash the beads with 1 mL ice-cold Immunoprecipitation buffer, resuspend them by pipetting up and down several times and place the tubes into a magnetic rack (*see* **Note 15**). Wait for 1 min to allow the beads to be captured by the magnet and remove the supernatant. Repeat the process two more times

for a total of three washes. After the last wash, resuspend the beads in the original volume of Immunoprecipitation buffer and keep them on ice.

10. Place the frozen leaves into a mortar, precooled with liquid N_2. Grind the sample with a pestle until the plant material becomes fine powder, and mix thoroughly with 0.2 g PVPP. Transfer the powder (ca. 1 mL) to a precooled 2 mL microcentrifuge tube.
11. Add 1 mL ice-cold Immunoprecipitation buffer, and mix until all the powder is completely dissolved. Centrifuge for 15 min at 4 °C at 20,000 × *g*. Transfer the supernatant into a new tube. If the supernatant contains debris, centrifuge again for 5 min under the same conditions.
12. Add 30–50 μL washed anti-GFP beads to 1000 μL of pre-cleared plant extract (*see* **Note 16**) and incubate for 30 min at 4 °C on a tube rotator or rocking platform.
13. Pellet the beads by placing the tubes into the magnetic rack. Wait for 1 min and pour down the supernatant.
14. Wash the beads with 500 μL ice-cold Immunoprecipitation buffer. Gently shake the tubes to suspend the beads and incubate for 5 min at 4 °C. Place them into the magnetic rack, wait for 1 min and remove the supernatant. Repeat two more times for a total of three washes.
15. After the last wash, proceed with the elution of the target protein. Add 80–100 μL 1× SDS PAGE Sample Buffer onto the pellet and mix by pipetting. Incubate the sample for 10 min at 95 °C.
16. Briefly spin the tubes and place them into the magnetic rack. Wait for 1 min and transfer the supernatant containing the isolated protein to a new tube.
17. Snap freeze the samples in liquid N_2 and store them at −80 °C. Thaw samples on ice before use and avoid multiple freeze–thaw cycles.

3.2 Extraction of SUMOylated phyB-GFP from Transgenic Arabidopsis Plants

1. Grow Arabidopsis seedlings expressing phyB-GFP at high levels (phyB-GFP expressed under the control of the constitutive viral *35S* promoter results in sufficient amounts of phyB-GFP for the presented assay.). Different growth conditions and transgenic plant lines can be used according to the planned experiment. Here we use 5-day-old Arabidopsis seedlings, grown on Murashige and Skoog medium under 12 h white light (100 μmol m^{-2} s^{-1})/12 h dark cycles at 22 °C.
2. Harvest 1 g of seedlings and snap freeze them in liquid N_2 (*see* **Note 17**).

3. Prepare fresh Immunoprecipitation buffer and wash the anti-GFP beads (*see* Subheading 3.1, **steps 8** and **9**).
4. Cool the mortar with liquid N_2. Grind the frozen Arabidopsis seedlings with a pestle to fine powder and transfer it to a pre-cooled 2 mL microcentrifuge tube.
5. Follow the immunoprecipitation protocol described in Subheading 3.1 (**steps 11–17**).

3.3 SDS-PAGE and Immunoblotting

1. Wash the Mini Protean glass plates thoroughly with household detergent, rinse with plenty of water, finally with ethanol and let them dry. Assemble the glass plate cassette according to the instructions of the manufacturer.
2. Prepare the resolving gel solution (*see* **Note 18**). Pipette the gel solution between the glass plates; leave about 1/4 of the space free for the stacking gel. Carefully pipette 2 mL 100% 2-propanol on top of the resolving gel, and wait until the resolving gel polymerizes (30–60 min).
3. When polymerization is complete, a clear line will appear between the gel surface and 2-propanol. Discard the 2-propanol and gently wash the upper surface of the gel with water. Prepare the stacking gel solution and carefully pipette it on the resolving gel; avoid formation of bubbles. Insert combs and allow the gel to polymerize for 30 min.
4. Remove combs carefully. Put the gel into the electrophoresis tank, fill the tank with fresh 1× Running buffer until it covers the wells of the gel. Rinse the wells by pipetting the running buffer up and down several times.
5. Load the protein molecular weight marker and the protein extracts to be tested (Results of Subheadings 3.1 and 3.2) into the wells of the gel. This work is easier using long and thin gel-loading micropipette tips. Depending on the expressed protein levels and the sensitivity of the detection, 5–10 μL or 25–30 μL anti-GFP precipitated protein extract is sufficient for anti-GFP or anti-SUMO1 western blot detection, respectively (*see* **Note 19**).
6. Set the applied voltage between 40 and 60 V during stacking. When the dye front reaches the resolving gel, increase the voltage to 80–120 V. The optimal duration of the gel run depends on the molecular weight of the protein of interest. In our case the ca. 170 kDa PHYB-GFP-SUMO1 chimeric protein can be visualized after a run of about 3 h.
7. After the electrophoresis, remove the gel sandwich from the running tank and open the glass plates gently using a spatula. Discard the stacking gel and immerse the resolving gel to a container filled with Transfer buffer. Rinse the gel for 10 min with gentle agitation.

8. Cut the PVDF membrane to the size of the gel. Activate the membrane by rinsing in methanol (5 min), washing with water (5 min) and equilibrating in Transfer buffer for 5 min.
9. Assemble the gel blotting sandwich in a flat container filled with cold (4 °C) Transfer buffer. Immerse a layer of sponge (5 mm thick) in the container, and place a sheet of 3MM Whatman filter paper on it. Next, gently lift the gel and place it on the wet filter paper and then place the activated and equilibrated PVDF membrane on the gel. Avoid bubbles and wrinkles. Put a wet filter paper on the PVDF membrane, followed by a next layer of sponge. Close the plastic holders placed on the upper and bottom side of the completed sandwich and insert it into the blotting tank (the membrane should face toward the positive cathode and the gel toward the negative anode), and fill up with cold transfer buffer. Immerse a cold (−20 °C) cooling unit into the tank, place the blotting apparatus at 4 °C and attach a power supply to the tank.
10. Apply 30–40 V for overnight (14–20 h) or 100 V for short (2–3 h) transfers (*see* **Note 20**).

3.4 Immunoblotting and Signal Detection

1. Remove the PVDF membrane from the blotting apparatus and rinse with TBST for 10 min with gentle agitation on a rocking platform. Also use the rocking platform for the subsequent washing and hybridization steps. Apply enough liquid to cover the membrane completely.
2. Block the membrane in TBST containing 5% milk powder at room temperature for 1–2 h.
3. Rinse the membrane with TBST (two times 5 min) to remove the milk powder.
4. Incubate the membrane with appropriate dilutions of the primary antibody in TBST buffer for 2 h (*see* **Note 21**).
5. Wash the membrane five times for 5 min with TBST.
6. Incubate the membrane in TBST containing the peroxidase-conjugated secondary antibody solution (in our case it was diluted 1:20000) for 1 h.
7. Wash the membrane five times for 5 min in TBST.
8. Apply the chemiluminescent substrate to the membrane according to the manufacturer's recommendation.
9. Acquire luminescent signals using a liquid nitrogen-cooled charge-coupled device camera (Micromax; Roper Scientific). Modern CCD camera systems provide convenience and high sensitivity, but using conventional film detection and developing combined with a simple flatbed scanner could also provide decent results.

4 Notes

1. Grow the tobacco plants in a growth chamber at constant 24 °C temperature and 16 h light–8 h dark cycles. Alternatively, a 26 °C and 14 h light/10 h dark light regime can also be used. It is important to have healthy leaves, and the plants should have no developed flower buds.
2. Throughout the whole protocol, whenever we refer to "water" we recommend to use purified water with a resistance of 18.2 MΩ·cm at 25 °C and a total organic content of less than five parts per billion (Millipore).
3. The expression of phyB-GFP can be controlled by either *35S* promoter or *PHYB* promoter, but it is worth to note that the application of the strong constitutive *35S* promoter results in higher phyB-GFP levels, which makes the detection of the SUMOylation easier. This is particularly important in those cases when only a small fraction of the examined protein pool is SUMOylated.
4. Always prepare fresh stock solution.
5. Always prepare fresh stock solution. Add 1 M NEM to the Immunoprecipitation buffer right before use, since it is not stable in aqueous solution. NEM blocks the activity of SUMO-proteases [31, 32], thus its presence is a key factor of the protein isolation procedure.
6. The 2× SDS PAGE Sample Buffer can be stored at −20 °C. Alternatively, commercially available version of this buffer also exist (Bio-Rad).
7. Alternatively, dissolve 29.2 g acrylamide and 0.2 g *N′N′*-bis-methylene-acrylamide in water, fill up to 100 mL and filter through 3MM Whatman filter paper. Store the acrylamide solution at 4 °C in the dark. Unpolymerized acrylamide is a potential neurotoxin, therefore be sure to apply the necessary precautions.
8. The applied acrylamide concentration in the resolving gel should fit the molecular weight of the examined proteins. In this protocol we use 8% acrylamide, which gives good resolution around 140–180 kDa, what is the approximate molecular weight range of phyB-GFP and its SUMOylated form.
9. Store TEMED stock solution at 4 °C in the dark.
10. It is convenient to prepare 10× Electrophoresis running buffer, to store at room temperature and to make 1× dilution prior to use.
11. Alternatively, use the same amount of the less toxic ethanol instead of methanol. Addition of 0.015% SDS may result in better transfer of the high molecular weight proteins.
12. It is convenient to prepare 10× TBST, and to make 1× dilution when needed.

13. Use 10 mM $MgCl_2$ as blank. The suspensions may be too dense for direct measurement in the spectrophotometer. Prepare 10× dilutions using 10 mM $MgCl_2$ ensuring that the resulting OD_{610} values are below 0.5.
14. It is advisable to collect noninjected leaves for negative controls. These samples will be able to show the possible nonspecific protein bands on the western blot.
15. Use low-retention plastic filter tips. Keep the beads in a homogenous suspension at all times when pipetting. Variation in the amount of beads can lead to lower reproducibility. Instead of using a magnetic rack, the anti-GFP beads can be pelleted by centrifuging the tubes at 1300 × *g* at 4 °C for 1 min.
16. At this point the input sample can be collected. Mix the cleared plant extract with an equal volume of 2× SDS PAGE Sample Buffer and incubate for 5 min at 95 °C. This sample can be tested using the western blot assay to show the amount of phyB-GFP expressed in the plant and taken into the forthcoming immunoprecipitation reaction.
17. Arabidopsis seedlings homozygous for the *35S:PHYB-GFP* transgene have higher relative phyB-GFP protein level than transiently transformed tobacco leaves. Along with this observation, note that collecting 1 g of Arabidopsis seedling is more difficult than harvesting 1.5 g tobacco leaves. It is worth to bulk up the Arabidopsis seed stocks before starting this kind of assay.
18. Calculate the necessary volumes of the resolving and stacking gels. It depends on the volume of the glass gel cassette and the number of gels casted at the same time. For a standard 80 × 83 × 1 mm cell cassette we calculate 4.5 mL resolving gel and 2 mL stacking gel volumes.
19. It is advisable to test what amount of PHYB-GFP was immunoprecipitated in the different samples. Perform the SDS-PAGE and western blot procedure with only a 3–5 μL of protein extracts and quantify the obtained GFP signals. Then, normalize the signal and load sufficient volumes of extract containing identical amount of GFP signal for the final anti-SUMO1 and anti-GFP blots.
20. These values work well with the Bio-Rad Mini Protean system. For further setting details consult the instruction manual of your blotting apparatus.
21. The dilution of the antibody depends on the recommendation of the antibody producer, the applied buffer system and the expression level of the examined proteins. During this protocol we applied anti-GFP antibody (1:10000), anti-HA antibody (1:5000), and anti-SUMO1 antibody (1:8000).

Acknowledgments

We wish to thank Prof. Ari Sadanandom for critical reading of the manuscript. This work in Szeged was supported by grants from the Gazdaságfejlesztési és Innovációs Operatív Program (GINOP -2.3.2-15-2016-00015 and GINOP-2.3.2-15-2016-00032).

References

1. Friso G, van Wijk KJ (2015) Posttranslational protein modifications in plant metabolism. Plant Physiol 169:1469–1487
2. Mahajan R, Delphin C, Guan T, Gerace L, Melchior F (1997) A small ubiquitin-related polypeptide involved in targeting RanGAP1 to nuclear pore complex protein RanBP2. Cell 88:97–107
3. Elrouby N, Coupland G (2010) Proteome-wide screens for small ubiquitin-like modifier (SUMO) substrates identify Arabidopsis proteins implicated in diverse biological processes. Proc Natl Acad Sci U S A 107:17415–17420
4. Miller MJ, Barrett-Wilt GA, Hua Z, Vierstra RD (2010) Proteomic analyses identify a diverse array of nuclear processes affected by small ubiquitin-like modifier conjugation in Arabidopsis. Proc Natl Acad Sci U S A 107:16512–16517
5. Lopez-Torrejon G, Guerra D, Catala R, Salinas J, del Pozo JC (2013) Identification of SUMO targets by a novel proteomic approach in plants. J Integr Plant Biol 55:96–107
6. Park HJ, Kim WY, Park HC, Lee SY, Bohnert HJ, Yun DJ (2011) SUMO and SUMOylation in plants. Mol Cells 32:305–316
7. Castro PH, Tavares RM, Bejarano ER, Azevedo H (2012) SUMO, a heavyweight player in plant abiotic stress responses. Cell Mol Life Sci 69:3269–3283
8. Elrouby N (2017) Regulation of plant cellular and organismal development by SUMO. Adv Exp Med Biol 963:227–247
9. Kurepa J, Walker JM, Smalle J, Gosink MM, Davis SJ, Durham TL, Sung DY, Vierstra RD (2003) The small ubiquitin-like modifier (SUMO) protein modification system in Arabidopsis. Accumulation of SUMO1 and −2 conjugates is increased by stress. J Biol Chem 278:6862–6872
10. Vierstra RD (2012) The expanding universe of ubiquitin and ubiquitin-like modifiers. Plant Physiol 160:2–14
11. Yates G, Srivastava AK, Sadanandom A (2016) SUMO proteases: uncovering the roles of deSUMOylation in plants. J Exp Bot 67:2541–2548
12. Sadanandom A, Adam E, Orosa B, Viczian A, Klose C, Zhang C, Josse EM, Kozma-Bognar L, Nagy F (2015) SUMOylation of phytochrome-B negatively regulates light-induced signaling in *Arabidopsis thaliana*. Proc Natl Acad Sci U S A 112:11108–11113
13. Lin XL, Niu D, Hu ZL, Kim DH, Jin YH, Cai B, Liu P, Miura K, Yun DJ, Kim WY, Lin R, Jin JB (2016) An arabidopsis SUMO E3 ligase, SIZ1, negatively regulates photomorphogenesis by promoting COP1 activity. PLoS Genet 12:e1006016
14. Rodriguez MS, Dargemont C, Hay RT (2001) SUMO-1 conjugation in vivo requires both a consensus modification motif and nuclear targeting. J Biol Chem 276:12654–12659
15. Xue Y, Zhou F, Fu C, Xu Y, Yao X (2006) SUMOsp: a web server for sumoylation site prediction. Nucleic Acids Res 34(Web Server):W254–W257
16. Ren J, Gao X, Jin C, Zhu M, Wang X, Shaw A, Wen L, Yao X, Xue Y (2009) Systematic study of protein sumoylation: development of a site-specific predictor of SUMOsp 2.0. Proteomics 9:3409–3412
17. Zhao Q, Xie Y, Zheng Y, Jiang S, Liu W, Mu W, Liu Z, Zhao Y, Xue Y, Ren J (2014) GPS-SUMO: a tool for the prediction of sumoylation sites and SUMO-interaction motifs. Nucleic Acids Res 42(Web Server issue):W325–W330
18. Colby T, Matthai A, Boeckelmann A, Stuible HP (2006) SUMO-conjugating and SUMO-deconjugating enzymes from Arabidopsis. Plant Physiol 142:318–332
19. Gali H, Juhasz S, Morocz M, Hajdu I, Fatyol K, Szukacsov V, Burkovics P, Haracska L (2012) Role of SUMO modification of human PCNA at stalled replication fork. Nucleic Acids Res 40:6049–6059
20. Burkovics P, Sebesta M, Sisakova A, Plault N, Szukacsov V, Robert T, Pinter L, Marini V, Kolesar P, Haracska L, Gangloff S, Krejci L (2013) Srs2 mediates PCNA-SUMO-

dependent inhibition of DNA repair synthesis. EMBO J 32:742–755

21. Weber AR, Schuermann D, Schar P (2014) Versatile recombinant SUMOylation system for the production of SUMO-modified protein. PLoS One 9:e102157
22. Okada S, Nagabuchi M, Takamura Y, Nakagawa T, Shinmyozu K, Nakayama J, Tanaka K (2009) Reconstitution of *Arabidopsis thaliana* SUMO pathways in *E. coli*: functional evaluation of SUMO machinery proteins and mapping of SUMOylation sites by mass spectrometry. Plant Cell Physiol 50:1049–1061
23. Iribarren PA, Berazategui MA, Cazzulo JJ, Alvarez VE (2015) Biosynthesis of SUMOylated proteins in bacteria using the *Trypanosoma brucei* enzymatic system. PLoS One 10:e0134950
24. Park HC, Choi W, Park HJ, Cheong MS, Koo YD, Shin G, Chung WS, Kim WY, Kim MG, Bressan RA, Bohnert HJ, Lee SY, Yun DJ (2011) Identification and molecular properties of SUMO-binding proteins in Arabidopsis. Mol Cells 32:143–151
25. Nukarinen E, Tomanov K, Ziba I, Weckwerth W, Bachmair A (2017) Protein sumoylation and phosphorylation intersect in Arabidopsis signaling. Plant J 9:505–517
26. Kircher S, Kozma-Bognar L, Kim L, Adam E, Harter K, Schafer E, Nagy F (1999) Light quality-dependent nuclear import of the plant photoreceptors phytochrome A and B. Plant Cell 11:1445–1456
27. Fernandez AP, Gil P, Valkai I, Nagy F, Schafer E (2005) Analysis of the function of the photoreceptors phytochrome B and phytochrome D in *Nicotiana plumbaginifolia* and *Arabidopsis thaliana*. Plant Cell Physiol 46: 790–796
28. Palagyi A, Terecskei K, Adam E, Kevei E, Kircher S, Merai Z, Schafer E, Nagy F, Kozma-Bognar L (2010) Functional analysis of amino-terminal domains of the photoreceptor phytochrome B. Plant Physiol 153: 1834–1845
29. Ewan R, Pangestuti R, Thornber S, Craig A, Carr C, O'Donnell L, Zhang C, Sadanandom A (2011) Deubiquitinating enzymes AtUBP12 and AtUBP13 and their tobacco homologue NtUBP12 are negative regulators of plant immunity. New Phytol 191:92–106
30. Lakatos L, Csorba T, Pantaleo V, Chapman EJ, Carrington JC, Liu YP, Dolja VV, Calvino LF, Lopez-Moya JJ, Burgyan J (2006) Small RNA binding is a common strategy to suppress RNA silencing by several viral suppressors. EMBO J 25:2768–2780
31. Lois LM, Lima CD, Chua NH (2003) Small ubiquitin-like modifier modulates abscisic acid signaling in Arabidopsis. Plant Cell 15:1347–1359
32. Li SJ, Hochstrasser M (2000) The yeast ULP2 (SMT4) gene encodes a novel protease specific for the ubiquitin-like Smt3 protein. Mol Cell Biol 20:2367–2377

Chapter 6

Approaches to Identify Protein Ubiquitination Sites in Plants

Feifei Yu and Qi Xie

Abstract

Identification of protein ubiquitination sites is very important for the functional analysis of a targeted protein. Sample preparation before LC-MS/MS assay is essential for this experimental process. Here, we describe two efficient methods for preparing samples for identifying ubiquitination sites in plant proteins by LC-MS/MS.

Key words Ubiquitination site, Protein expression, Plant protein extraction, Immunoprecipitation, Cell-free assay, LC-MS/MS

1 Introduction

Ubiquitination is an important posttranslational modification (PTM) and functions mainly in determining the degradative fate of regulatory and abnormal proteins [1, 2]. Proteins encoded by approximately 5% of the genes in *Arabidopsis thaliana* are related to the ubiquitin–proteasome system (UPS), and these proteins function at diverse plant growth stages and in response to various environmental stresses [3, 4]. The ubiquitination reaction on specific target proteins needs a three-step cascaded catalytic reaction accomplished by three enzymes, ubiquitin-activating enzyme (E1), which activates ubiquitin molecules in an ATP-dependent manner; ubiquitin conjugation enzyme (E2), which can accept the activated ubiquitin molecules on cysteine from the E1 to form an E2–ubiquitin intermediate via a thioester bond; and ubiquitin ligase (E3), which facilitates the transfer of ubiquitin from E2–ubiquitin intermediate to targeted substrates by forming an isopeptide bond between a glycine residue of ubiquitin and a lysine residue of the target protein.

Due to the large number of E3s in plants, E3 ligases are well recognized to be responsible for the specificity of ubiquitination by recruiting appropriate target proteins. For instance, there are over

Andreas Hiltbrunner (ed.), *Phytochromes: Methods and Protocols*, Methods in Molecular Biology, vol. 2026,
https://doi.org/10.1007/978-1-4939-9612-4_6, © Springer Science+Business Media, LLC, part of Springer Nature 2019

1500 different E3s in *Arabidopsis* [5]. On the basis of the mechanism of action and subunit composition, E3s are classified into four types: homologous to the E6-AP carboxyl terminus (HECT), really interesting new gene (RING), U-box, and cullin-RING ligases (CRLs) [6]. In the past few years, many researches focused on the identification of the specific substrates of E3 ubiquitin ligases to uncover the diverse function of E3s in different biological processes. At the same time, many approaches to identify substrates of E3s have been developed, such as yeast two-hybrid, global protein stability profiling, and in vitro ubiquitination screens [7].

As we know, monoubiquitin or polyubiquitin chains often are attached to a lysine residue of the modified protein, and the mutation of this position can significantly affect protein stability, subcellular localization, and biological functions [8, 9]. However, the methods of identification of ubiquitination sites in substrates are still much limited. The approach involving mutagenesis followed by a degradation assay is not only laborious but also time-consuming, especially when the ubiquitin is conjugated to a non-lysine type amino acid. How to identify ubiquitination site(s) of proteins turns to be an important unsolved question. Fortunately, LC-MS/MS has become more efficient in recent years. In this chapter, we will introduce two simple approaches to identifying ubiquitination sites by LC-MS/MS analysis.

2 Materials

2.1 Plant Materials and Growth Condition

1. *Arabidopsis* plants (Col-0) expressing a GFP- or Myc-tagged version of the protein, for which ubiquitination should be investigated.
2. Liquid 1/2 MS medium (pH 5.8, autoclaved) in a 500 mL flask.
3. Tissue culture room with 16-h light–8-h dark cycles.
4. Constant shaking incubator placed in the tissue culture room.

2.2 Plant Protein Extraction

1. MG132 is added to the culture medium 8–12 h before sample collection to enrich the ubiquitinated proteins.
2. 1 M Tris–MES (pH 8.0): dissolve 27.86 g Tris in 230 mL ddH_2O, and then add about 33.76 g MES to adjust pH to 8.0.
3. 1 M dithiothreitol (DTT): dissolve 15.425 g DTT in 100 mL ddH_2O, filter with a 0.45 μm sterile membrane, aliquot, and store at −20 °C.
4. 0.1 M phenylmethanesulfonyl fluoride (PMSF): dissolve 1.74 g PMSF in 100 mL methyl alcohol, aliquot and store at −20 °C.

5. Native buffer: 50 mM Tris–MES (pH 8.0), 0.5 M sucrose, 1 mM $MgCl_2$, 10 mM EDTA; sterilize by filtration and store at 4 °C. Add 5 mM DTT, 1 mM PMSF, protease inhibitor cocktail Complete Mini tablets (one tablet per 10 mL) (Roche), and 0.2% NP40 before use (*see* **Note 1**).

2.3 Protein Expression in E. coli and Protein Purification

1. *Escherichia coli* strain BL21 (DE3) (*see* **Note 2**).
2. Plasmid for expression of GST-tagged protein in *E. coli*: pGEX-6P-1 (*see* **Note 3**).
3. LB medium: dissolve 5 g yeast extract, 10 g tryptone, and 10 g NaCl in 1 L dH_2O. Autoclave and store at room temperature.
4. Isopropylthio-β-galactoside (IPTG): 100 mM stock. Dissolve in dH_2O, and filter with a 0.45 μm sterile membrane. Store at −20 °C.
5. 40% glucose: dissolve 40 g glucose in 100 mL dH_2O, filter with a 0.45 μm sterile membrane, and store at 4 °C (*see* **Note 4**).
6. GST binding buffer (GBB): 50 mM Tris–HCl (pH 8.0), 120 mM NaCl, 0.5% NP40. Add 1 mM DTT and 1 mM PMSF before use (*see* **Note 1**).
7. Beads for GST fusion protein purification: Glutathione Sepharose 4B beads (GE Healthcare).
8. Pressure machine: JN-mini (JNBIO Company).

2.4 Sodium Dodecyl Sulfate–Polyacrylamide Gel Electrophoresis (SDS-PAGE)

1. 30% acrylamide–bis-acrylamide (29:1) solution (Genestar); store at 4 °C.
2. 10% ammonium persulfate (APS): dissolve 1 g APS in 10 mL ddH_2O, and store at 4 °C.
3. *N,N,N′,N′*-tetramethylethylenediamine (TEMED) (AMRESCO); store at 4 °C.
4. 10% SDS: dissolve 1 g SDS in 10 mL ddH_2O, and store at room temperature.
5. 1.5 M Tris–HCl (pH 8.8): dissolve 181.6 g Tris base in 800 mL ddH_2O, adjust pH to 8.8 with HCl, and then fix the volume to 1 L with additional ddH_2O. Autoclave and store it at room temperature.
6. 1 M Tris–HCl (pH 6.8): dissolve 121.2 g Tris base in 800 mL ddH_2O, adjust pH to 6.8 with HCl, and then fix the volume to 1 L with additional ddH_2O. Autoclave and store it at room temperature.
7. 5 mL SDS-PAGE separating gel (10%): 1.9 mL ddH_2O, 1.7 mL 30% acrylamide–bis-acrylamide (29:1) solution, 1.3 mL 1.5 M Tris–HCl (pH 8.8), 50 μL 10% SDS, 50 μL 10% APS, 2 μL TEMED.

8. 2 mL SDS-PAGE stacking gel: 1.4 mL ddH_2O, 0.33 mL 30% acrylamide–bis-acrylamide (29:1) solution, 0.33 mL 1 M Tris–HCl (pH 6.8), 20 μL 10% SDS, 20 μL 10% APS, 2 μL TEMED.

2.5 Immuno precipitation

1. 10 mM MG132: dissolve 5 mg MG132 powder in 2.1 mL DMSO. Aliquot it in appropriate volume, store at −80 °C (*see* **Note 5**).
2. 0.45 μm sterile membrane.
3. Anti-Myc-tag mAb-Agrose (MBL), and GFP-Trap_MA (Dynabeads coated with anti-GFP antibody) (Chromotek).
4. A shaker to invert the samples in tubes.
5. A cold room at 4 °C.

2.6 In-Gel Digestion

1. 25 mM ammonium bicarbonate–50% acetonitrile buffer.
2. 50 mM ammonium bicarbonate.
3. 55 mM iodoacetamide in 50 mM ammonium bicarbonate.
4. Trypsin (Sigma).
5. 5% trifluoroacetic acid and 50% acetonitrile.
6. 0.1% formic acid.
7. 0.45 μm centrifugal filter.

3 Methods

Here, two strategies for identifying ubiquitination sites are introduced, including immunoprecipitation from transgenic plants and a cell-free assay combined with prokaryotic fusion protein expression system. The immunoprecipitation assay with transgenic *Arabidopsis* can mimic the event in vivo, while the cell-free assay is time-saving. The samples are separated on an SDS-PAGE gel before LC-MS/MS assay, since it can remove most chemical impurity from the samples. Both methods need LC-MS/MS assay to identify ubiquitination sites.

3.1 Plant Protein Expression and Extraction

1. 0.04 g seeds of the transgenic *Arabidopsis* plants expressing a tagged version (e.g., GFP, Myc) of the protein for which ubiquitination is investigated are sterilized, sowed in liquid 1/2 MS medium in a 500 mL flask, and cultured at 22 °C in 16-h light–8-h dark cycles under constant shaking; alternatively, they can be grown on 1/2 MS plates. Ten-day-old seedlings are collected and frozen in liquid nitrogen (*see* **Note 6**).
2. Seedlings are ground into fine powder in liquid nitrogen using mortar and pestle, and native extraction buffer is added at 2 mL/g tissue powder. Mix well and keep on ice for 5 min.

3. The plant extracts are centrifuged at 12,000 × *g* for 30 min at 4 °C. The supernatant is collected (*see* **Note 7**).
4. Filter the protein supernatant with a 0.45 μm sterile membrane into a precooled tube; the filtered supernatant is used for the following immunoprecipitation assay (*see* **Note 8**).

3.2 Immuno precipitation

1. Wash 40 μL agarose beads or Dynabeads coated with specific antibody (here, anti-Myc or anti-GFP antibody) three times with precooled PBS buffer supplemented with 1 mM PMSF (*see* **Note 9**).
2. Add the prepared beads into the plant extract from Subheading 3.1, and add MG132 to a final concentration of 50 μM (*see* **Note 5**).
3. Keep the samples gently rotating at 4 °C in order to facilitate binding of target protein by the beads. Rotate the samples at room temperature for 0.5 h and then transfer them to 4 °C and continue rotating for 4 h (*see* **Note 10**).
4. Collect the beads by centrifugation or using a magnetic rack, and wash the beads for six times (10 min each time) in pre-cooled PBS buffer with additional 150 mM NaCl (*see* **Note 11**).
5. Discard the buffer completely after the last wash; proteins are then eluted in 40 μL 0.5% SDS, 10 mM Tris–HCl (pH 6.8) at 95 °C for 5 min and diluted into 1 mL PBS buffer (*see* **Note 12**).
6. To enrich the specific target protein, reimmunoprecipitate the target protein with the same antibody as in **step 1** at room temperature (*see* **Note 12**).
7. Collect and wash the beads as in **step 4**; discard the buffer completely after the last wash: resuspend the beads in 30 μL 2×SDS protein loading buffer, and heat at 95 °C for 5 min.
8. Separate the samples on an SDS-PAGE gel.
9. The part of the lanes on the SDS-PAGE gel running above the target protein band (i.e., where ubiquitinated target protein is expected to run) is collected; proceed with in-gel digest and LC-MS/MS as described in Subheading 3.5 (*see* **Note 13**).

3.3 Expression of Proteins in Bacteria and Purification of Proteins

1. The plasmid for recombinant expression of GST-tagged target protein is transformed into *E. coli* strain BL21 (DE3). A single colony is chosen to inoculate 5 mL liquid LB medium; the culture is incubated at 37 °C overnight (*see* **Note 2**).
2. The overnight culture is diluted into 300 mL LB liquid medium to adjust the concentration to OD_{600} = 0.1; the 300 mL culture is incubated at 18 °C under constant shaking.
3. After about 6 h, check OD_{600}. When OD_{600} reached to 0.6–0.8, 0.8 mM IPTG is added to induce protein expression in bacteria; at this stage, also add glucose to a final concentration of 0.2%. The culture is kept shaking at 18 °C (*see* **Note 14**).

4. The bacteria cells are collected after 12–16 h by centrifugation at 6,000 × *g* for 10 min at 4 °C. Resuspend in 20 mL GBB buffer and break up the cells by a pressure machine or ultrasonic pulverizer. The supernatant is collected after centrifugation at 12,000 × *g* for 30 min at 4 °C, and separated into two new 15 mL tubes.
5. Wash 300 μL Glutathione Sepharose 4B beads three times with GBB, and dilute them into the 20 mL protein supernatant. After binding for 2 h at 4 °C, collect the beads by centrifugation at 6,000 × *g* for 30 s and discard the supernatant.
6. Wash the beads with GST-fused proteins with GBB three times for 10 min by rotating. After the last wash, discard the supernatant completely.
7. Resuspend an aliquot of the beads in 2×SDS protein loading buffer, and heat at 95 °C for 5 min; separate the sample on an SDS-PAGE and include bovine serum albumin (BSA) as reference.
8. Stain the SDS-PAGE gel by colloidal Coomassie brilliant blue (CBB) and quantify the GST-tagged protein by comparing with BSA as reference (*see* **Note 15**).
9. Aliquot the beads with GST-fused proteins in 10% glycerol and store at −80 °C.

3.4 Cell Free Assay

1. 200 ng GST-fused protein is added to 200 μL plant extracts (containing about 500 μg total proteins) and incubated for 30 min at room temperature with addition of MG132 at a final concentration of 50 μM, and then transferred to 4 °C for 4 h. The plant extracts are obtained from seedlings grown on 1/2 MS plates for 10 days as described in Subheading 3.1.
2. Wash Glutathione Sepharose 4B beads coated with target protein six times (10 min each time) by GBB with additional 150 mM NaCl (*see* **Note 11**).
3. Discard the buffer completely after the last wash; resuspend the beads in 30 μL 2×SDS protein loading buffer, and heat at 95 °C for 5 min.
4. Run the samples on an SDS-PAGE gel.
5. The part of the lanes on the SDS-PAGE gel running above the target protein band (i.e., where ubiquitinated target protein is expected to run) is collected for in-gel digestion (*see* **Note 13**).

3.5 In-Gel Digestion and LC-MS/MS Analysis

1. The gel sample containing ubiquitinated target protein from Subheadings 3.2 or 3.4 is cut into 2–3 mm^2 pieces; work in a dust-free environment.
2. The gel pieces are destained in 25 mM ammonium bicarbonate–50% acetonitrile buffer.

3. Incubate the gel pieces with 10 mM DTT in 50 mM ammonium bicarbonate at 56 °C for 1 h.
4. Incubate the gel pieces with 55 mM iodoacetamide in 50 mM ammonium bicarbonate in the dark for 45 min at room temperature.
5. Digest the gel pieces with trypsin (enzyme-to-substrate ratio 1:50) in 25 mM ammonium bicarbonate at 37 °C overnight.
6. Extract the peptides from the gel pieces with buffer containing 5% trifluoroacetic acid and 50% acetonitrile by ultrasonication (twice for 15 min).
7. The peptide-containing extract is freeze-dried in a SpeedVac concentrator; the peptides are redissolved in 0.1% formic acid and filtered by a 0.45 μm centrifugal filter before LC-MS/MS assay.
8. Analyze the samples by LC-MS/MS (*see* **Note 16**).

4 Notes

1. Stocks of DTT (1 M) and PMSF (100 mM) are unstable at 4 °C and room temperature, thus store them in small aliquots at −20 °C and avoid repeated freeze-thawing. They should be added just before the assay.
2. *Transetta* (DE3) is more efficient than standard expression strains for protein expression from genes that contain rare codons (including AUA, AGG, AGA, CUA, CCC, and GGA).
3. Other kinds of expression plasmids also can be used. It is important to ensure that the fusion protein can be efficiently expressed in bacteria and can function normally.
4. 40% glucose cannot be autoclaved, so it should be sterilized by filtering with a 0.45 μm sterile membrane.
5. MG132 is not stable at 4 °C and −20 °C; thus, aliquot it into small aliquots and store at −80 °C. In order to maintain full activity of MG132, MG132 should not be frozen and thawed more than twice.
6. Since agar can reduce the efficiency of protein extraction in native buffer, it is suggested to sow the *Arabidopsis* seeds in 1/2 MS liquid medium or on filter membranes plated on 1/2 MS agar plates.
7. As control, always include extracts from plants transformed with the empty vector. The control sample is necessary for immunoprecipitation assay.

8. To remove the debris, the extracted total proteins should be filtered with a 0.45 μm sterile membrane. This can reduce background in the immunoprecipitation assay.
9. Since the bead slurries often contain ethyl alcohol or sodium azide, they should be washed three times before use. The buffer should be discarded completely after the first wash in order to completely remove the ethyl alcohol or sodium azide.
10. To reduce protein degradation during immunoprecipitation, the assay is always performed at 4 °C. The incubation of total proteins at room temperature can facilitate the ubiquitination reaction, and increase the abundance of ubiquitinated target proteins. If the target protein is degraded rapidly at room temperature, it is suggested to incubate only at 4 °C.
11. To reduce the unspecific binding of protein, additional NaCl can be added in the wash buffer. The concentration of NaCl in the wash buffer can reach to 600 mM.
12. As we know, if proteins interacting with the target protein can be ubiquitinated, they will affect the LC-MS/MS analysis of the ubiquitination sites of the target protein. The SDS treatment followed by reprecipitation can remove the interacting proteins [10].
13. It is better to separate the samples on an SDS-PAGE gel and subject them to in-gel digestion than directly performing in-solution digestion. The SDS-PAGE process can remove most impurities in the IP samples. The samples should not be separated too much in the SDS-PAGE gel assay. The shorter the gel run, the less trypsin will be needed for the digest.
14. The IPTG working concentration is dependent on the specific plasmid and protein, and can range from 0.1 to 1 mM. It might be necessary to test different temperatures for expression to find the most appropriate condition.
15. Because there are always truncated target proteins and/or contaminating proteins, the best method of rough quantifying the concentration of full-length target protein is comparing the main band of target protein with the band of a known amount of BSA by colloidal CBB.
16. Ubiquitin chains cannot only be attached to lysine but also serine, threonine, and cysteine. Thus, it is important to consider this when analyzing the raw data from Q Exactive™ mass spectrometer.

Acknowledgments

This work was supported by grant 2016YFA0500500 from the National Key R&D Program of China.

References

1. Vierstra RD (2012) The expanding universe of ubiquitin and ubiquitin-like modifiers. Plant Physiol 160:2–14
2. Proietto M, Bianchi MM, Ballario P, Brenna A (2015) Epigenetic and posttranslational modifications in light signal transduction and the circadian clock in *Neurospora crassa*. Int J Mol Sci 16:15347–15383
3. Vierstra RD (2003) The ubiquitin/26S proteasome pathway, the complex last chapter in the life of many plant proteins. Trends Plant Sci 8:135–142
4. Smalle J, Vierstra RD (2004) The ubiquitin 26S proteasome proteolytic pathway. Annu Rev Plant Biol 55:555–590
5. Hua ZH, Vierstra RD (2011) The cullin-RING ubiquitin-protein ligases. Annu Rev Plant Biol 62:299–334
6. Vierstra RD (2009) The ubiquitin-26S proteasome system at the nexus of plant biology. Nat Rev Mol Cell Biol 10:385–397
7. Iconomou M, Saunders DN (2016) Systematic approaches to identify E3 ligase substrates. Biochem J 473:4083–4101
8. Mukhopadhyay D, Riezman H (2007) Proteasome-independent functions of ubiquitin in endocytosis and signaling. Science 315:201–205
9. Pickart CM, Eddins MJ (2004) Ubiquitin: structures, functions, mechanisms. BBA-Mol Cell Res 1695:55–72
10. Chen Q, Yang X, Xie Q (2016) Plant proteostasis: Approaches to determine protein ubiquitination residue types. Springer, New York. https://doi.org/10.1007/978-1-4939-3759-2

Chapter 7

Expression, Purification, and Spectral Characterization of Phytochromes

Yun-Jeong Han, Jae-Yong Cho, and Jeong-Il Kim

Abstract

Expression and purification of recombinant proteins are important for the structure–function study of phytochromes. However, it is difficult to purify phytochrome proteins from natural sources or using a bacterial expression system, due to the presence of multiple phytochrome species and low expression and solubility, respectively. Here we describe the expression of recombinant full-length plant phytochromes in the yeast *Pichia pastoris*, and the spectral analysis of chromophore-assembled phytochromes before and after the purification by streptavidin affinity chromatography.

Key words Phytochromes, Recombinant protein, *Pichia pastoris*, Streptavidin affinity chromatography, Difference spectrum

1 Introduction

Phytochromes are dimeric proteins with each monomer (120–130 kDa) possessing a covalently linked chromophore, and there are two spectrally distinct red-absorbing Pr and far-red-absorbing Pfr forms [1]. They are synthesized as the Pr form in the dark, and phototransformed into the Pfr form upon exposure to red light. Conversely, the Pfr form can be converted to the Pr form by the absorption of far-red light. The photoconversion between the Pr and Pfr forms is a unique spectral feature of phytochromes, which is known to be important for the regulation of light signaling in plants [2, 3].

Isolation and purification of a protein is important for its biochemical and functional analyses. Native phytochromes can be purified directly from plants [4]. However, the amount of phytochromes in plant tissues is low and the purification procedures are known to be difficult and something of an art. Especially, plants have multiple phytochrome species, which makes it difficult to purify a specific phytochrome species from natural resources. Thus, individual phytochrome species have

Andreas Hiltbrunner (ed.), *Phytochromes: Methods and Protocols*, Methods in Molecular Biology, vol. 2026,
https://doi.org/10.1007/978-1-4939-9612-4_7,

been expressed as recombinant forms in a few microorganisms such as *Escherichia coli*, *Saccharomyces cerevisiae*, and *Hansenula polymorpha* [5–8]. However, the expression level of recombinant phytochromes in those organisms is low. Especially in *E. coli*, plant phytochromes are expressed mostly in inclusion bodies (i.e., proteins in an insoluble and inactive form), although some phytochrome fragments have been obtained [9]. In contrast, full-length plant phytochromes have been successfully expressed in the methylotrophic yeast *Pichia pastoris* [10–13]. Here we describe the detailed procedures for the expression and purification of recombinant full-length plant phytochromes using the *Pichia* expression system.

P. pastoris can use methanol as carbon and energy sources [14]. However, alcohol oxidase (AOX) for the metabolism of methanol in *P. pastoris* is not so efficient (i.e., poor affinity to O_2) which the cell compensates by generating large amounts of AOX to survive in a methanol-enriched environment. Thus, the promoter for *AOX* transcription in *P. pastoris* is extraordinarily strong, which can make up to 30% of total soluble proteins in cells grown with methanol only. Thus, phytochrome coding sequences need to be cloned into a *Pichia* expression vector pPIC3.5K under the control of the *AOX* promoter. Then, the vector harboring a phytochrome coding sequence with an affinity tag for purification, such as a ten-amino acid streptavidin affinity-tag (*Strep-tag*) at the C-terminal end, is transformed into *P. pastoris* strain GS115. After selection of transformants with multicopy integration, recombinant phytochrome proteins can be expressed in minimal methanol media. Harvested cells are then broken using a homogenizer in liquid nitrogen (Fig. 1), and the phytochrome-containing protein fraction is obtained by ammonium sulfate precipitation. After in vitro reconstitution with a chromophore, phycocyanobilin (PCB), the recombinant phytochromes are purified by streptavidin affinity chromatography. During the purification steps, we analyzed spectral properties of the obtained recombinant phytochromes, such as absorption and difference spectra (Figs. 2, 3, and 4). Later, the chromophore ligation in recombinant phytochrome proteins was confirmed by a zinc fluorescence assay after SDS-PAGE (sodium dodecyl sulfate–polyacrylamide gel electrophoresis) (Fig. 5). Using the methods described here, we could successfully express and spectrally analyze the recombinant proteins of phytochrome A (phyA) from *Avena sativa* (oat), *Pisum sativum* (pea), and *Brachypodium distachyon* (a monocot model plant), and those of type II phytochromes such as *Arabidopsis thaliana* phytochromes B and D.

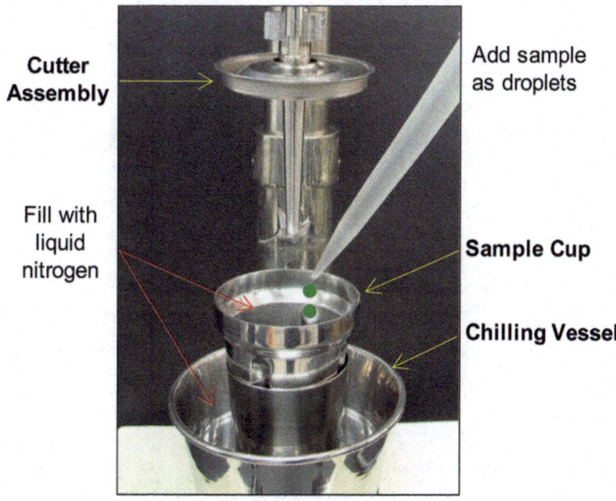

Fig. 1 A homogenizer to break the *Pichia* cells. The *Pichia* cells were broken in liquid nitrogen to prevent heat generation during homogenization. For this, the sample cup and chilling vessel were filled with liquid nitrogen, and the cell suspension was put into the sample cup as droplets (shown in green dots) using a 10 mL pipetman

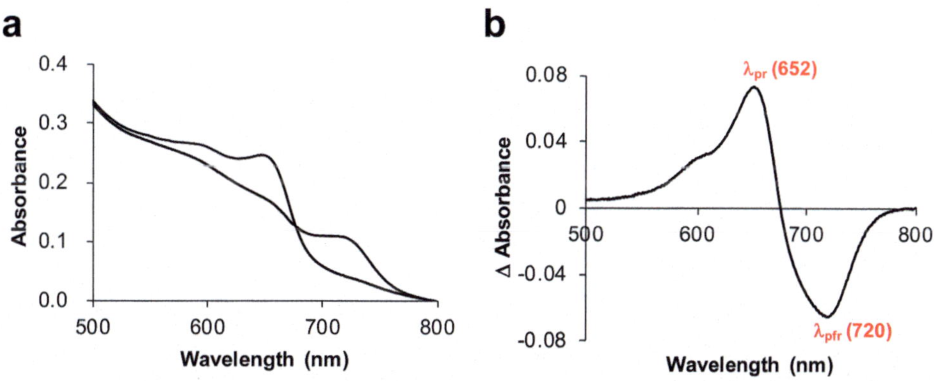

Fig. 2 Spectral analysis of ammonium sulfate-precipitated proteins assembled with PCB. (**a**) Absorption spectra of Pr and Pfr forms. After in vitro reconstitution and dialysis under green safety light, the Pr absorption spectrum was measured (solid line). After illumination with red light, the Pfr spectrum (dotted line) was obtained. (**b**) A difference spectra using the absorption spectra in (**a**). The difference spectrum (ΔAbsorbance) was obtained by subtracting the Pfr spectrum from the Pr spectrum. Maximum absorption peaks of Pr (λ_{Pr}) and Pfr (λ_{Pfr}) are recorded at 652 nm and 720 nm, respectively. The ratio of the heights (A_{min}/A_{max}) is ~0.96, and the difference of inter-peak distance (Δ of maxima) is 68 nm

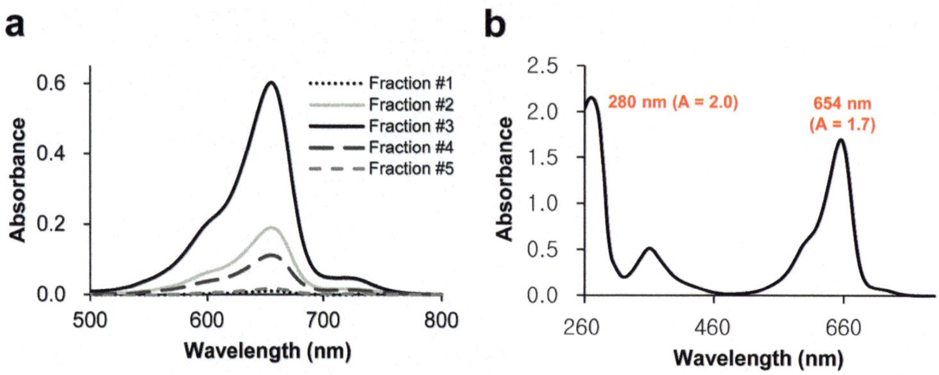

Fig. 3 Spectral analysis of purified recombinant phytochromes. (**a**) Absorption spectra of elution fractions. After collecting elution fractions (0.5 mL each), absorption spectra of each elution fraction were obtained. Fractions #2–4 usually contained recombinant phytochromes, with concentrations being the highest in fraction #3. (**b**) A representative absorption spectrum of an elution fraction with high concentration of *Avena sativa* phyA (AsphyA). The specific absorbance ratio (SAR) of this protein sample is 0.85 (= 1.7/2.0). If it is necessary to obtain phytochromes with a higher purify, gel filtration can be performed in the next purification step

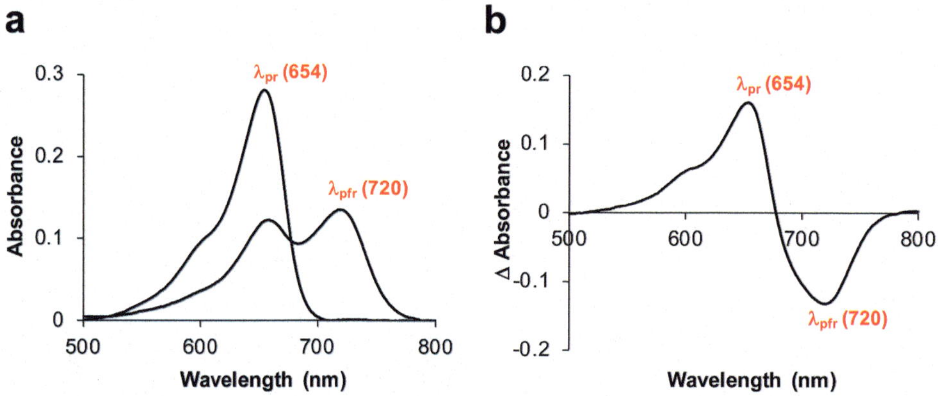

Fig. 4 Absorption and difference spectra of purified recombinant phytochromes. (**a**) Absorption spectra of the Pr and Pfr forms of purified phytochromes. Maximum absorption peaks of Pr (solid line, λ_{Pr}) and Pfr (dotted line, λ_{Pfr}) are recorded at 654 nm and 720 nm, respectively. The ratio of the peak maxima (A_{Pfr}/A_{Pr}) is 0.48. (**b**) A difference spectrum using the absorption spectra in (**a**). Maximum absorption peaks are same as those in absorption spectra

2 Materials

Prepare all solutions using ultrapure deionized water (DW; 18 MΩ·cm at 25 °C) and analytical or tissue culture-grade reagents.

2.1 Linearization of Plasmid DNA

1. pPIC3.5K (Invitrogen) plasmid harboring a phytochrome coding sequence (at least 10 μg).
2. Restriction enzyme for linearization: *Pme*I for *AtPHYB*.

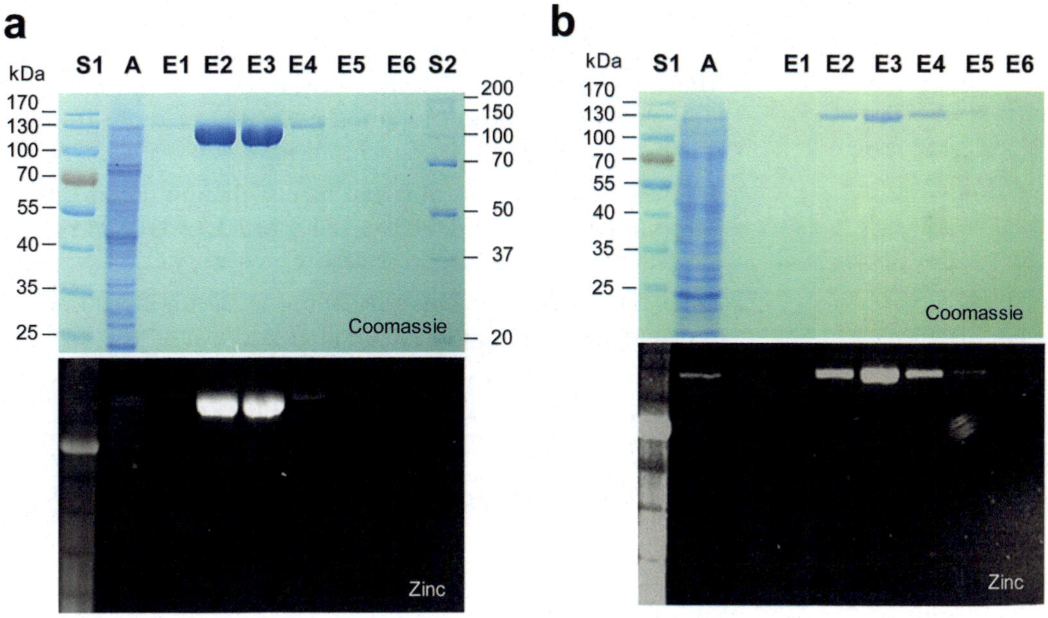

Fig. 5 Zinc fluorescence analysis of recombinant phytochromes. Coomassie blue stained gel and zinc fluorescence of AsphyA (**a**) and AtphyB (**b**) proteins. S1 and S2, protein size standards from Thermo Fisher Scientific and Bio-Rad, respectively; A, ammonium sulfate-precipitated proteins; E1–E6, elution fractions from streptavidin affinity chromatography. In the presence of zinc acetate, phytochromes form a complex that emits fluorescence in the orange region of the spectrum when viewed under ultraviolet light. Thus, chromophore ligation in recombinant phytochromes can be confirmed by observing zinc fluorescence

3. 5 M NaCl (146.1 g/500 mL in DW), absolute ethanol, and 70% ethanol.

2.2 Growth of Pichia Cells for Electroporation

1. *P. pastoris* strain GS115 (Invitrogen).
2. 20% Dextrose: dissolve 200 g dextrose in 1 L of DW, and autoclave for 15 min.
3. YPD (yeast extract–peptone–dextrose) medium: Dissolve 10 g yeast extract and 20 g of peptone in 900 mL of water. After autoclaving, add 100 mL of 20% dextrose.
4. YPD plate: Dissolve 10 g yeast extract, 20 g of peptone, and 15 g agar in 900 mL of water. After autoclaving, add 100 mL of 20% dextrose and pour into petri dishes.
5. 1 M sorbitol: dissolve 91.085 g D-sorbitol in 500 mL of DW, and autoclave.

2.3 Electroporation and Selection of Pichia Cells with Multicopy Inserts

1. An electroporator, and 0.2 cm electroporation cuvettes.
2. MD (Minimal Dextrose)/−His plate (MD plate without histidine): 26.7 g Minimal SD (synthetic defined) Base, 0.77 g −His DO (dropout) Supplement, and 1.5% agar in 1 L DW. Autoclave and pour into petri dishes.

3. YPD-G418 (Geneticin) plate: Dissolve G418 disulfate (Duchefa Biochemie) in DW to a final concentration of 100 mg/mL and sterilize the solution by 0.20 μm filtration. Add the G418 solution to autoclaved YPD medium to make 4 mg/mL or 2 mg/mL concentration.
4. 10× YNB (13.4% yeast nitrogen base with ammonium sulfate without amino acids): Dissolve 134 g in 1 L of DW and sterilize using 0.20 μm filter. Store at 4 °C.
5. 500× Biotin (0.02%): Dissolve 20 mg D-Biotin in 100 mL of DW and sterilize using 0.20 μm filter. Store at 4 °C.
6. MM (Minimal Methanol) medium: For 10 mL, mix 1 mL of 10× YNB, 20 μL of 500× D-biotin, 100 μL of methanol, and 8.88 mL of autoclaved DW.
7. 30% glycerol: Mix 60 mL of glycerol with 140 mL of water and autoclave.

2.4 Induction of Recombinant Phytochrome Expression in *P. pastoris*

1. 20× glycerol (20%): Mix 40 mL of glycerol with 160 mL of water, and autoclave.
2. MGY (Minimal Glycerol) medium: For 200 mL, combine aseptically 169.6 mL of autoclaved DW in a 1 L baffled flask with 20 mL of 10× YNB, 0.4 mL of 500× biotin, and 10 mL of 20× glycerol.
3. MM medium for 500 mL: In 444 mL of autoclaved DW in a 2 L baffled flask, add 50 mL of 10× YNB, 1 mL of 500× biotin, and 5 mL of methanol.

2.5 Preparation of Recombinant Phytochrome Proteins from *P. pastoris* cells

1. A homogenizer with a sample cup and a chilling vessel (*see* Fig. 1 and **Note 1**).
2. TE buffer: 100 mM Tris (pH 8.0) and 1 mM EDTA.
3. Stock solutions of protease inhibitors: 0.1 M PMSF (phenylmethylsulfonyl fluoride) in isopropanol, 10 mg/mL leupeptin in DW, and 1 mg/mL pepstatin in methanol.

2.6 Purification of Phytochromes by Streptavidin Affinity Chromatography

1. Green safety light: fluorescent lamps equipped with plastic filters having a maximal transmittance at 500 nm.
2. Chromophore for in vitro reconstitution to generate holophytochrome: PCB (phycocyanobilin; *see* **Note 2**).
3. Dialysis membrane: Molecular weight cutoff (MWCO) = 12,000–14,000 Da.
4. Dialysis buffer: 20 mM Tris (pH 8.0) and 0.5 mM EDTA.
5. Gravity flow Strep-Tactin® Sepharose Column (ready-to-use, 1 mL; IBA Lifesciences).
6. Elution buffer: 20 mM Tris (pH 8.0), 0.5 mM EDTA, and 5 mM desthiobiotin (Sigma-Aldrich).

2.7 Spectral Analysis and Zinc Fluorescence Assay

1. A double beam UV-Vis spectrophotometer, and two semi-micro quartz cuvettes.
2. A fiber optic illuminator system equipped with 666 and 750 nm interference filters: The light intensities from this system are 8 W/m^2 for red light and 6 W/m^2 for far-red light.
3. Zinc solution: 150 mM Tris–HCl (pH 7.0) and 20 mM Zinc acetate.
4. Coomassie gel staining solution: Dissolve 2 g of Coomassie brilliant blue in the solution containing 450 mL of DW, 450 mL of methanol, and 100 mL of acetic acid.
5. Destaining solution: Mix 500 mL of DW, 400 mL of methanol, and 100 mL of acetic acid.

3 Methods

Here, we used the nomenclature system of phytochromes as described previously [15]. For example, we designated the chromophore-assembled holophytochrome B as phyB, and the phytochrome B gene/coding sequence as *PHYB*. The methods described here begin with the transformation of pPIC3.5K vector carrying a phytochrome coding sequence into *P. pastoris* cells. This is because the cloning strategy into pPIC3.5K will vary depending on the phytochrome to be expressed. As an example, the cloning strategy for *A. thaliana PHYB* coding sequence (*AtPHYB*) with *Strep-tag* was summarized in **Note 3**.

3.1 Linearization of Plasmid DNA

1. Digest approximately 10 μg of pPIC3.5K plasmid harboring the coding sequence of a phytochrome with a restriction enzyme cutting in the vector backbone. For example, digest pPIC3.5K carrying *AtPHYB* with *Pme*I (*see* **Note 4**).
2. Precipitate the linearized plasmid DNA by adding 10% (v/v) of 5 M NaCl and 2 volumes of absolute ethanol and incubating for 20 min at −20 °C. Harvest DNA pellet by centrifugation at 15,000 × *g* for 10 min at 4 °C.
3. Remove the supernatant and wash with 1 mL of 70% ethanol. Centrifuge at 15,000 × *g* for 5 min at 4 °C and remove all of the supernatant. Store the open tube at room temperature until the ethanol has evaporated (usually for 5 min).
4. Dissolve the DNA pellet with 10 μL of sterile DW, and store at −20 °C for the following *Pichia* transformation.

3.2 Growth of Pichia Cells for Electroporation

1. Streak *P. pastoris* strain GS115 (hereafter, GS115) onto an YPD plate and incubate the plate at 30 °C for 3 days (*see* **Note 5**).

2. Inoculate 10 mL of YPD medium in a 50 mL sterile conical tube with a single colony of GS115, and grow overnight at 30 °C with vigorous shaking (250–300 rpm) (*see* **Note 6**).
3. Inoculate 400 mL of fresh YPD medium in a 2 L baffled cell culture flask with 400 μL of cells from the culture made in **step 2**, and incubate at 30 °C overnight (usually 13–15 h) with vigorous shaking (*see* **Note 7**).
4. Using two sterile 250 mL centrifuge bottles, harvest the cells from the culture that has an OD_{600} (i.e., absorbance at 600 nm) between 1.2 and 1.4 by centrifugation at 3000 × *g* for 5 min at 4 °C. Discard the supernatant and resuspend cell pellet with 400 mL of ice-cold, sterile DW (*see* **Note 8**).
5. Pellet the cells by centrifugation as in **step 4**. Discard the supernatant and resuspend cell pellet with 200 mL of ice-cold, sterile DW.
6. Centrifuge as in **step 4**. Discard the supernatant and resuspend cell pellet with 16 mL of ice-cold, 1 M sorbitol.
7. Centrifuge as in **step 4**. Discard the supernatant completely, and resuspend cell pellet with 0.8 mL of ice-cold, 1 M sorbitol.
8. Use the electrocompetent cells for electroporation immediately (*see* **Note 9**).

3.3 Electroporation and Selection of Pichia Cells with Multicopy Inserts

1. Put 80 μL of the electrocompetent cells into a 0.2 cm electroporation cuvette that has been chilled on ice, and add 10 μL of the linearized plasmid DNA. Mix the suspension by tapping, and stand the cuvette on ice for 5 min.
2. Set an electoroporator, such as MicroPulser™ Electoroporator (Bio-Rad), for *Pichia* transformation (voltage = 2.0 kV). Place the cuvette in the chamber and pulse once.
3. Remove the cuvette from the chamber and immediately add 0.9 mL of ice-cold, 1 M sorbitol to the cuvette. Mix gently.
4. Spread the cells in the cuvette on three MD/-His plates (~330 μL/plate), and incubate at 30 °C for 2 days (*see* **Note 10**).
5. Using a sterile spreader and a pipetman, collect the colonies from the plates using 5 mL of sterile DW (repeat one more; i.e., collect cells with 10 mL of DW), and transfer the cell suspension into a sterile 50 mL conical tube (*see* **Note 11**).
6. Vortex for 1 min, and spread 50–100 μL of cell suspension on YPD plates containing 4 mg/mL of G418 (*see* **Note 12**).
7. Incubate plates at 30 °C and check daily. G418-resistant colonies will take 2–5 days to appear. Then, streak early-appearing G418-resistant cells (at least four cells) onto YPD plates con-

taining 2 mg/mL of G418, and incubate the plates at 30 °C for 2–4 days (*see* **Note 13**).

8. Select single colonies from each streaked plate and inoculate 10 mL of MM medium in a sterile 50 mL conical tube. Grow them at 30 °C for 2–3 days with vigorous shaking (*see* **Note 14**).
9. Harvest the cells by centrifugation at 3000 × *g* for 5 min at 4 °C, and resuspend cell pellet with 0.5 mL of 30% glycerol (for the 10 mL culture of MM medium). Freeze the cells in ~200 μL aliquots in liquid nitrogen, and store at −80 °C until use for protein expression.

3.4 Induction of Recombinant Phytochrome Expression in P. pastoris

1. Thaw the cells stored in −80 °C on ice and inoculate 10 mL of MM medium in a 50 mL sterile conical tube with 10 μL of the thawed cells. Incubate the cells in MM medium overnight (not more than 24 h) at 30 °C with vigorous shaking (*see* **Note 15**).
2. Inoculate 200 mL of MGY medium in a 1 L baffled cell culture flask with cells from the culture made in **step 1** (usually, make two cultures), and incubate at 30 °C with vigorous shaking until culture reaches an OD_{600} = 2–6 (approximately 16–18 h) (*see* **Note 16**).
3. Using sterile 250 mL centrifuge bottles, harvest the cells from the culture that has an OD_{600} between 2 and 6 by centrifugation at 3000 × *g* for 5 min at room temperature. Decant carefully and remove the supernatant completely, and resuspend cell pellet with 500 mL of MM medium in a 2 L baffled cell culture flask (usually, make two 500 mL cultures of MM medium from the 200 mL culture of MGY medium). During the suspension, adjust OD_{600} = 0.8–1.0. Cover the flask with two layers of sterile cheesecloth for aeration and return to incubator to continue growth (*see* **Note 17**).
4. After exactly 24 h of induction in the MM medium, harvest the cells with 500 mL centrifuge bottles by centrifugation at 3000 × *g* for 5 min at 4 °C (two centrifugation, ~250 mL each). Remove the supernatant carefully, and resuspend cell pellets with 40 mL of ice-cold DW. Transfer the resuspended cells to 50 mL conical tubes, and centrifuge at 3000 × *g* for 5 min at 4 °C (*see* **Note 18**).
5. Remove the supernatant completely, and store the harvested cells in 50 mL conical tubes at −80 °C before use for protein preparation (*see* **Note 19**).

3.5 Preparation of Recombinant Phytochrome Proteins from P. pastoris Cells

1. Thaw the harvested cells stored at −80 °C on ice (usually two conical tubes, each containing the harvested cells from 500 mL culture of MM medium). Meanwhile, set up a homogenizer

for cell break and cool it with liquid nitrogen (Fig. 1) (*see* **Note 20**).

2. Add 10 mL of TE buffer to each conical tube and resuspend the cells thoroughly by vortexing.
3. Add protease inhibitors (100 μL of 0.1 M PMSF, 10 μL of 10 mg/mL leupeptin, 10 μL of 1 mg/mL pepstatin) to one conical tube (~10 mL of cell suspension), and put the suspension as droplets using a 10 mL pipetman into the sample cup of the homogenizer.
4. Break the *Pichia* cells by homogenization at 5,000 × *g* for 5 min in liquid nitrogen, and thaw the homogenized cells quickly at 40 °C in a water bath circulator.
5. Add 100 μL of 0.1 M PMSF, and put the cell suspension as droplets to the sample cup as in **step 3**.
6. Homogenize as in **step 4**. Thaw and transfer the homogenized cells to a 50 mL centrifuge tube. Store the tube on ice for **step 8**.
7. Repeat **steps 3–6** with another conical tube containing cell suspension.
8. Obtain crude cell extracts by centrifugation at 27,000 × *g* for 30 min at 4 °C.
9. Measure the volume of the supernatant from two centrifuge tubes by transferring to a new 50 mL centrifuge tube and add ammonium sulfate (0.23 g/mL of the supernatant). Dissolve ammonium sulfate completely by gentle mixing and store in 4 °C until use for spectral analysis or protein purification (*see* **Note 21**).

3.6 Purification of Phytochromes by Streptavidin Affinity Chromatography

1. Harvest the ammonium sulfate-precipitated proteins by centrifugation at 15,000 × *g* for 5 min at 4 °C.
2. Remove the supernatant completely, and dissolve protein pellet with 4 mL of TE buffer (for the protein sample from 1 L culture of MM media) on ice using a 1 mL pipetman (*see* **Note 22**).
3. Centrifuge at 15,000 × *g* for 10 min at 4 °C, and measure the volume of the supernatant by transferring to a 15 mL conical tube on ice. From this step, experiments should be carried out under green safety light.
4. Add chromophore (i.e., PCB) to a final concentration of 10 μM for in vitro reconstitution, and incubate for 1 h on ice (*see* **Note 23**).
5. Carry out dialysis with 2 L of dialysis buffer for 2 h (*see* **Note 24**).

6. Filtrate the dialyzed solution with 0.45 μm filter to a new 15 mL conical tube. If necessary, save some amount for SDS-PAGE and Zinc fluorescence analysis (*see* **Note 25**).
7. Load the filtrated solution (usually ~4.5 mL) to a 1 mL ready-to-use Strep-Tactin® Sepharose Column that is equilibrated with TE buffer (*see* **Note 26**).
8. After loading the entire protein sample, wash the column thoroughly with TE buffer until the absorbance at 280 nm is less than 0.01.
9. Elute the bound proteins with 0.5 mL of elution buffer. Collect 0.5 mL of the eluate in at least six 1.5 mL tubes. If necessary, save part of each fraction for SDS-PAGE and zinc fluorescence analysis.
10. Scan the absorbance of each fraction, and if necessary, also scan the Pr and Pfr spectra. Freeze the elution fractions in liquid nitrogen and store in −80 °C.

3.7 Spectral Analysis of Recombinant Phytochrome Proteins

1. With the ammonium sulfate-precipitated proteins incubated with PCB or the elution fractions from streptavidin affinity chromatography, the absorption spectrum is obtained by scanning 800–500 nm or 800–260 nm using a UV-Vis spectrophotometer. This spectrum will represent the Pr form of phytochromes, if the samples have not been exposed to light before the spectral measurement (Figs. 2 and 3).
2. Illuminate with red light for 15 min on ice, and obtain the absorbance spectrum as in **step 1**. The spectrum obtained here represents the Pfr form of phytochromes (Fig. 4a).
3. Illuminate with far-red light for 5 min on ice, and obtain the absorbance spectrum as in **step 1**. The spectrum obtained represents the Pr form of phytochromes (Fig. 4a) (*see* **Note 27**).
4. From Pr and Pfr absorption spectra, a difference spectrum (i.e., ΔAbsorbance) is obtained by subtracting the Pfr spectrum from the Pr spectrum (Figs. 2b and 4b).
5. From the absorption and difference spectra of Pr and Pfr forms, spectral characteristics of phytochromes can be analyzed, which includes maximum absorption peaks (λ_{Pr} and λ_{Pfr}), the ratio of the heights in difference spectra (A_{min}/A_{max}), difference of interpeak distance ($\lambda_{Pfr} - \lambda_{Pr}$; Δ of maxima), the ratio of the peak maxima on the absorbance spectra (A_{Pfr}/A_{Pr}), and specific absorbance ratio (SAR; A_{Pr}/A280, ratio of the absorbance at Pr peak to absorbance at 280 nm) (Figs. 2-4).

3.8 Zinc Fluorescence Assay to Assess Chromophore Ligation in Phytochromes

1. Prepare the samples (usually, ammonium sulfate-precipitated proteins and elution fractions) for SDS-PAGE by adding 5× SDS loading buffer. Boil the samples at 100 °C for 3–5 min, and spin down the samples.
2. Prepare for 10% resolving gel and 4% stacking gel. Load the boiled samples onto the SDS-PAGE gel and run until the dye front reached the bottom of the gel (*see* **Note 28**).
3. Following the SDS-PAGE, transfer the gel into zinc solution and incubate for 5–10 min at room temperature with gentle shaking (*see* **Note 29**).
4. Visualize zinc fluorescence under UV light (312 nm) using Polaroid camera with a red filter or using a gel imaging instrument such as Gel Doc™ EZ Imager (Bio-Rad) (Fig. 5).
5. Wash the gel with DW and stain with Coomassie Brilliant Blue for 30 min at room temperature with gentle shaking.
6. Destain the gel by soaking it in the destaining solution at room temperature with gentle shaking.
7. Take a picture with the destained gel or scan after drying the gel.

4 Notes

1. When *P. pastoris* cells are grown in methanol media, it is difficult to break the cells efficiently using routine methods (*see* also **Note 20**). We found that a physical breaking method in liquid nitrogen using a homogenizer (Model no. AM-3 from Nihonseiki Kaisha Ltd.) is effective to obtain the crude extract of *P. pastoris* cells containing recombinant phytochrome proteins, although it is not so efficient. If this kind of homogenizer is not available, grinding the droplets of cell suspension in liquid nitrogen using mortar and pestle is an alternative method.
2. Phycocyanobilin (PCB) was prepared from lyophilized *Spirulina platensis* by methanolysis, as previously described [16]. Alternatively, it can be purchased from Frontier Scientific, Inc.
3. For expression and purification of recombinant proteins of AtphyB, we first cloned *ATPHYB* coding sequence (3519 bp) into pASK75 (3266 bp) to attach *Strep-tag* at the C-terminal end using *Sma*I and *Eco*47III, and then cloned it into pPIC3.5K (9.0 kb) using *Sna*BI and *Not*I. The *Strep-tag* is composed of ten amino acids (Ser-Ala-Trp-Arg-His-Pro-Gln-Phe-Gly-Gly) that can bind to streptavidin. Thus, a protein with *Strep-tag* can be purified by affinity chromatography using streptavidin-bound resin. The final construct of pPIC3.5K harboring

AtPHYB has the following sequence, 5′-*Sna*BI-ACCATGG-*AtPHYB*-**TAT***AGCGCTTGGCGTCACCCGCAGTTCGGTGGT*TAA-*Not*I-3′. The start and stop codons are underlined, and the last amino acid (tyrosine) of AtphyB is shown in bold. A Kozak sequence including the start codon (ACCATGG) is necessary for protein expression in the eukaryotic system, and the *Strep-tag* sequence is shown in italics. Of note, the pASK75 vector containing the *Strep-tag* was originally purchased from Biometra (Göttingen, Germany), which is no longer to be purchased. Vectors with *Strep-tag* can now be purchased from IBA Lifesciences: for example, pASG and pPSG vectors. Alternatively, the *Strep-tag* can be added to a gene by PCR using primers that contain the coding sequence of the tag.

4. Linearization of plasmid DNA is necessary to stimulate homologous recombination between the transforming DNA and regions of homology within the *Pichia* genome. Due to the limitation of available restriction enzyme sites in phytochrome genes, *Pme*I was chosen to make single cut in the 5′*AOX* promoter for homologous recombination. To obtain sufficient number of *Pichia* transformants, more than 10 μg of plasmid DNA should be linearized.
5. Incubation period is usually 2–4 days, depending on the status of the GS115 cell used. Once single colony was obtained, move to the next step.
6. For aeration, do not close the cap of the conical tube tightly. Use tape to fix the cap loosely closed.
7. The use of baffled flask is recommended for better aeration during the culture.
8. The cell density (OD_{600} = 1.2–1.4) is important for transformation efficiency. For electroporation, a cell density of 5–7 × 10^7 cells/mL is recommended. Considering that OD_{600} = 1 is approximately 5 × 10^7 cells/mL in *P. pastoris*, OD_{600} = 1.2–1.4 corresponds to 6.0–7.0 × 10^7 cells/mL.
9. Sometimes, it is recommended to freeze the electrocompetent cells in 80 μL aliquots. However, the transformation efficiency will decrease significantly. Thus, only fresh electrocompetent cells should be used for the multiple copy integration of phytochrome expression constructs into the *Pichia* genome.
10. Incubation period can be varied (usually 1.5–2 days). The important thing is that colonies with tiny sizes (usually less than 0.2 mm) should be collected for next steps. In addition, there should be thousands of colonies on the three plates in total. Thus, check the growing colonies frequently, and move to the next step when tiny colonies have appeared.

11. The collected cells can be stored at 4 °C for a month and used again for further selection of the *Pichia* cells with multicopy inserts, if necessary.
12. The volume varies depending on cell density. It is recommended to spread ~10^5 cells/plate for selection (for example, 100 μL of cells with 0.1 of OD_{600}). Usually, three G418-containing plates are used for selection of *Pichia* transformants with each construct. It is known that the level of Geneticin (G418) resistance depends on the number of vectors integrated. A single copy of pPIC3.5K integrated into the *Pichia* genome confers resistance to G418 to a level of ~0.25 mg/mL, and multiple integrated copies can increase the G418 resistance level from 0.5 mg/mL (1–2 copies) up to 4 mg/mL (7–12 copies). To obtain sufficient amount of recombinant phytochromes, the *Pichia* cells with multicopy inserts are essential, so only transformants growing on YPD plates containing 4 mg/mL of G418 are selected for phytochrome expression.
13. For saving time, single colony isolation can be performed using YPD plates containing 2 mg/mL of G418, instead of 4 mg/mL. Incubation period varies depending on the status of cells. Once single colony was obtained, move to the next step.
14. The growth period varies depending on the status of cells. For making a cell stock, cells in log-phase (OD_{600} = 2–6) should be harvested. Check the cell growth frequently and harvest when the cells reach at the log-phase.
15. It is better to finish the seed culture within 24 h. If the growth time of cells to reach the log-phase is over 24 h, the expression level of recombinant phytochromes can be decreased. If a fresh cell stock is used, 10 μL of inoculation is enough for the cells to reach the log-phase within 24 h. However, if the cell stock is old (usually stored for several months), the volume of inoculation can be increased up to 30 μL. For better aeration, close the cap loosely and use tape at one side to fix the cap to the conical tube.
16. The cells in the log-phase are best for induced expression of recombinant phytochromes. The growth phase can be estimated by measuring OD_{600}. If OD_{600} is between 2 and 6, the cells are in the middle of the log-phase growth.
17. The cell density is important for high induction of recombinant phytochromes. The final OD_{600} should be approximately 0.8 (not more than 1). If the OD_{600} of cells in 200 mL of the MGY medium is 4, the OD_{600} of cells in two 500 mL cultures of MM medium (= 1000 mL in total) will be approximately 0.8.

18. The induction time is important for the lysis of induced cells. If the *Pichia* cells are grown in MM media for more than 24 h, it is hard to break the cells, although the expression level of phytochrome protein might increase. Thus, harvest the cells at or before 24 h after the induction of recombinant proteins in MM media.

19. If the protein induction procedures are done without problems, the wet volume of the harvested cells from the 500 mL culture of MM medium is approximately 5 mL. Sometimes, the harvested cell volume is over 6 mL, but in this case the expression level of recombinant phytochromes is usually reduced.

20. It is not easy to break the *Pichia* cells efficiently. It might be due to the cell wall structure of the cells grown with methanol as a carbon source. Thus, the *Pichia* cells cannot be broken with traditional methods, including freeze/thaw cycles, vortexing with glass beads, digestion of cell wall with Zymolyase in combination with ultrasonication, French press and Dounce homogenizer, and so on. To break the *Pichia* cells, a method combining mechanical disruption in liquid nitrogen with freeze/thaw cycles is suitable using a homogenizer which contains a cup for sample and a chilling vessel for cooling with liquid nitrogen (*see* Fig. 1). In this method, cell suspension can be added as small droplets to the cup containing liquid nitrogen, which make the cell suspension as solid spheres. In addition, homogenization can be carried out without heat generation, because the sample cup and chilling vessel are filled with liquid nitrogen. Therefore, the *Pichia* cell break is performed as follows: freeze (cells stored at −80 °C)—thaw (for cell suspension)—freeze (added into liquid nitrogen)—homogenization (i.e., mechanical disruption)—thaw (for cell suspension)—freeze (added into liquid nitrogen)—homogenization—thaw.

21. The incubation time for protein precipitation with ammonium sulfate is usually for 2 h or overnight. It is recommended to use the phytochrome proteins precipitated with ammonium sulfate within 3 days. For purification of recombinant proteins, the ammonium sulfate-precipitated proteins from at least 1 L culture of MM medium are necessary to obtain absorption and difference spectra of purified phytochromes (Figs. 3–4). However, for spectral characterization to compare wild-type and mutant phytochromes, the ammonium sulfate-precipitated proteins from 500 mL culture of MM medium are enough to obtain difference spectra (Fig. 2).

22. Usually, 4 mL of TE buffer is added to dissolve the ammonium sulfate-precipitated proteins from 1 L culture of MM media for the purification of phytochromes. However, to get absorption spectra of crude phytochrome samples (i.e., not for the purification purpose), we recommend to prepare ammonium sulfate-precipitated proteins from 500 mL culture of MM media and add 1 mL of TE buffer to dissolve the proteins. Then, transfer to a 1.5 mL tube, centrifuge at 15,000 × *g* for 10 min at 4 °C and use the supernatant (i.e., ammonium sulfate fraction) for spectral analysis after chromophore ligation.
23. PCB is light-sensitive, so it should be wrapped with foil for storage at −80 °C. Addition of the chromophore to the ammonium sulfate fraction is optimal for reconstitution and free chromophore is removed during purification, which is advantageous compared to the addition of the chromophore to purified apophytochromes.
24. For spectral characterization of wild-type and mutant phytochromes by measuring difference spectra, the dialysis step is not necessary. Ammonium sulfate-precipitated proteins incubated with PCB can be used directly for the spectral analysis.
25. It is recommended to scan the absorbance of the dialyzed solution (from 800 to 500 nm) (Fig. 2). This is necessary to confirm the presence of chromophore-assembled phytochrome before the purification step by affinity chromatography.
26. It is better to load ~1 mL of protein sample to the column at once, and continue the loading until all the filtrated solution is used. Although other streptavidin-conjugated resins (e.g., streptavidin–agarose) can be used, phytochrome proteins with high purity could be obtained when the prepacked Strep-Tactin® Sepharose Column (1 mL) was used.
27. It is known that the Pfr forms are less stable than the Pr forms. Thus, phytochrome samples should be stored as Pr after this step.
28. SDS-PAGE is a standard method and therefore no detailed procedure is described here.
29. Zinc fluorescence assay was used to confirm the PCB ligation in the recombinant phytochrome proteins. In the presence of zinc acetate, bilin-containing proteins (such as PCB-assembled phytochromes) can form a complex of the bilin chromophore and the zinc ion, which fluoresces in the orange region of the spectrum when viewed under UV light. The fluorescence can be visualized in SDS-PAGE gels to confirm ligation of the bilin chromophore by the phytochrome proteins.

Acknowledgments

This research was supported by Basic Science Research Program through the National Research Foundation of Korea (NRF) funded by the Ministry of Education, Science and Technology (grant no. 2017R1A2B4010349) and in part by Next-Generation BioGreen 21 Program from Rural Development Administration, Republic of Korea (grant no. PJ01332701).

References

1. Rockwell NC, Su YS, Lagarias JC (2006) Phytochrome structure and signaling mechanisms. Annu Rev Plant Biol 57:837–858
2. Li J, Li G, Wang H, Wang Deng X (2011) Phytochrome signaling mechanisms. Arabidopsis Book 9:e0148
3. Xu X, Paik I, Zhu L, Huq E (2015) Illuminating progress in phytochrome-mediated light signaling pathways. Trends Plant Sci 20:641–650
4. Lapko VN, Song PS (1995) A simple and improved method of isolation and purification for native oat phytochrome. Photochem Photobiol 62:194–198
5. Hill C, Gartner W, Towner P, Braslavsky SE, Schaffner K (1994) Expression of phytochrome apoprotein from *Avena sativa* in Escherichia coli and formation of photoactive chromoproteins by assembly with phycocyanobilin. Eur J Biochem 223:69–77
6. Mozley D, Remberg A, Gartner W (1997) Large-scale generation of affinity-purified recombinant phytochrome chromopeptide. Photochem Photobiol 66:710–715
7. Murphy JT, Lagarias JC (1997) Purification and characterization of recombinant affinity peptide-tagged oat phytochrome A. Photochem Photobiol 65:750–758
8. Gartner W, Hill C, Worm K, Braslavsky SE, Schaffner K (1996) Influence of expression system on chromophore binding and preservation of spectral properties in recombinant phytochrome A. Eur J Biochem 236:978–983
9. Mukougawa K, Kanamoto H, Kobayashi T, Yokota A, Kohchi T (2006) Metabolic engineering to produce phytochromes with phytochromobilin, phycocyanobilin, or phycoerythrobilin chromophore in *Escherichia coli*. FEBS Lett 580:1333–1338
10. Wu SH, Lagarias JC (1996) The methylotrophic yeast *Pichia pastoris* synthesizes a functionally active chromophore precursor of the plant photoreceptor phytochrome. Proc Natl Acad Sci U S A 93:8989–8994
11. Shin AY, Han YJ, Song PS, Kim JI (2014) Expression of recombinant full-length plant phytochromes assembled with phytochromobilin in *Pichia pastoris*. FEBS Lett 588:2964–2970
12. Shin AY, Han YJ, Baek A, Ahn T, Kim SY, Nguyen TS, Son M, Lee KW, Shen Y, Song PS, Kim JI (2016) Evidence that phytochrome functions as a protein kinase in plant light signalling. Nat Commun 7:11545
13. Jeong AR, Lee SS, Han YJ, Shin AY, Baek A, Ahn T, Kim MG, Kim YS, Lee KW, Nagatani A, Kim JI (2016) New constitutively active phytochromes exhibit light-independent signaling activity. Plant Physiol 171:2826–2840
14. Romanos M (1995) Advances in the use of *Pichia pastoris* for high-level gene expression. Curr Opin Biotechnol 6:527–533
15. Quail PH, Briggs WR, Chory J, Hangarter RP, Harberd NP, Kendrick RE, Koornneef M, Parks B, Sharrock RA, Schafer E, Thompson WF, Whitelam GC (1994) Spotlight on phytochrome nomenclature. Plant Cell 6:468–471
16. Terry MJ, Maines MD, Lagarias JC (1993) Inactivation of phytochrome- and phycobiliprotein-chromophore precursors by rat liver biliverdin reductase. J Biol Chem 268:26099–26106

Chapter 8

In Vivo Spectroscopy

Cornelia Klose

Abstract

In vivo spectroscopy is used to directly assay phytochromes in intact plant material. The method is depending on the photoreversibility of phytochromes displaying light induced absorbance changes in response to actinic irradiation. Dual-wavelength ratio spectrophotometers (ratiospects) are the instruments successfully used for assaying phytochromes in highly scattering plant material. In the present chapter I describe the general instrument setup of an automated ratiospect and explain the measuring procedure and data calculation required to determine the total amount of photoreversible phytochromes in a sample as well as the proportion of phytochrome present in the Pfr conformation.

Key words Phytochrome, Photoreversibility, Ratiospectrophotometer, Ratiospect, In vivo spectroscopy, Photoconversion

1 Introduction

Phytochromes are photoreversible pigments; they interconvert between the red light (R) absorbing Pr form and the far-red light (FR) absorbing Pfr form. This phytochrome photointerconversion is repeatedly reversible and can be utilized to measure the concentration of phytochromes in living plant material using spectrophotometric assays. The change in absorbance between Pr (λ_{max} = 665 nm) and Pfr (λ_{max} = 730 nm) can be directly correlated to the phytochrome concentration in the sample. The main difficulty in doing spectrophotometric measurements in plant material is the inhomogeneous nature of the sample: plant tissues are highly scattering, exhibit heterogeneous pigment distribution and may contain other light absorbing pigments (e.g., chlorophylls) that potentially interfere with the measurement. Absorption assays in light scattering materials is a complex subject not appropriate to discuss within the scope of this chapter. For a detailed discussion of the theoretical basis of light absorption spectroscopy in light-scattering samples and potential artifacts that may lead to incorrect data interpretation we refer to Gross et al. [1]. It is important to

Andreas Hiltbrunner (ed.), *Phytochromes: Methods and Protocols*, Methods in Molecular Biology, vol. 2026, https://doi.org/10.1007/978-1-4939-9612-4_8,

note that light scattering within the material is even advantageous for in vivo spectroscopy assays since it causes extension of the measuring light path and thereby amplification of light absorption in the sample. Because of the abovementioned complications, in vivo spectroscopy requires a spectrophotometer that is capable of accurately measuring small absorbance differences in highly scattering samples with strong background absorption [2, 3]. Dual-wavelength ratiospectrophotometers (ratiospects) have been successfully used for this and several custom built, automated ratiospects have been developed and built in the past [4–6].

Different wavelength pairs can be used for in vivo spectroscopic phytochrome assays. Due to the light absorption changes between Pr and Pfr a combination of R (665 nm) and FR (730 nm) measuring light beams would give the strongest signal. However, a combination of FR (730 nm) and long-wavelength FR (around 780 nm) might be preferred in order to minimize interference by chlorophylls that, even in very small amounts, absorb strongly in the red light range. The presence of chlorophylls should be generally avoided for in vivo spectroscopic measurements and the use of etiolated plant material as well as short irradiation conditions not inducing chlorophyll accumulation is highly recommended. Although chlorophyll absorbs only weakly in FR, it has been shown that it can interfere with phytochrome assays at 730 nm even in only partially greened plants [7].

Another limitation is the abundance of the phytochrome species that is to be examined in the plant tissue. For example, in *Arabidopsis thaliana* phytochrome A (phyA) is highly abundant and well measurable in etiolated seedlings, but the other phytochromes phyB, C, D, or E are not detectable at endogenous levels. For measuring those phytochromes in vivo, transgenic lines have to be used that overexpress the phytochrome of interest preferentially in a *phyA* mutant background. Alternatively the light labile phyA can be depleted by a 2–3 h red light preirradiation.

Here, the general setup and operation mode of a custom-built, fully automated dual-wavelength ratiospect is presented, a typical measuring procedure is described and data calculation is discussed. In vivo spectroscopy using dual wavelength ratiospects is generally applicable for measuring phytochrome not only in intact plant tissue but also recombinantly expressed phytochrome reconstituted with chromophore in cell cultures from different species including yeast, bacteria or mammalian cells. In addition the setup can also be used to measure (partially) purified phytochrome in vitro. However, in the present chapter we focus on phytochrome measurements in intact *Arabidopsis thaliana* seedlings. The technique not only allows to detect total amounts of photoreversible phytochromes (Ptot) but also to determine the proportion of phytochrome that is in the active Pfr or inactive Pr conformation. Possible applications include measurements of Ptot degradation and Pfr specific degradation, Pfr thermal reversion (dark reversion) and the detection of Pfr levels established in light in a wavelength- and light intensity-dependent manner.

2 Materials

2.1 Ratiospect

Described here is a custom built, computer controlled, fully automated dual wavelength ratiospect comparable to the one described in Lamparter et al. [6], with slight modifications (Fig. 1). The system requires two types of light sources: measuring light for absorbance measurements as well as actinic light to execute photoconversion of phytochrome in the sample. Measuring light is provided by two light emitting diodes (LEDs) or laser diodes (LDs) emitting 730 and 787 nm light (*see* **Note 1**). Importantly, measuring light intensity should be sufficiently low to avoid any photoconversion occurring during the measurement. Actinic light is generated by tungsten lamps in combination with either R (λ_{max} = 660 nm) or long-wavelength FR (λ_{max} = 776 nm) filters.

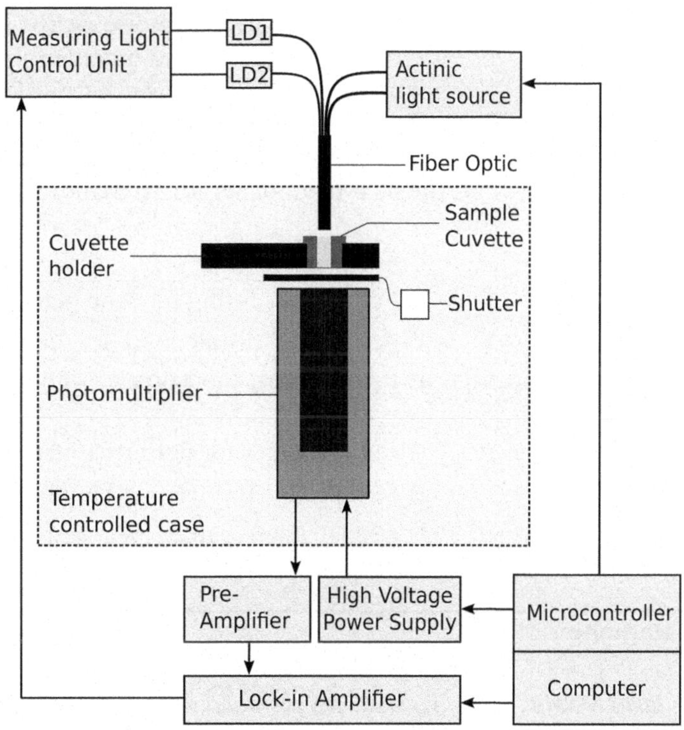

Fig. 1 General setup of the dual-wavelength ratiospectrophotometer (ratiospect). Monochromatic measuring light (730/787 nm) as well as actinic light (660/776 nm) is guided through fiber optics and passes vertically down through the sample containing cuvette. During actinic irradiation the photomultiplier (PM) is protected by a shutter. The light signal detected by the PM is converted via a logarithmic preamplifier and then perceived by the lock-in amplifier. An internal oscillator at the lock-in amplifier controls alternating light emission of the two measuring light beams with a high frequency (e.g., 1 kHz) and provides the reference signal for recovery of the received signals. Details are described in Subheading 2.1 in the text

Light from each light source is guided through separate arms of a fiber optic merging into a common fiber optic arm on top of the cuvette holder at the measuring position. Light passes vertically down through the sample cuvette reaching the photomultiplier (PM). A shutter between cuvette and PM protects the PM from light during actinic irradiation. For additional protection, high voltage at the PM is reduced during the actinic light irradiation. Cuvette holder and PM are arranged in a temperature controlled case and temperature is adjusted to 4 °C in order to minimize noise at the PM. The signal output of the PM is converted by a logarithmic preamplifier which in turn is connected with the signal input of a lock-in amplifier. The lock-in amplifier is essential to detect and extract very small signals in the presence of high background noise. An internal oscillator at the lock-in amplifier controls the two LD/LEDs resulting in alternating light emission of the two measuring light beams with a high frequency (e.g., 1 kHz). It further provides the reference signal that is needed to recover received signals at low signal-to-noise ratios. Prior to measurements, zero adjustment at the lock-in amplifier is performed automatically by adjusting the light intensity of the LD/LED and by attenuating the measuring light beams using variable combinations of neutral filters located in front of the LD/LEDs. The sensitivity range of the lock-in amplifier is automatically adjusted stepwise.

2.2 Plant Growth

1. Petri dishes.
2. Filter paper fitting into the petri dishes.
3. Autoclaved distilled water.
4. Ethanol 70% and 100%.

2.3 Sample Preparation

1. Light sources for light treatments as required.
2. Sample cuvettes wrapped in film and precooled on ice.
3. Scalpel, forceps, tissue, metal or glass rod fitting into the cuvette.

3 Methods

3.1 Plant Growth

To obtain 4 days old etiolated *Arabidopsis thaliana* seedlings:

1. Surface-sterilize seeds: shake seeds in 1 ml 70% EtOH for 5 min, discard supernatant, add 1 ml 100% EtOH and shake for 2 min, transfer seeds onto a filter paper in a hood and dry them completely.
2. Sow seeds in petri dishes containing four layers of filter paper wetted with autoclaved distilled water and keep them in a light-tight box.
3. Stratify the seeds for at least 2 days at 4 °C in darkness.

4. To induce germination expose the plates for 8 h to white light. Put them back into light-tight black boxes after induction of germination.
5. Incubate the plates for 4 days in darkness at 22 °C.

3.2 Sample Preparation

1. Use 4-day-old etiolated *Arabidopsis thaliana* seedlings to perform light treatments as required, do not irradiate seedlings longer than 4 h with continuous red, blue, or white light to prevent chlorophyll accumulation.
2. Under dim green-safelight: carefully scrape the seedlings from the filter paper and collect them on a tissue (*see* **Note 2**).
3. Under dim green-safelight: thoroughly pat the seedlings dry.
4. Under dim green-safelight: transfer the seedlings into precooled sample cuvettes and determine the fresh weight.
5. Carefully compress the seedlings in the cuvette without scrunching them using a precooled metal rod in order to obtain an even sample geometry.
6. Measure the samples in the ratiospect.

3.3 Measuring Procedure

A typical measuring procedure is described here, but single steps (number of measuring cycles, duration of actinic light irradiation, measuring time) can be defined variably by the user in the computer program. After starting the measuring procedure, the program automatically initiates zero adjustment and sensitivity range tuning of the lock-in amplifier. Whenever during the measuring procedure the detected signals overshoot the selected sensitivity range, it is readjusted automatically. A typical measuring protocol comprises six measuring cycles, each cycle consisting of a 25 s actinic FR irradiation followed by a 25 s absorbance measurement and a subsequent 25 s actinic R irradiation again followed by a 25 s absorbance measurement. From these measuring cycles the total amount of photoreversible phytochrome (Ptot) in the sample can be calculated taking into account the photoequilibrium according to Mancinelli [8].

In order to determine the initial Pfr amount present in the sample at the beginning of the measuring procedure careful absorbance baseline detection is essential (*see* **Note 3**). Then, the actinic FR irradiation within the first measuring cycle photoconverts all initial Pfr in the sample back to Pr. The absorbance change initiated by this first FR treatment compared to the baseline absorbance can be used to calculate the initial Pfr amount (*see* **Note 4**) relative to the total amount of photoreversible phytochrome (Pfr/Ptot).

3.4 Calculations

Signals on the *y*-axis of the photoreversibility plot (Fig. 2) represent voltage output of the lock-in amplifier and can be converted into absorbance differences ΔA using a calibration factor deter-

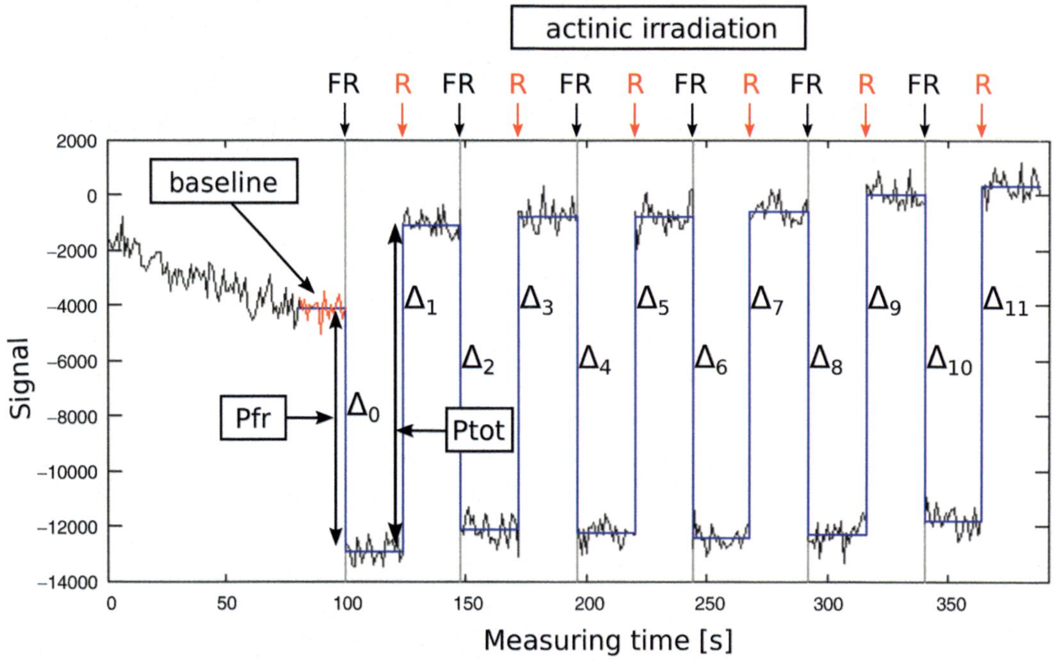

Fig. 2 Graphical display of an example measurement. A typical measuring protocol comprises six measuring cycles: 25 s actinic FR irradiation followed by a 25 s absorbance measurement and subsequent 25 s actinic R irradiation followed by 25 s absorbance measurement, each. The photoreversibility plot shows voltage signals derived from the lock-in amplifier on the *y*-axis. Only measuring time periods are shown on the *x*-axis, intermittent actinic R and FR light treatments are indicated with red and black arrows. Continuous blue lines depict averaged signals for each absorbance measurement. Red labelled measured signals indicate the data range used for baseline detection. Pfr and Ptot calculation is explained in Subheading 3.3 in the text

mined utilizing specific calibration glasses. ΔA is the difference in absorbance A at the two measuring light wavelengths 730 and 787 nm.

$$\Delta A = A_{730} - A_{787}$$

The phytochrome quantity in the sample is linearly proportional to the photoreversibility value $\Delta\Delta A$ that is given by the difference between the absorption difference measurements ΔA after actinic R (ΔA^{R}) or FR (ΔA^{FR}) irradiation (*see* **Note 5**).

$$\Delta\Delta A = \left|\Delta A^{\mathrm{R}} - \Delta A^{\mathrm{FR}}\right|$$

In order to enhance the accuracy of the $\Delta\Delta A$ determination several measuring cycles (typically 6) are performed. $\Delta_1 - \Delta_{11}$ represent the individual $\Delta\Delta A$ measurements during the repeated cycles (Fig. 2). The total amount of photoreversible phytochrome Ptot is calculated from the average $\overline{\varnothing}$ of the absorbance differences

$\Delta_1 - \Delta_{11}$ induced by the actinic light treatments and normalized to seedling fresh weight (fw). The correction factor 1.15 has to be included to account for the fact that even by irradiation with R only 87% of the phytochrome can be photoconverted to the Pfr form [8]. Ptot is given in $\Delta(\Delta A)$/mg fw.

$$\text{Ptot} = \frac{\bar{\Delta} \times 1.15}{fw}$$

The relative Pfr/Ptot ratio φ is calculated from the absorbance difference Δ_0 between the first actinic FR treatment and the baseline:

$$\varphi = \frac{\Delta_0}{\bar{\Delta} \times 1.15}$$

4 Notes

1. As mentioned in the introduction, although the use of 660–670 nm light as measuring wavelength would give higher sensitivity due to stronger absorbance changes following actinic light irradiation, 730 nm (maximum wavelength of the Pfr absorbance spectrum) was chosen in order to minimize interference caused by chlorophylls. The second measuring light at 787 nm serves as a reference wavelength where no phytochrome absorbance change occurs upon actinic light irradiation.
2. Work as fast and consistent as possible when preparing the samples. Cool down harvested seedlings immediately to avoid any unintended changes in Pfr/Ptot ratios.
3. Signals might be drifting because of small changes in light path or temperature in living material occurring after sample preparation. Therefore, baseline is measured over 100 s to allow for stabilization of the signal.
4. Reliable Pfr determination is only feasible using FR as first actinic light treatment. Initial actinic R irradiation photoconverts protochlorophyllide to chlorophyllide in samples that have not been pretreated with strong red light or have been kept for longer time periods in darkness allowing reaccumulation of protochlorophyllide. Protochlorophyllide to chlorophyllide transformation changes the absorbance spectrum of the sample and would therefore interfere with the Pfr determination. The respective measuring cycles (usually after the first actinic R treatment) should also be omitted for Ptot calculation. Subsequent measuring cycles used for Ptot determination

are not affected when protochlorophyllide to chlorophyllide transformation by the first actinic R irradiation was saturating.

5. $\Delta\Delta A$ measurements of different samples are only comparable to each other as long as sample structure is consistent and the effects of light scattering on the measuring light pathlength are the same in individual samples. Therefore, uniform sample preparation is absolutely essential. Linearity of the proportional relation between phytochrome amount and $\Delta\Delta A$ has to be tested for each sample type. For the described setup linearity was assured for etiolated *Arabidopsis thaliana* seedlings in a range from 80 mg to 140 mg fresh weight.

References

1. Gross J, Seyfried M, Fukshansky L et al (1984) In: Smith H, Holmes MG (eds) In vivo spectrophotometry. Methods in photomorphogenesis, biological technique series. Academic Press, London, pp 131–157
2. Butler WL (1962) Absorption of light by turbid materials. J Opt Soc Am 52:292–299
3. Butler WL (1964) Absorption spectroscopy in vivo theory and application. Annu Rev Plant Physiol 15:451–460
4. Pratt LH, Wampler JE, Rich ES (1984) An automated dual-wavelength spectrophotometer optimized for Phytochrome assay. Instrum Sci Technol 13:269–287
5. Butler WL, Norris KH, Siegelman HW et al (1959) Detection, assay, and preliminary purification of the pigment controlling photoresponsive development of plants. Proc Natl Acad Sci U S A 45:1703–1708
6. Lamparter T, Hughes J, Hartmann E (1994) A fully automated dual-wavelength photometer for phytochrome measurements and its application to phytochrome from chlorophyll-containing extracts. Photochem Photobiol 60:179–183
7. Grill R (1977) Influence of chlorophyll content on phytochrome measurements in turnip cotyledons. Planta 134:11–16
8. Mancinelli AL (1994) The physiology of phytochrome action. In: Kendrick RE, Kronenberg GHM (eds) Photomorphogenesis in plants. Springer, Dordrecht, pp 211–269

Chapter 9

Basic Phytochrome B Calculations

Robert W. Smith and Christian Fleck

Abstract

Mathematical models are important tools in helping us to understand complex biological systems. Models of phytochrome-regulated systems in *Arabidopsis thaliana* have shown the importance of dimerization, nuclear transport, and thermal/dark reversion in mediating phytochrome activity and plant development. Here we go through the steps required to calculate the steady-state amounts of phytochrome subspecies relative to the total phytochrome molecule population. Starting from a simplified two-state system we expand and apply the technique to the extended phytochrome dimer model. Additionally, we provide a Python package that can automatically calculate the proportion of phytochrome B in a particular state given specific experimental conditions.

Key words Mathematical modeling, Ordinary differential equations, Phytochromes, Computer programming

1 Introduction

Plant phytochromes switch between two states (inactive P_r and active P_{fr}) in a light-dependent manner to regulate plant physiology. Phytochromes are synthesized in the P_r state within the cytosol of plant cells [1–3]. Upon exposure to red light, the phytochromes switch from P_r to P_{fr}, enter the nuclei of cells whereby they interact with a range of other proteins, form nuclear speckles/bodies, and regulate transcription [4–8]. When exposed to far-red light, the active P_{fr} conformers are switched back to the P_r state and the system is reset. As well as the light-regulated reactions, P_{fr} molecules are also constantly relaxing back to the P_r state through thermal reversion [9–12]. However, given that this process is slow, it is often only important for regulating phytochrome dynamics in darkness or under low intensity light [12, 13]. The reactions shown in Fig. 1a form the basic two-state phytochrome system [3, 13, 14].

In the presence of synthesis and degradation reactions, experimentally measuring the proportions of phytochrome populations within specific states, and how fast they conformationally

Andreas Hiltbrunner (ed.), *Phytochromes: Methods and Protocols*, Methods in Molecular Biology, vol. 2026,
https://doi.org/10.1007/978-1-4939-9612-4_9,

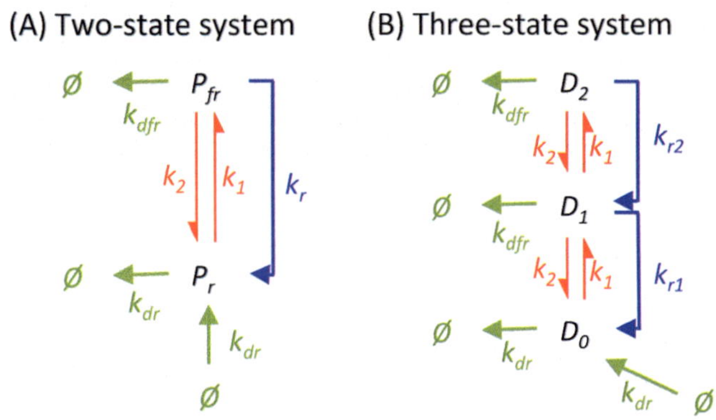

Fig. 1 Phytochrome B dynamics in plants. Simplified schematic representations of phytochrome B dynamics in plants. (**a**) The two-state system whereby phyB switches between P_r and P_{fr} states. (**b**) The three-state system whereby dimers of phyB P_r and P_{fr} are formed. Here, $D_0 = P_r - P_r$, $D_1 = P_r - P_{fr}$, and $D_2 = P_{fr} - P_{fr}$ dimers. Parameters: k_1, k_2=light-regulated photoconversion rates; k_{di} = degradation rates; k_{ri} = thermal reversion rate of P_{fr} to P_r state. Light-regulated reactions are colored red, thermal reversion in blue, and degradation or synthesis reactions in green

change from one state to another, is often difficult. Consequently, mathematical models and computational optimization techniques have been utilized to provide a better understanding as to how phytochrome subpopulations regulate plant development [3, 13, 15–17]. These models have helped explain how phytochrome B (phyB) dynamics relate to hypocotyl elongation of *Arabidopsis thaliana* and why phytochrome A shows activity under far-red light [3, 12, 14].

To highlight the utility of mathematical models, the two-state system of phyB was constructed mathematically and matched to data showing thermal reversion of P_{fr}, degradation kinetics, nuclei speckle formation, and, ultimately, hypocotyl elongation all under a range of different light conditions (Fig. 1a; [3]). However, this model was unable to explain why hypocotyl elongation is not strongly inhibited under high intensity ~700 nm light [12]. The two-state model was extended to incorporate phyB dimerization—a phenomenon that had been previously observed experimentally [18]. The resulting three-state model (P_r–P_r, P_r–P_{fr} and P_{fr}–P_{fr} dimers rather than P_r and P_{fr} molecules in the two-state model) correctly matched fluence-response curves of hypocotyl elongation under high intensity 690–716 nm light (Fig. 1b; [12]).

Here, we discuss the two-state and three-state models, showing that the steady-state (after prolonged periods of light exposure) proportions of phytochrome species can be directly calculated given the experimental conditions. We then introduce a Python script that automatically performs these calculations that we hope the photobiology community will find of use in the future.

2 Methods

2.1 Models

2.1.1 Original Models and Simplifying Assumptions

Both of the original models proposed by Rausenberger et al. and Klose et al. are too complex to be analyzed at steady state without numerical simulations [3, 12]. Here, we shall quickly review the two systems and the simplifying assumptions we have made to aid the analytical calculations introduced in the following sections. We refer readers to the original publications for further details and model equations.

As well as photoconversion between different phyB states, the two models also include a range of other biological processes (Fig. 2): intracellular transport from cytoplasm to nucleus (defined by parameter k_{in} in Fig. 2) and, ultimately, into nuclear bodies (denoted 'NB' in Fig. 2, where k_3 and k_4 are the P_{fr}-dependent NB association and dissociation rates, and k_5 is the P_{r}-dependent NB dissociation rate [12]). Within each compartment, the degrada-

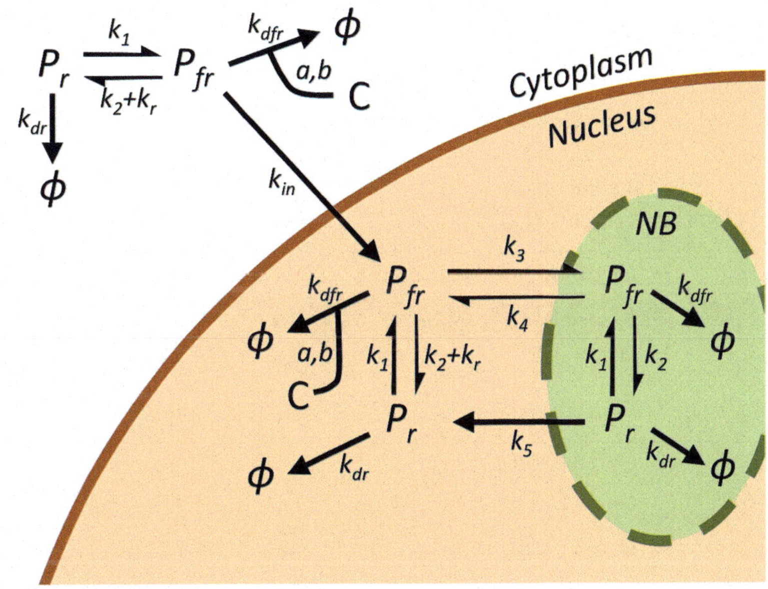

Fig. 2 Phytochrome B dynamics captured in full models. Schematic of all the biological processes in the full models of phyB dynamics [3, 12]. As in Fig. 1, phyB undergoes photoconversion, thermal reversion and degradation. However, in the full models, phyB is also transported between cellular compartments, including the cytoplasm, nucleus, and nuclear speckles/bodies (NB). Within the NBs there is no thermal reversion process. Degradation processes also depend on compartment, whereby an interacting protein, C, enhances P_{fr} degradation in the cytoplasm and nucleus. The parameter k_{in} represents nuclear import of phyB, k_3 and k_4 are the P_{fr}-dependent NB association and dissociation rates, k_5 is the P_{r}-dependent NB dissociation rate, and a and b are indicative of the sensitivity of P_{fr} to C and the enhancement of degradation by C, respectively

tion and thermal reversion of phyB differs. As observed experimentally, thermal reversion is strongly suppressed within NBs due to their prolonged visibility after long periods of darkness [3]. Degradation rates of P_r and P_{fr} also differ due to the experimental observation that phyB degrades faster under red light when compared to levels of phyB in etiolated seedlings [3]. This suggests that P_{fr} degrades faster than P_r. Further iterations of the model went on to include compartment-dependent degradation mechanisms of P_{fr} whereby degradation is enhanced by phyB-interacting proteins in the cytoplasm and nucleus but not in NBs [12]. These degradation mechanisms are controlled by two parameters (a and b in Fig. 2 and Table 1) that determine the affinity between P_{fr} and an interacting protein (a) and the enhancement of degradation due to the interaction (b).

Whilst these models, when simulated numerically, are more accurate at describing phyB dynamics, they are intractable to obtain analytic *estimates* of species concentrations in the phyB system. In order to perform the analytical derivations below we make two simplifying assumptions. First, P_r, P_{fr}, and their respective dimers degrade at the same rate in *all* cellular compartments. Second, we

Table 1
Format for input file detailing simulated parameter values

Parameter	Biological interpretation	Units	Value 1 (default values)	···	Value X (simulating no dark reversion)
k_{dr}	Phytochrome degradation	min^{-1}	0.00061	···	0.00061
k_{dfr}	P_{fr}-specific degradation	min^{-1}	49.57	···	49.57
k_{r1}	D_1 dark reversion	min^{-1}	0.497	···	0
k_{r2}	D_2 dark reversion	min^{-1}	0.0051	···	0
k_{31}	D_1 NB formation	min^{-1}	1.36	···	1.36
k_{32}	D_2 NB formation	min^{-1}	4.43	···	4.43
k_{41}	D_1 NB breakdown	min^{-1}	4.1	···	4.1
k_{42}	D_2 NB breakdown	min^{-1}	0.13	···	0.13
k_5	D_0 NB breakdown	min^{-1}	0.03	···	0.03
k_{in}	P_{fr} nuclear import	min^{-1}	58.4	···	58.4
M	Number of NBs	–	6	···	6
a	Sensitivity of P_{fr}-specific degradation	–	0.56	···	0.56
b	Strength of P_{fr}-specific degradation	–	9	···	9

Up to X different parameter sets can be simulated. *See* Fig. 2 for a schematic illustration of parameters biological interpretation. The first numeric column shows the default values. See text for details about parameter notation allow for thermal reversion within NBs. These two assumptions (although biologically inaccurate) allow us to ignore cellular compartments in our model and construct simpler, mathematically tractable systems with which to estimate the proportions of P_r, P_{fr}, D_0, D_1, and D_2 within cells.

2.1.2 Two-State Model

The two-state model incorporates dynamics for the P_r and P_{fr} concentrations within a cell. The ordinary differential equations (ODEs) read as follows:

$$\begin{aligned} \frac{dP_r}{dt} &= k_{dr}(1-P_r)+(k_2+k_r)P_{fr}-k_1P_r \\ \frac{dP_{fr}}{dt} &= k_1P_r-(k_2+k_r+k_{dfr})P_{fr}, \end{aligned} \quad (1)$$

where dx/dt is the change in concentration x over a time-step dt, k_r, is the thermal reversion rate, k_{dr} and k_{dfr} are the conformation-dependent degradation rates, and k_1 and k_2 are the light-regulated photoconversion rates. We normalized the equations such that the amount of P_r before light exposure is equal to 1, that is, 100% of the phyB population is in the P_r state. The light-regulated photoconversion rates are given by $k_i = \sum_\lambda N_\lambda \sigma_i^\lambda$, where N_λ is the wavelength-dependent light intensity of the experimental conditions in μmol/m^{2}s and σ^λ is the photoconversion cross-section at wavelength λ between the two states in m^2/mol (*see* **Note 1**; [13]).

2.1.3 Three-State Model

Similarly to the two-state model, we can simplify the model proposed by Klose et al. to read as follows:

$$\begin{aligned} \frac{dD_0}{dt} &= k_{dr}(1-D_0)+(k_2+k_{r1})D_1-2k_1D_0 \\ \frac{dD_1}{dt} &= 2k_1D_0+2(k_2+k_{r2})D_2-(k_1+k_2+k_{r1}+k_{dfr})D_1 \\ \frac{dD_2}{dt} &= k_1D_1-\left[2(k_2+k_{r2})+k_{dfr}\right]D_2 \end{aligned} \quad (2)$$

The states of the dimer are denoted by $D_0 = P_r - P_r$, $D_1 = P_r - P_{fr}$ and $D_2 = P_{fr} - P_{fr}$ [12]. The thermal reversion of D_2 is slower than D_1 such that $k_{r2} < k_{r1}$. Again the model is normalized such that the amount of D_0 before light exposure is equal to 1. Note that, in this model, the multiplication of conformation switching from the homodimers to the heterodimer by 2 represents the fact that the heterodimer can take two different heterodimer forms, $P_r - P_{fr}$ and $P_{fr} - P_r$, such that D_1 is the total concentration of these two forms.

2.2 Steady-State Analysis

Here we will go through the steps required to estimate the proportion of the phyB molecular population in a given state after prolonged experimental conditions. To do this we assume that the system has reached steady state.

2.2.1 Definition: Steady State

When we refer to solving a mathematical model at *steady state* what we are implying is that the concentration of each system component does not change over time, that is, they are constant. Thus, all equations of the system can be set $dx/dt = 0$. In relation to the phytochrome system this means that the sample has been exposed to a given light condition for a long enough period of time that the concentrations of P_r, P_{fr}, and their respective dimers are stable.

2.2.2 Two-State Model

By setting $dP_{fr}/dt = 0$ in Eq. 1, we get

$$k_1 P_r - \left(k_2 + k_r + k_{dfr}\right) P_{fr} = 0.$$

From this, it is clear to see that

$$k_1\left(P_r + P_{fr}\right) - \left(k_1 + k_2 + k_r + k_{dfr}\right) P_{fr} = 0,$$

is also true as the added $k_1 P_{fr}$ terms cancel with each other.

Thus, using $P_{tot} = P_r + P_{fr}$, rearrangement leads to

$$\begin{aligned} \frac{P_{fr}}{P_{tot}} &= \frac{k_1}{k_1 + k_2 + k_r + k_{dfr}} \\ \frac{P_r}{P_{tot}} &= 1 - \frac{P_{fr}}{P_{tot}} = \frac{k_2 + k_r + k_{dfr}}{k_1 + k_2 + k_r + k_{dfr}}. \end{aligned} \tag{3}$$

Using these expressions, we can also go one step further and ask what the effect of light intensity is on the system. By substituting $k_i = \sum_\lambda N_\lambda \sigma_i^\lambda$ into the above expressions we get

$$\frac{P_{fr}}{P_{tot}} = \frac{\sum_\lambda N_\lambda \sigma_r^\lambda}{\sum_\lambda N_\lambda \left(\sigma_r^\lambda + \sigma_{fr}^\lambda\right) + k_r + k_{dfr}}. \tag{4}$$

From this, as $N_\lambda \to \infty$, the effects of dark reversion and degradation of P_{fr} become negligible compared to the photoconversion reactions. Hence, the proportion of P_{fr} in the system within this limit is dominated by the photoconversion spectra or light-regulated reactions

$$\lim_{N_\lambda \to \infty} \frac{P_{fr}}{P_{tot}} = \frac{\sum_\lambda N_\lambda \sigma_r^\lambda}{\sum_\lambda N_\lambda \left(\sigma_r^\lambda + \sigma_{fr}^\lambda\right)}. \tag{5}$$

2.2.3 Three-State Model

In order to obtain expressions for D_0, D_1, D_2, and P_{fr}/P_{tot} from the three-state model, we need to simplify the system. We do this in two ways. First, we reduce the system from three variables to two by using $D_{tot} = D_0 + D_1 + D_2$, such that $D_0 = D_{tot} - D_1 - D_2$. This means that we go from having three equations and three unknown variables to having three equations and two unknowns (D_1/D_{tot} and D_2/D_{tot}). Second, we set $k_{dr} = 0$, implying that synthesis or degradation of phyB D_0 is very slow relative to other reactions in the system [3]. This removes the constant term from the equation for dD_0/dt and allows the system to be rewritten into new variables. Thus, upon setting the time derivatives to zero we can rewrite the system as

$$
\begin{aligned}
0 &= (2k_1 + k_2 + k_{r1})\frac{D_1}{D_{tot}} - 2k_1\left(1 - \frac{D_2}{D_{tot}}\right) \\
0 &= 2k_1 + 2(k_2 + k_{r2} - k_1)\frac{D_2}{D_{tot}} - (3k_1 + k_2 + k_{r1} + k_{dfr})\frac{D_1}{D_{tot}} \\
0 &= k_1\frac{D_1}{D_{tot}} - \left[2(k_2 + k_{r2}) + k_{dfr}\right]\frac{D_2}{D_{tot}},
\end{aligned}
\tag{6}
$$

Substituting the third expression of Eq. 6 into the first to remove the D_1/D_{tot} terms yields the following:

$$
\begin{aligned}
0 &= (2k_1 + k_2 + k_{r1})\frac{2(k_2 + k_{r2}) + k_{dfr}}{k_1}\frac{D_2}{D_{tot}} + 2k_1\left(1 - \frac{D_2}{D_{tot}}\right) \\
\frac{D_2}{D_{tot}} &= \frac{2k_1^2}{2k_1^2 + (2k_1 + k_2 + k_{r1})\left[2(k_2 + k_{r2}) + k_{dfr}\right]}
\end{aligned}
\tag{7}
$$

Further substitution of this term into the second expression of Eq. 6 and solving for D_1/D_{tot} gives

$$
\begin{aligned}
\frac{D_1}{D_{tot}} &= \frac{1}{3k_1 + k_2 + k_{r1} + k_{dfr}}\left[2k_1 + \frac{4k_1^2(k_2 + k_{r2} - k_1)}{2k_1^2 + (2k_1 + k_2 + k_{r1})\left[2(k_2 + k_{r2}) + k_{dfr}\right]}\right] \\
&= \frac{2k_1\left\{2k_1^2 + (2k_1 + k_2 + k_{r1})\left[2(k_2 + k_{r2}) + k_{dfr}\right]\right\} + 4k_1^2(k_2 + k_{r2} - k_1)}{(3k_1 + k_2 + k_{r1} + k_{dfr})\left\{2k_1^2 + (2k_1 + k_2 + k_{r1})\left[2(k_2 + k_{r2}) + k_{dfr}\right]\right\}}.
\end{aligned}
\tag{8}
$$

Finally, using the fact that $D_{tot} = D_0 + D_1 + D_2$, we obtain

$$
\begin{aligned}
\frac{D_0}{D_{tot}} &= 1 - \frac{D_1}{D_{tot}} - \frac{D_2}{D_{tot}} \\
&= \frac{(k_1 + k_2 + k_{r1} + k_{dfr})\left\{2k_1^2 + (2k_1 + k_2 + k_{r1})\left[2(k_2 + k_{r2}) + k_{dfr}\right]\right\}}{(3k_1 + k_2 + k_{r1} + k_{dfr})\left\{2k_1^2 + (2k_1 + k_2 + k_{r1})\left[2(k_2 + k_{r2}) + k_{dfr}\right]\right\}} \\
&= -\frac{4k_1^2(k_2 + k_{r2} - k_1) + 2k_1^2}{(3k_1 + k_2 + k_{r1} + k_{dfr})\left\{2k_1^2 + (2k_1 + k_2 + k_{r1})\left[2(k_2 + k_{r2}) + k_{dfr}\right]\right\}}.
\end{aligned}
\tag{9}
$$

We can now use these expressions to calculate the total amount of P_r and P_{fr} in the system at steady state. Recall that in D_2 we have all the phyB molecules in the P_{fr} state, whilst in D_1 half the molecules are in the P_{fr} state. This means that

$$\frac{P_{\text{fr}}}{P_{\text{tot}}} = \frac{D_2 + D_1/2}{D_{\text{tot}}} \tag{10}$$

and

$$\frac{P_{\text{r}}}{P_{\text{tot}}} = 1 - \frac{P_{\text{fr}}}{P_{\text{tot}}} = \frac{D_0 + D_1/2}{D_{\text{tot}}}. \tag{11}$$

We shall not provide the exact expressions here, but one should note that as $N_\lambda \to \infty$ (or equivalently $\{k_{\text{r1}}, k_{\text{r2}}, k_{\text{dfr}}\} \to 0$) the resulting pool size of P_{fr} is given by

$$\lim_{N_\lambda \to \infty} \frac{P_{\text{fr}}}{P_{\text{tot}}} = \frac{\sum_\lambda N_\lambda \sigma_{\text{r}}^\lambda}{\sum_\lambda N_\lambda \left(\sigma_{\text{r}}^\lambda + \sigma_{\text{fr}}^\lambda\right)}.$$

as is the case for the simpler two-state model in Eq. 5.

2.3 Computational Tool

We hope, from the above derivations, that we have shown how the mathematical models of phytochrome activity can be analyzed to estimate the steady-state percentage of P_{fr} in the system. To aid the photobiology community, we have developed a computational tool that numerically simulates the full model proposed by Klose et al. [12] or analytically solves the simplified three-state model given a set of input conditions. In Fig. 3, we provide a workflow as to how the algorithm functions and what the output could look like. These input conditions can be formatted within text files or can be submitted to the tool from the command-line interface. The calculator can be downloaded from http://gitlab.com/

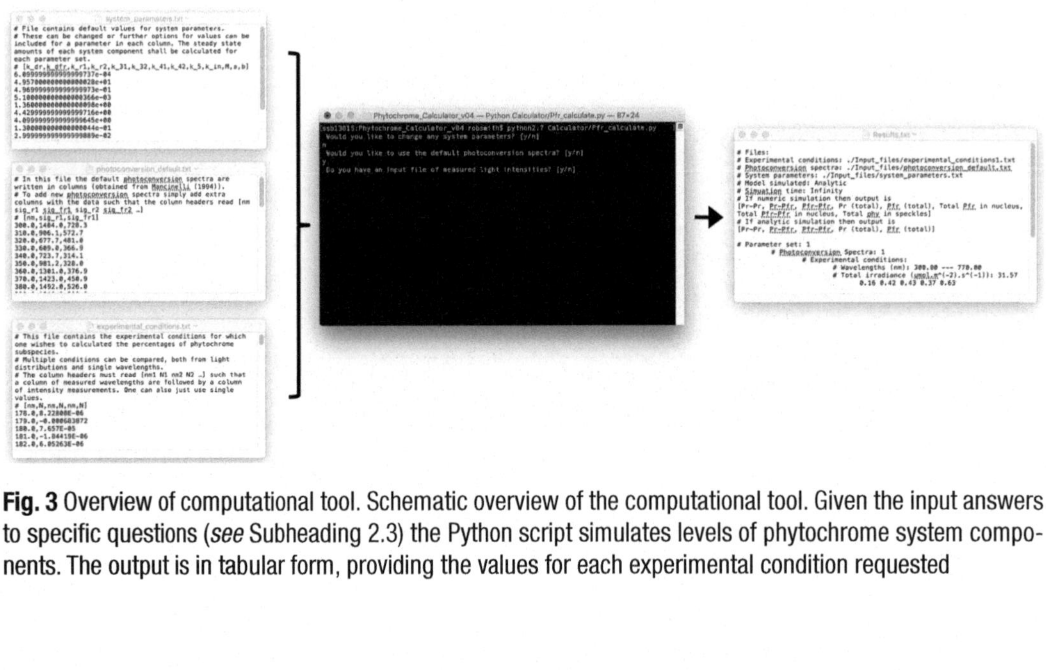

Fig. 3 Overview of computational tool. Schematic overview of the computational tool. Given the input answers to specific questions (*see* Subheading 2.3) the Python script simulates levels of phytochrome system components. The output is in tabular form, providing the values for each experimental condition requested

wurssb/Phytochrome_Calculator (*see* **Note 2**). Examples of input files required to use the calculator can also be found here. To calculate the relevant values the following questions need to be answered.

2.3.1 Would You Like to Change Any Kinetic Rates Within the phyB System?

The amount of P_{fr} within the phytochrome system depends on the rates at which phyB switches between different states, thermal reversion and degradation (Eqs. 4 and 10). Consequently, users can either use the published values of the degradation and thermal reversion rates (as default) or change them to any value they desire. This could be advantageous if, for example, one wished to know how much P_{fr} exists in a molecule population of mutated phyB whereby the thermal reversion rate (given by the k_{r*} parameters in Figs. 1 and 2 and Table 1) is altered, as in [10, 11]. The altered parameter values can be input into the calculator from the command line or using a text file containing a table similar to Table 1 where the first numerical column shows the default values published by [12].

2.3.2 Would You Like to Use the Default Photoconversion Spectra?

As discussed above, P_{fr} levels are dependent on the speed of light-regulated reactions (k_1 and k_2 above) as determined by the photoconversion spectra (σ_r^λ and σ_{fr}^λ; *see* Eqs. 4 and 10). By default, our calculator uses the values published for full-length phytochromes [13]. However, one may wish to alter the light-regulated photoconversion rates in the phyB system or measure new photoconversion spectra. For example, the N-terminal 650 amino acid truncated form of the phyB protein used in synthetic biology has recently been shown to respond to light more slowly in vitro than the full-length protein in vivo [16]. Thus, if a user had measured or knew that a particular phyB variant had altered photoconversion spectra compared to the full-length protein, then the user can calculate P_{fr} levels with these new values. The measured photoconversion spectra should be stored in a text file following the table structure shown in Table 2.

Table 2
Format for input file allowing multiple photoconversion spectra to be simulated

Wavelength, λ (nm)	σ_r^1 (μmol/m²s)	σ_{fr}^1 (μmol/m²s)	σ_r^2 (μmol/m²s)	σ_{fr}^2 (μmol/m²s)	...	σ_r^X (μmol/m²s)	σ_{fr}^X (μmol/m²s)
300	1404.0	728.3	67.12	71.52	...	4.953	0.468
310	906.1	572.7	79.26	67.39	...	5.798	0.352
320	677.7	481.0	101.0	64.71	...	6.957	0.468
⋮	⋮	⋮	⋮	⋮	⋱	⋮	⋮

Up to *X* different photoconversion spectra can be simulated. See text for details about parameter notation

Table 3
Format for input file containing experimental conditions

λ_1	N_{λ_1}	λ_2	N_{λ_2}	⋯	λ_X	N_{λ_X}
640	0.0834	665	5.0	⋯	740	1.0
641	0.0824			⋯		
642	0.0829			⋯		
⋮	⋮	⋮	⋮	⋱	⋮	⋮

Up to X experimental conditions can be simulated, including light distributions. Here λ_i is the wavelength of the light source (either a single value or a distribution) and N_{λ_i} is the corresponding light intensity

2.3.3 Do You Have an Input File of Measured Light Intensities?

In order to calculate the amount of P_{fr}, one needs to specify the experimental conditions that determine k_1 and k_2. Users must provide a wavelength (or wavelengths) and their respective light intensities at which to calculate steady-state P_{fr} levels. Importantly, users can either calculate these values at a single wavelength with a single intensity (as would be the case if one would use idealized lasers experimentally), across a distribution of wavelengths (as is the case when using LEDs experimentally) or across a complete light spectrum (as is the case when measuring, for example, natural sunlight). The measured light spectrum can be input into the calculator from the screen or with a text file using the table format shown in Table 3. This provides users with the freedom to simulate any experimental condition currently found in the published literature.

2.3.4 Would You Like to Solve the System Analytically or Numerically?

In the above derivations, we have shown how the full dynamic system of phyB activity can be reduced under steady-state conditions using simplifying assumptions to a single equation determining the percentage of P_{fr} in the molecule population. Here we allow the users to choose whether they wish to calculate the percentage of P_{fr} using the analytical steady-state equations or the full dynamic system schematically illustrated in Fig. 2 [3, 12]. In **Note 3** we discuss the conditions under which the two solutions differ.

2.3.5 Enter Path for Results File

Finally, we require the user to specify where they would like their results to be saved. The results are printed to the screen and a text file is automatically created containing different amounts of information depending on whether the system has been simulated numerically or analytically. If the system has been simulated numerically then information with regards to the relative levels of each dimer form, the total amount of P_{r} and P_{fr} in the system, how much P_{fr} is in the nucleus, how much D_2 is in the nucleus, and how much phyB is in nuclear speckles can be obtained. However, if the system is simulated analytically then one can only obtain informa-

Table 4
Format of output file

Simulation	Total D_0	Total D_1	Total D_2	Total P_r	Total P_{fr}	Total nuclear P_{fr}	Total nuclear D_2	Total phy in NBs
Analytic	0.98	0.02	0.00	0.99	0.01	–	–	–
Numeric	0.98	0.02	0.00	0.99	0.01	0.01	0.00	0.17

Example of which information is contained in the output file given that the phytochrome system is simulated either analytically or numerically

tion about the relative levels of each dimer form, the total amount of P_r and P_{fr} in the system since the analytic calculation does not take into account different cellular compartments (*see* Subheading 2.2.2 above). From the current published literature, this information should be sufficient to relate phyB activity to physiological responses such as hypocotyl elongation [12]. The output file is organized to provide this information in tabular form (as seen in Table 4) for all the requested sets of kinetic rates, photoconversion spectra, and experimental conditions specified above.

3 Notes

1. An important aspect of mathematical modeling is ensuring that the equations have consistent dimensions. For example, all the k_i's in the models outlined above are in units min^{-1} such that $dx/dt = [\text{concentration}] \cdot [\text{min}]^{-1}$. However,

$$k_1 = \sum_{\lambda_{\max}}^{\lambda=\lambda_{\min}} N_\lambda \sigma_r^\lambda = \frac{\mu mol}{m^2 s} \times \frac{m^2}{mol} \neq \frac{1}{\text{min}}.$$

Thus, the light-regulated parameters need to be rescaled by

$$\alpha \sim \frac{mol}{\mu mol} \times \frac{s}{\text{min}} = 6 \times 10^{-5} min^{-1},$$

such that $k_1 \rightarrow \alpha k_1$. The same needs to be applied to k_2 to ensure that one calculates the correct dynamics of the phyB system and the P_{fr} levels.

In these expressions, $\lambda_{\min}$ and $\lambda_{\max}$ represent the minimum and maximum wavelengths for which the photoconversion spectra σ^λ is measured. Thus, even if light intensity is measured at other wavelengths, the k_i's can only be calculated if the photoconversion spectra are known. As default in our computational tool, we use the photoconversion spectra measured by [13] where $\lambda_{\min} = 300$ nm and $\lambda_{\max} = 770$ nm. Further methods have also been developed to calculate the photoconversion spectra for phytochrome proteins [16].

2. The code was written using Python version 2.7 with NumPy version 1.8 and SciPy version 0.13. We recommend users to maintain use of Python version 2.7 when running the calculator, but our current implementation should be compatible with newer versions of NumPy and SciPy. Detailed information about the tool's usage, as discussed above, can also be found in the README files and we encourage users to read these before using the program. All input files need to be generated as comma-delimited tables with the columns shown in Tables 1, 2, and 3. Example files are provided with the Python scripts.
3. Within the computational tool, we provide the user with the option of analyzing the phyB system numerically or analytically. The choice of technique used depends on what the user wishes to analyze. For example, solving the full system numerically allows the user to observe how altering speckle formation kinetics (the parameters k_{3*}, k_{4*}, or k_5 in Table 1, *see* Subheading 2.1.1) leads to changes in P_{fr}/P_{tot}. However, whilst solving the system analytically restricts the number of parameters that can be changed in the system, it does allow the user to study a wider range of experimental conditions easily. In calculating the analytical expressions, we used the steady-state assumption and simplified models (*see* Subheading 2). However, in our numerical solver, the user can define how long they wish to simulate the phyB system for. This means that, under certain conditions, the analytical and numerical solutions of the phyB system will differ. We have already shown above that when light intensity is high, the amount of P_{fr} in the system at steady state tends to the ratio of photoconversion spectra. This is similarly the case for the numerical solution. However, when light intensity is low (i.e., when $N_\lambda < 0.1$ μmol/m²s), the dynamics of the phyB system are slower, meaning that the steady-state level of P_{fr} will not be reached and will require longer simulation time to simulate. Thus, the numerical solution after short time-periods shall not match the steady-state level under these conditions. Users are advised to take this into consideration when performing their analyses.

Acknowledgments

We would like to thank Andreas Hiltbrunner and Rik van Rosmalen for testing the tool and for providing valuable feedback as to its usage and how it can be made easier for users. RWS is funded by FP7 Marie Curie Initial Training Network grant agreement number

316723 and EU Horizon 2020 grant agreement number 634942. CF is funded by HFSP Research grant RGP0025/2013.

References

1. Rüdiger W, Thümmler F, Cmiel E et al (1983) Chromophore structure of the physiologically active form (Pfr) of phytochrome. Proc Natl Acad Sci U S A 80:6244–6248
2. Li L, Lagarias JC (1992) Phytochrome assembly. J Biol Chem 267:19204–19210
3. Rausenberger J, Hussong A, Kircher S et al (2010) An integrative model for phytochrome B mediated photomorphogenesis: from protein dynamics to physiology. PLoS One 5:e10721
4. Chen M, Schwab R, Chory J (2003) Characterization of the requirements for localization of phytochrome B to nuclear bodies. Proc Natl Acad Sci U S A 100:14493–14498
5. Tepperman JM, Hudson ME, Khanna R et al (2004) Expression profiling of phyB mutant demonstrates substantial contribution of other phytochromes to red-light-regulated gene expression during seedling de-etiolation. Plant J 38:725–739
6. Al-Sady B, Kikis EA, Monte E et al (2008) Mechanistic duality of transcription factor function in phytochrome signaling. Proc Natl Acad Sci U S A 105:2232–2237
7. Park E, Park J, Kim J et al (2012) Phytochrome B inhibits binding of phytochrome-interacting factors to their target promoters. Plant J 72:537–546
8. Pfeiffer A, Nagel MK, Popp C et al (2012) Interaction with plant transcription factors can mediate nuclear import of phytochrome B. Proc Natl Acad Sci U S A 109:5892–5897
9. Schäfer E, Schmidt W (1974) Temperature dependence of phytochrome dark reactions. Planta 116:257–266
10. Medzihradszky M, Bindics J, Ádám É et al (2013) Phosphorylation of phytochrome B inhibits light-induced signaling via accelerated dark reversion in Arabidopsis. Plant Cell 25:535–544
11. Burgie ES, Bussell AN, Walker JM et al (2014) Crystal structure of the photosensing module from a red/far-red light-absorbing plant phytochrome. Proc Natl Acad Sci U S A 111:10179–10184
12. Klose C, Venezia F, Hussong A et al (2015) Systematic analysis of how phytochrome B dimerization determines its specificity. Nat Plants 1:15090
13. Mancinelli AL (1994) The physiology of phytochrome action. In: Kendrick RE, Kronenberg GMH (eds) Photomorphogenesis in plants. Kluwer Academic Publishers, Dordrecht
14. Rausenberger J, Tscheuschler A, Nordmeier W et al (2011) Photoconversion and nuclear trafficking cycles determine phytochrome A's response profile to far-red light. Cell 146:813–825
15. Hennig L, Büche C, Schäfer E (2000) Degradation of phytochrome A and the high irradiance response in Arabidopsis: a kinetic analysis. Plant Cell Environ 23:727–734
16. Smith RW, Helwig B, Westphal AH et al (2016) Unearthing the transition rates between pho toreceptor conformers. BMC Syst Biol 10:110
17. Smith RW, Helwig B, Westphal AH et al (2017) Interactions between phyB and PIF proteins alter thermal reversion reactions in vitro. Photochem Photobiol 93:1525–1531
18. Li H, Zhang J, Vierstra RD et al (2010) Quaternary organization of a phytochrome dimer as revealed by cryoelectron microscopy. Proc Natl Acad Sci U S A 107:10872–10877

Chapter 10

Quantitative Analysis of Photobodies

Chan Yul Yoo, Desiree Williams, and Meng Chen

Abstract

Photobodies are membraneless subnuclear organelles that contain the red and far-red photoreceptors, phytochromes. Photobody biogenesis has been postulated to play important roles in early light signaling events. The size and number of photobodies are highly dynamic in response to the quality and quantity of light and correlated tightly with phytochrome-mediated seedling morphogenesis. Here, we provide a detailed protocol for characterization of the three-dimensional morphology of photobodies, including sample preparation, fluorescence microscopy, and image analysis. Although this method was developed initially for characterizing photobodies, it can be adopted to analyze other membraneless or membrane-bound subcellular organelles.

Key words Photobody, Phytochrome, Nuclear body, Photomorphogenesis, Subnuclear organization

1 Introduction

Photobodies are membraneless photoreceptor-containing, light-sensory subnuclear domains in plants [1]. In Arabidopsis, photobodies contain the red and far-red light photoreceptors phytochrome A to E (phyA-E) [2, 3], the blue light photoreceptor CRY2 (cryptochrome 2) and possibly CRY1 [4–9], the UV-B receptor UVR8 (UV RESISTANCE LOCUS 8) [10], and a battery of light signaling components [11]. Although the interactions among the photoreceptors on photobodies have not been systematically investigated, the dynamic changes in photobody morphology have been well studied using Arabidopsis transgenic lines expressing fluorescent protein-tagged phytochrome B (phyB)—a photostable and the most prominent phytochrome in the light [2, 12, 13]. Phytochromes are bilin-containing proteins that exist in two relatively stable and photoconvertible forms, a red light-absorbing biologically inactive Pr form and a far-red light-absorbing biologically active Pfr form [14, 15]. Besides photoconversion, Pfr can also thermodynamically convert back to Pr in a process called dark reversion or thermoreversion; because the rate of

Andreas Hiltbrunner (ed.), *Phytochromes: Methods and Protocols*, Methods in Molecular Biology, vol. 2026,
https://doi.org/10.1007/978-1-4939-9612-4_10,

thermoreversion of phyB is accelerated with an increase in temperature, phyB is also considered a thermosensor [16, 17]. One of the earliest light responses at the cellular level is the translocation of photoactivated phytochromes from the cytoplasm to the nucleus and their subsequent localization to photobodies [1, 2, 12]. The biogenesis of photobodies is promoted by the Pfr form of phyB and therefore is highly dynamic in response to changes in light quality and quantity as well as temperature [13, 17, 18].

The steady-state patterns of phyB-FP photobodies correlate with early light signaling events as well as downstream phytochrome-mediated morphological responses. Formation of a few large phyB-FP photobodies with diameters between 1 and 2 μm under high intensity of red light correlates with degradation of the hypocotyl growth-promoting transcription factor PIF3 (PHYTOCHROME-INTERACTING FACTOR 3) and phyB-mediated inhibition of hypocotyl growth [13, 19–21]. By contrast, disassembly of large phyB-FP photobodies into tens to hundreds of small photobodies by inactivation of phyB or in mutants of phyB signaling correlates with the accumulation of PIF3 and promotion of hypocotyl growth [18, 19, 22–24]. These results, combined with the data that PIF3 is localized to photobodies before its degradation during the dark-to-light transition [25, 26], support the model that photobodies play important roles in PIF3 degradation and PIF3-dependent regulation of light-responsive genes [23]. Similarly, the dynamics of phyB photobodies is also associated with the hypocotyl response to changes in temperature [17]. Taken together, although the precise function of photobodies is still poorly understood, the steady-state pattern of photobodies can be used as a readout for early light signaling events and even genetic screens to identify components involved in early steps of phytochrome signaling [13, 19, 21].

The dynamic nature of photobodies presents two major challenges in characterization of photobody morphology. The first challenge is that exposure of living plant cells to the excitation light for FPs in fluorescence or confocal microscopy will alter the steady-state pattern of photobodies, and therefore, introduce experimental errors into photobody characterization. Based on our experience, this is particularly true for plant materials grown in the dark or dim light, as the strong excitation light (e.g., 488 nm excitation light for GFP) will promote rapid formation of small photobodies. The second challenge lies in the difficulty to quantitatively measure the number and volume of various sized photobodies. Here we describe an improved photobody characterization protocol that circumvents these obstacles by fixing plant materials without denaturing the fluorescent protein and analyzing the three-dimensional morphology of photobodies using Huygens Essential software. This method has been applied to characterize photobodies under a variety of light conditions in both wild-type and mutant backgrounds [21, 22, 24].

2 Materials

2.1 Plant Materials, Plant Growth Supplies, and Equipment

1. *PBG* (*phyB-GFP*), Arabidopsis transgenic line expressing phyB-GFP in *phyB-5* background [2].
2. ½ Murashige and Skoog (½ MS) with Gamborg's vitamins (Caisson Labs, MSP06) and 0.8% phyto agar.
3. LED chamber (Percival Scientific, Perry, IA).
4. Spectroradiometer (model PS-200, Apogee Instruments Inc., Logan, UT).

2.2 Reagents

1. 1% (v/v) paraformaldehyde (Electron Microscopy Sciences, 15710) (*see* **Note 1**).
2. 50 mM NH_4Cl in PBS.
3. 0.2% (v/v) Triton X-100 in PBS.
4. 300 nM DAPI in PBS.
5. ProLong Diamond Antifade Mountant (Thermo Fisher Scientific, P36965).
6. Superfrost Plus Micro Slide.
7. Micro cover glasses, rectangular, 22 × 40 mm, No. 1.5.

2.3 Imaging and Image Analysis

1. Confocal microscope or fluorescence microscope with deconvolution software (*see* **Note 2**).
2. Huygens Essential software (Scientific Volume Imaging, The Netherlands).

3 Methods

3.1 Preparation of Plant Materials

1. Sterilize *PBG* seeds and sow on ½ MS plate.
2. Stratify the seeds in the dark at 4 °C for 5 days.
3. Grow *PBG* seedlings under monochromatic 10 μmol m^{-2} s^{-1} red light at 21 °C for 4 days (*see* **Note 3**) or adjust the light/temperature conditions according to your specific experiments.

3.2 Fixation and Mounting

1. Fix seedlings with 1% paraformaldehyde in PBS under vacuum for 10 min under the respective growth conditions (*see* **Note 4**).
2. Wash seedlings 3× with 50 mM NH_4Cl in PBS for 5 min each to quench the residual amount of fixative.
3. Permeabilize seedlings with 0.2% Triton X-100 in PBS for 5 min.
4. Stain nuclei with 300 nM DAPI in PBS for 10 min.
5. Wash seedlings 3× with PBS for 5 min each (*see* **Note 5**).

6. Mount a few seedlings in ProLong Diamond Antifade on a slide with a coverslip and incubate overnight in the dark to cure the sample (*see* **Note 6**).
7. Seal the edges of the coverslip with nail polish. The slide can be imaged immediately or stored at 4 °C in the dark (*see* **Note 7**).

3.3 Fluorescence Imaging and Photobody Quantification

1. Select a specific cell type to characterize photobody morphology. We found that the morphology of photobodies varies dramatically in different cell types and even in the same cell type under different modes of growth. To demonstrate a correlation between photobody biogenesis and the regulation of hypocotyl growth by light, In our previous studies we have focused on the epidermal cells in the upper one third of the hypocotyl [13, 19, 21, 22], because these cells are in the growing zone along the hypocotyl on the fourth day after germination when the seedlings are imaged [27, 28].
2. PhyB-GFP photobodies can be imaged with either a confocal microscope or a fluorescence microscope with a 63× or 100× oil immersion objective (*see* **Note 8**). Three-dimensional image stacks of individual nuclei containing photobodies are acquired. For confocal microscopy, we take optical sections with a Z step of 0.70 μm. For fluorescence microscopy, we take optical sections with a Z step of 0.24 μm. For statistical analysis, we collect photobody images from at least 50 nuclei from three biological replicates.
3. Load image stacks into Huygens Essential. Three-dimensional stacks taken by a fluorescence microscope can be deconvolved by Huygens Essential software or other deconvolution software before imported into Huygens Essential.
4. Analyze the number and volume of photobodies using Object Analyzer (Fig. 1). The threshold and seeds are set as default (*see* **Note 9**).
5. Export the data and calculate the number of large or small photobodies for each nuclei using Excel (*see* **Note 10**).

4 Notes

1. This paraformaldehyde solution is methanol-free, which is important to prevent the fluorescent protein from denaturation during fixation [29].
2. We used a Zeiss LSM510 inverted confocal microscope equipped with a Plan-Apochromat 100×/1.4 Oil DIC objective and a Zeiss Axio Observer Z1 inverted fluorescence microscope equipped with a Plan-Apochromat 100×/1.4 Oil

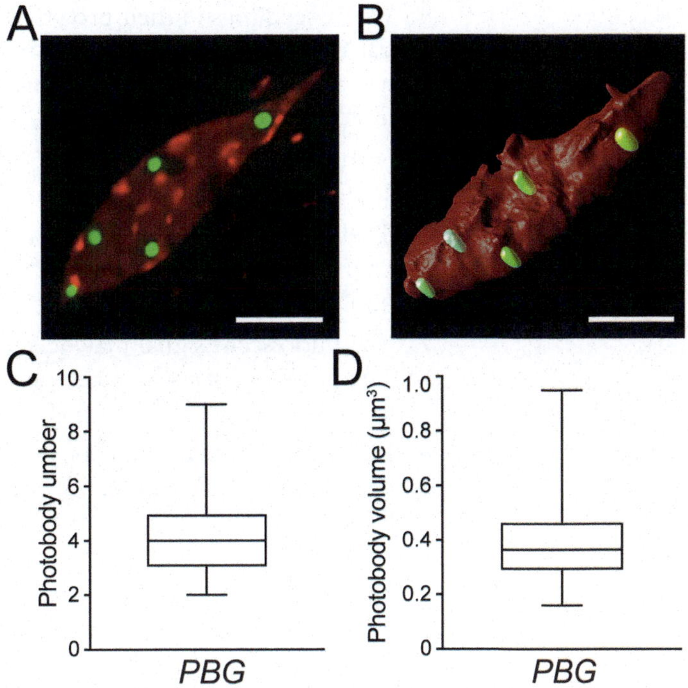

Fig. 1 Quantitative analysis of photobody morphology. Photobodies were analyzed in nuclei of epidermal cells in the upper 1/3 region of hypocotyls of 4-day-old ***PBG*** seedlings grown under 10 μmol m^{-2} s^{-1} red light. Three-dimensional image stacks of individual nuclei were acquired at room temperature with a Zeiss Axio Observer Z1 fluorescence microscope equipped with a Plan-Apochromat 100×/1.4 Oil DIC objective and an Axiocam 506 mono CCD camera (Carl Zeiss, Thornwood, NY). Filters used for GFP were exciter, 470/40 nm/nm; emitter, 525/50 nm/nm; and FT 495 nm beamsplitter (Zeiss Filter Set 38 HE). Filters used for DAPI were exciter, G 365 nm; emitter, 445/50 nm/nm; FT 395 beamsplitter (Zeiss Filter Set 49). The image stacks of nuclei with a ***Z*** step size of 0.24 μm were subjected to Quick Maximum Likelihood Estimation deconvolution by using Huygens Essential software (Scientific Volume Imaging) on a HP Z840 Workstation (HP, Palo Alto, CA) (**a**) Representative maximum projection image of a DAPI-stained (red) nucleus with photobodies (green). Scale bar equals to 5 μm (**b**) Three-dimensional image of the nucleus shown in (**a**) reconstructed by using Huygens Essential Object Analyzer (**c**, **d**) Box and whisker plots showing the number (**c**) and volume (**d**) of the photobodies described above. The boxes represent from 25th to 75th percentile; the bars equal to the median values. The data were from total 66 nuclei

DIC objective and an Axiocam 506 mono CCD camera (Carl Zeiss, Thornwood, NY).

3. This is the standard condition where phyB-GFP is localized to only large photobodies [13]. When characterizing a new light signaling mutant in the *PBG* background, we look at its steady-state photobody morphology under this condition first, and then alter either the light intensity or the temperature.

4. We fix seedlings in their growth condition to minimize potential changes in photobody morphology during fixation. Do not fix samples on ice because temperature may affect the morphology of photobodies.
5. Seedlings can be stored in PBS at 4 °C. We normally mount and image the seedlings within a week.
6. Seedlings can also be mounted with PBS for imaging. However, ProLong Diamond Antifade is recommended to reduce photobleaching.
7. The fluorescent signal may decrease slowly at 4 °C over time. Therefore, we usually image photobodies within a week after fixation.
8. Either a confocal microscope or a fluorescence microscope can be used for imaging nuclei of epidermal cells. However, a confocal microscope is recommended for imaging nuclei of inner cells.
9. Threshold and seeds are automatically calculated from the image. Do not change threshold and seeds, which will affect the size of photobodies.
10. PhyB-GFP is localized only to large photobodies under 10 μmol m^{-2} s^{-1} red light. We therefore define large photobodies based on the volumes of photobodies under 10 μmol m^{-2} s^{-1} red light.

Acknowledgments

This work was supported by National Institute of Health grant R01GM087388 to M.C.

References

1. Chen M, Chory J (2011) Phytochrome signaling mechanisms and the control of plant development. Trends Cell Biol 21:664–671
2. Yamaguchi R, Nakamura M, Mochizuki N, Kay SA, Nagatani A (1999) Light-dependent translocation of a phytochrome B-GFP fusion protein to the nucleus in transgenic *Arabidopsis*. J Cell Biol 145:437–445
3. Kircher S, Gil P, Kozma-Bognár L, Fejes E, Speth V, Husselstein-Müller T, Bauer D, Ádám É, Schäfer E, Nagy F (2002) Nucleocytoplasmic partitioning of the plant photoreceptors phytochrome A, B, C, D, and E is regulated differentially by light and exhibits a diurnal rhythm. Plant Cell 14:1541–1555
4. Kleiner O, Kircher S, Harter K, Batschauer A (1999) Nuclear localization of the Arabidopsis blue light receptor cryptochrome 2. Plant J 19:289–296
5. Yu X, Sayegh R, Maymon M, Warpeha K, Klejnot J, Yang H, Huang J, Lee J, Kaufman L, Lin C (2009) Formation of nuclear bodies of Arabidopsis CRY2 in response to blue light is associated with its blue light–dependent degradation. Plant Cell 21:118–130
6. Gu N-N, Zhang Y-C, Yang H-Q (2012) Substitution of a conserved glycine in the PHR domain of *Arabidopsis* CRYPTOCHROME 1 confers a constitutive light response. Mol Plant 5:85–97

7. Lian H-L, He S-B, Zhang Y-C, Zhu D-M, Zhang J-Y, Jia K-P, Sun S-X, Li L, Yang H-Q (2011) Blue-light-dependent interaction of cryptochrome 1 with SPA1 defines a dynamic signaling mechanism. Genes Dev 25:1023–1028
8. Liu B, Zuo Z, Liu H, Liu X, Lin C (2011) *Arabidopsis* cryptochrome 1 interacts with SPA1 to suppress COP1 activity in response to blue light. Genes Dev 25:1029–1034
9. Más P, Devlin PF, Panda S, Kay SA (2000) Functional interaction of phytochrome B and cryptochrome 2. Nature 408:207–211
10. Favory J-J, Stec A, Gruber H et al (2009) Interaction of COP1 and UVR8 regulates UV-B-induced photomorphogenesis and stress acclimation in *Arabidopsis*. EMBO J 28:591–601
11. Van Buskirk EK, Decker PV, Chen M (2012) Photobodies in light signaling. Plant Physiol 158:52–60
12. Kircher S, Kozma-Bognar L, Kim L, Adam E, Harter K, Schäfer E, Nagy F (1999) Light quality-dependent nuclear import of the plant photoreceptors phytochrome A and B. Plant Cell 11:1445–1456
13. Chen M, Schwab R, Chory J (2003) Characterization of the requirements for localization of phytochrome B to nuclear bodies. Proc Natl Acad Sci U S A 100:14493–14498
14. Rockwell NC, Su Y-S, Lagarias JC (2006) Phytochrome structure and signaling mechanisms. Annu Rev Plant Biol 57:837–858
15. Burgie ES, Vierstra RD (2014) Phytochromes: an atomic perspective on photoactivation and signaling. Plant Cell 26:4568–4583
16. Jung J-H, Domijan M, Klose C et al (2016) Phytochromes function as thermosensors in *Arabidopsis*. Science 354:886–889
17. Legris M, Klose C, Burgie ES, Rojas CCR, Neme M, Hiltbrunner A, Wigge PA, Schäfer E, Vierstra RD, Casal JJ (2016) Phytochrome B integrates light and temperature signals in *Arabidopsis*. Science 354:897–900
18. Klose C, Venezia F, Hussong A, Kircher S, Schäfer E, Fleck C (2015) Systematic analysis of how phytochrome B dimerization determines its specificity. Nat Plants 1:15090
19. Chen M, Galvão RM, Li M, Burger B, Bugea J, Bolado J, Chory J (2010) *Arabidopsis* HEMERA/pTAC12 initiates photomorphogenesis by phytochromes. Cell 141:1230–1240
20. Galvão RM, Li M, Kothadia SM, Haskel JD, Decker PV, Van Buskirk EK, Chen M (2012) Photoactivated phytochromes interact with HEMERA and promote its accumulation to establish photomorphogenesis in *Arabidopsis*. Genes Dev 26:1851–1863
21. Huang H, Yoo CY, Bindbeutel R, Goldsworthy J, Tielking A, Alvarez S, Naldrett MJ, Evans BS, Chen M, Nusinow DA (2016) PCH1 integrates circadian and light-signaling pathways to control photoperiod-responsive growth in *Arabidopsis*. elife 5:e13292
22. Van Buskirk EK, Reddy AK, Nagatani A, Chen M (2014) Photobody localization of phytochrome B is tightly correlated with prolonged and light-dependent inhibition of hypocotyl elongation in the dark. Plant Physiol 165:595–607
23. Qiu Y, Li M, Pasoreck EK et al (2015) HEMERA couples the proteolysis and transcriptional activity of PHYTOCHROME INTERACTING FACTORs in Arabidopsis photomorphogenesis. Plant Cell 27:1409–1427
24. Qiu Y, Pasoreck EK, Reddy AK, Nagatani A, Ma W, Chory J, Chen M (2017) Mechanism of early light signaling by the carboxy-terminal output module of *Arabidopsis* phytochrome B. Nat Commun 8:1905
25. Bauer D, Viczián A, Kircher S et al (2004) CONSTITUTIVE PHOTOMORPHOGENESIS 1 and multiple photoreceptors control degradation of PHYTOCHROME INTERACTING FACTOR 3, a transcription factor required for light signaling in Arabidopsis. Plant Cell 16:1433–1445
26. Al-Sady B, Ni W, Kircher S, Schäfer E, Quail PH (2006) Photoactivated phytochrome induces rapid PIF3 phosphorylation prior to proteasome-mediated degradation. Mol Cell 23:439–446
27. Gendreau E, Traas J, Desnos T, Grandjean O, Caboche M, Höfte H (1997) Cellular basis of hypocotyl growth in *Arabidopsis thaliana*. Plant Physiol 114:295–305
28. Wang Z-Y, Nakano T, Gendron J et al (2002) Nuclear-localized BZR1 mediates brassinosteroid-induced growth and feedback suppression of brassinosteroid biosynthesis. Dev Cell 2:505–513
29. Nybo K (2012) GFP imaging in fixed cells. BioTechniques 52:359–360

Chapter 11

Regulation of Alternative Splicing by Phytochrome

Tomonao Matsushita

Abstract

Recently, phytochrome has been shown to regulate not only transcription but also alternative splicing at a similar genomic scale in *Arabidopsis*. Here I describe the protocols for detecting phytochrome-mediated light-responsive alterations in alternative splicing in *Arabidopsis*, using RT-PCR technique followed by quantification with the Bioanalyzer.

Key words Phytochrome, Alternative splicing, Regulation of gene expression

1 Introduction

Plants adapt to their fluctuating environment by monitoring surrounding light conditions through several informational photoreceptors, among which phytochrome is the most important and the best characterized. Phytochrome can be reversibly activated and inactivated by absorbing red and far-red light, respectively, which enables plants to sense red/far-red light ratio in the environments [1–3]. Upon absorption of red light, activated phytochromes are translocated from the cytoplasm to the nucleus, where they inhibit several transcription factors to induce genome-wide changes in mRNA levels of the target genes, thereby mediating various light responses in plants [4, 5].

Accumulating evidence suggests that light regulates not only transcription but also other aspects of gene expression [6]. It has long been reported that light affects the alternative splicing of several genes in plants [7–9]. However, the identity of the photoreceptors responsible for these light-regulated alternative splicing events remained elusive. Through an extensive forward genetic screen, RRC1, a novel SR protein involved in alternative splicing control, was identified as a positive regulator of phytochrome signaling [10]. This finding suggested that the regulation of alternative splicing plays an important role in phytochrome signal transduction. Then recently, mRNA sequencing (mRNA-seq) analysis of red

Andreas Hiltbrunner (ed.), *Phytochromes: Methods and Protocols*, Methods in Molecular Biology, vol. 2026,
https://doi.org/10.1007/978-1-4939-9612-4_11,

light-responsive and phytochrome-dependent transcriptomic changes during seedling de-etiolation revealed that phytochrome controls not only transcription, but also alternative splicing, at a similar genomic scale to mediate light responses in *Arabidopsis* [11]. Coincidentally, through a similar mRNA-seq analysis, phytochrome-regulated alternative splicing was also independently confirmed in the moss *Physcomitrella* [12], strongly suggesting that this mechanism is widely conserved in plants. Here I describe detailed methods for detecting and quantifying phytochrome signal-induced changes in alternative splicing patterns in *Arabidopsis.*

2 Materials

1. *Arabidopsis thaliana* seeds of wild-type Columbia and phytochrome-deficient *phyA-211 phyB-9* double mutants [13, 14].
2. RNeasy Plant Mini Kit (Qiagen).
3. RNase-free DNase set (Qiagen).
4. RNase-free micropipette filter tips.
5. RNase & DNase-free 2.0 mL microcentrifuge tubes (Cat.# 508-GRD-Q, QSP, San Diego, USA).
6. 6 mm zirconia beads, dry-heat sterilized.
7. Tissue homogenizer 2010 Geno/Grinder (SPEX SamplePrep, USA).
8. SuperScript III First-Strand Synthesis System for RT-PCR (Invitrogen).
9. *Taq* DNA polymerase Ex Taq (TaKaRa).
10. *2100 Bioanalyzer (Agilent).*
11. *DNA 7500 Kit (Agilent).*
12. 0.6% agar plates containing 1× Murashige and Skoog medium without sucrose.
13. Light-emitting diodes with emission maximum at 634 nm.

3 Methods

Genome-wide analysis of phytochrome-controlled alternative splicing by mRNA sequencing is described elsewhere [11]. Here, an RT-PCR based procedure is described to validate phytochrome-controlled alternative splicing identified by genome-wide approaches and to investigate phytochrome-regulated alternative splicing of candidate transcripts. Since multiple splice variants cannot be quantified at the same time with a single primer set,

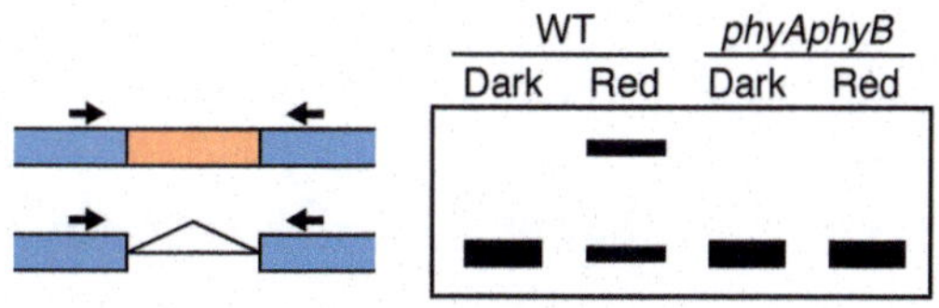

Fig. 1 Schematic of typical RT-PCR gel image of phytochrome-controlled alternative splicing (right). Gene-specific primers are designed as indicated by arrows to detect red light-responsive and phytochrome-dependent intron retention in the target gene (left). Blue, exon; orange, intron; black arrows, primers

quantitative RT-PCR is inappropriate for analyzing alternative splicing in an unbiased manner; however, it is very useful for verification of data from genome-wide approaches and for the investigation of specific splicing events of candidate transcripts. The primers for detection of a specific splicing event should be designed as shown in Fig. 1.

Perform all the procedures described here at room temperature, unless otherwise stated. Use RNase-free micropipette filter tips from total RNA extraction (Subheading 3.1, **step 5**) until cDNA synthesis (Subheading 3.2, **step 1**) to avoid contamination with RNase.

3.1 Sample Preparation and Total RNA Extraction

1. Grow 4-day-old etiolated *Arabidopsis* wild-type and *phyA phyB* double mutant seedling (*see* **Notes 1** and **2**) on 0.6% agar plates containing Murashige–Skoog salt mixture without sucrose (*see* **Note 3**), and expose them to continuous red light (8–10 μmol m^{-2} s^{-1}) for 0 or 1 h (*see* **Note 4**).
2. Put one zirconia bead into an empty 2-mL microcentrifuge tube.
3. Harvest the seedlings with forceps, and gently press them onto a soft paper towel to remove excess water (*see* **Note 5**). Then, put the sample into the 2-mL microcentrifuge tube containing a zirconia bead, firmly close the cap, and immediately place it in liquid nitrogen. These procedures should be carried out as quickly as possible (*see* **Note 6**) under dim green safe light in a dark room. At this stage, samples can be stored at −80 °C for several years.
4. Set the frozen microcentrifuge tubes in the tissue homogenizer 2010 Geno/Grinder, disrupt the samples at 1500 strokes/min for 20 s, and freeze the samples again in liquid nitrogen. Repeat this process twice.
5. Purify total RNA using an RNeasy Plant Mini Kit (Qiagen) and treat with RNase-free DNase (Qiagen) on the column, according to the manufacturer's instructions.
6. Measure the concentration and A_{260}/A_{280} value of the purified total RNA using a spectrophotometer, and confirm that A_{260}/A_{280} value is not less than two.

7. Add RNase-free water or freshly obtained ultrapure water to dilute the total RNA solution to the concentration of 0.5 μg/μL. If necessary, RNA samples can be stored at −80 °C for years.
8. For quality check, electrophorese 0.25–0.5 μg of the purified total RNA on 1–2% standard agarose gel for DNA (*see* **Note 7**).

3.2 Complementary DNA Synthesis and Reverse Transcription PCR

1. Use 1 μg of the total RNA obtained and synthesize complementary DNA (cDNA) using SuperScript III First-Strand Synthesis System for RT-PCR (Invitrogen) with oligo $(dT)_{20}$ primers, according to the manufacturer's instructions.
2. Design gene-specific primers for reverse transcription PCR (RT-PCR) to detect phytochrome-controlled alternative splicing in the target genes (*see* Fig. 1 and **Note 8**).
3. For each target gene, perform preliminary PCRs at different cycles (24, 28, 32, and 36 cycles) to optimize the number of PCR cycles (*see* **Note 9**).
4. Use cDNA synthesized that is equivalent to 50 ng of total RNA and perform RT-PCR with gene-specific primers in a 20-μL reaction volume. The basic PCR condition is as follows: 96 °C for 1 min; 96 °C for 30 s, 56 °C for 30 s, and 72 °C for 1 min for cycles determined in **step 3**; and 4 °C forever. If necessary, the annealing temperature should be modified depending on the primers used.
5. Electrophorese 5 μL of the 20-μL PCR reaction solution on 1–2% agarose gel to check the pattern of amplified bands.
6. The specific bands should be confirmed by subcloning and sequencing all the bands detected in each RT-PCR reaction.
7. Quantify the concentrations of amplified DNA fragments corresponding to splice variants, using the Agilent DNA 7500 Kit and 2100 Bioanalyzer (Agilent) or any other capillary electrophoresis devices such as MultiNA (SHIMADZU) and Genetic Analyzers (Applied Biosystems), according to the manufacturer's instructions (*see* **Note 10**).
8. Red light-responsive and phytochrome-dependent changes in the ratio among splice variants are regarded as being phytochrome-controlled alternative splicing (Fig. 1).

4 Notes

1. In this condition, approximately 20 μg of total RNA is obtained from 4-day-old etiolated (i.e., dark-grown) *Arabidopsis* seedlings derived from 200 seeds. Total RNA yield is much higher in light-grown seedlings or dark-grown seedlings of constitutively photomorphogenic mutants.

2. I use *phyA phyB* double mutants because phyA and phyB are the dominant members of the phytochrome family in *Arabidopsis* that function in a partially redundant manner in seedling de-etiolation [2].
3. It is well known that sucrose strongly affects both phytochrome signaling [15] and alternative splicing in many genes [16]. Therefore, sucrose should not be added to the medium, in order to more specifically detect phytochrome-controlled alternative splicing.
4. In order to analyze direct effects of phytochrome signals on the transcriptome, seedling samples with no difference in morphology should be compared, because morphological changes alone induce global alterations in alternative splicing patterns. Neither phytochrome mutation nor 1-h irradiation with continuous red light at fluence rates of 8–10 μmol m^{-2} s^{-1} cause morphological changes in etiolated seedlings. Higher fluence rates of red light should be avoided, not only because they cause morphological changes, but also because stronger red light induces stress responses and produces more sucrose through photosynthesis, both of which affect alternative splicing in a number of genes.
5. Excess water from the medium is frozen into hard ice, which makes it difficult to achieve thorough disruption of the sample.
6. It has been reported that some stress-responsive genes show very rapid increase in mRNA levels as early as less than 1 min after stress treatment [17, 18].
7. Loading dye containing SDS should be used to avoid degradation of RNA samples by RNase.
8. Amplified DNA fragments should be as small as possible but more than approximately 100 bp, which increases both sensitivity and resolution in detecting splice variants with 2100 Bioanalyzer.
9. In order to avoid heteroduplex formation or biased amplification among different fragments, the number of PCR cycles should be as few as possible within the linear range of amplification.
10. Detection resolution gets lower when DNA fragments are more than 2000 bp. For detecting DNA fragments less than 1200 bp, Agilent DNA 1000 Kit (Agilent) can also be used. However, even in that size range, I usually obtained better results with Agilent DNA 7500 Kit, in terms of both sensitivity and resolution. Agilent DNA 7500 Kit is especially useful when the size of PCR products is not yet determined, as I often detect unexpectedly large fragments in this RT-PCR analysis. In those cases, the extension time at 72 °C should be prolonged to 1.5 min in the PCR program (*see* Subheading 3.2, **step 4**).

Acknowledgments

I thank Drs. Tomokazu Ushijima (Kyushu University, Japan) and Hiromasa Shikata (Nagoya University, Japan) for helping to develop these methods. This study was supported by JST PRESTO Grant Number JPMJPR13B9; JSPS KAKENHI Grant Numbers JP25120720, JP16H01470, JP16H04808, JP16H06279, JP17H05727, JP18H04784, and JP18H04001; and a research grant from Takeda Science Foundation.

References

1. Galvão VC, Fankhauser C (2015) Sensing the light environment in plants: photoreceptors and early signaling steps. Curr Opin Neurobiol 34:46–53
2. Franklin KA, Quail PH (2010) Phytochrome functions in Arabidopsis development. J Exp Bot 61:11–24
3. Casal JJ (2013) Photoreceptor signaling networks in plant responses to shade. Annu Rev Plant Biol 64:403–427
4. Leivar P, Monte E (2014) PIFs: systems integrators in plant development. Plant Cell 26:56–78
5. Menon C, Sheerin DJ, Hiltbrunner A (2016) SPA proteins: SPAnning the gap between visible light and gene expression. Planta 244:297–312
6. Wu SH (2014) Gene expression regulation in photomorphogenesis from the perspective of the central dogma. Annu Rev Plant Biol 65:311–333
7. Mano S, Hayashi M, Nishimura M (1999) Light regulates alternative splicing of hydroxypyruvate reductase in pumpkin. Plant J 17:309–320
8. Simpson CG, Fuller J, Maronova M, Kalyna M, Davidson D, McNicol J et al (2008) Monitoring changes in alternative precursor messenger RNA splicing in multiple gene transcripts. Plant J 53:1035–1048
9. Jung KH, Bartley LE, Cao P, Canlas PE, Ronald PC (2009) Analysis of alternatively spliced rice transcripts using microarray data. Rice 2:44–55
10. Shikata H, Shibata M, Ushijima T, Nakashima M, Kong SG, Matsuoka K et al (2012) The RS domain of Arabidopsis splicing factor RRC1 is required for phytochrome B signal transduction. Plant J 70:727–738
11. Shikata H, Hanada K, Ushijima T, Nakashima M, Suzuki Y, Matsushita T (2014) Phytochrome controls alternative splicing to mediate light responses in Arabidopsis. Proc Natl Acad Sci U S A 111:18781–18786
12. Wu HP, Su YS, Chen HC, Chen YR, Wu CC, Lin WD et al (2014) Genome-wide analysis of light-regulated alternative splicing mediated by photoreceptors in *Physcomitrella patens*. Genome Biol 15:R10
13. Reed JW, Nagatani A, Elich TD, Fagan M, Chory J (1994) Phytochrome A and phytochrome B have overlapping but distinct functions in Arabidopsis development. Plant Physiol 104:1139–1149
14. Reed JW, Nagpal P, Poole DS, Furuya M, Chory J (1993) Mutations in the gene for the red/far-red light receptor phytochrome B alter cell elongation and physiological responses throughout Arabidopsis development. Plant Cell 5:147–157
15. Dijkwel PP, Huijser C, Weisbeek PJ, Chua NH, Smeekens SC (1997) Sucrose control of phytochrome A signaling in Arabidopsis. Plant Cell 9:583–595
16. Hartmann L, Drewe-Boß P, Wießner T, Wagner G, Geue S, Lee HC et al (2016) Alternative splicing substantially diversifies the transcriptome during early photomorphogenesis and correlates with the energy availability in Arabidopsis. Plant Cell 28:2715–2734
17. Suzuki N, Devireddy AR, Inupakutika MA, Baxter A, Miller G, Song L et al (2015) Ultrafast alterations in mRNA levels uncover multiple players in light stress acclimation in plants. Plant J 84:760–772
18. Moore M, Vogel M, Dietz K (2014) The acclimation response to high light is initiated within seconds as indicated by upregulation of AP2/ERF transcription factor network in *Arabidopsis thaliana*. Plant Signal Behav 9:976479

Chapter 12

Phytochrome Regulation of Seed Germination

Kijong Song and Giltsu Choi

Abstract

Seed germination assays consist of counting the number of germinated seeds, defined as seeds in which the radicle has ruptured the endosperm and emerged from the seed coat. In Arabidopsis seed germination assays, Arabidopsis seeds are surface-sterilized, plated on agar plates containing test compounds, and incubated at specific temperatures under specific light conditions, after which the germinated seeds are counted, either with the naked eye or under a microscope. This chapter describes step-by-step protocols for Arabidopsis seed germination assays under phytochrome-dependent conditions.

Key words Seed germination, Seed dormancy, Phytochrome, phyA, phyB, Light

1 Introduction

Seed germination determines the season of seedling emergence and the physical site of plants. To germinate under optimal conditions, plant seeds must carefully monitor environmental conditions and germinate only when these conditions are favorable for growth. Environmental conditions regulating seed germination include temperature, light, water, oxygen, and nutrients. Seed germination, however, is not controlled solely by environmental conditions but also by an internal condition called seed dormancy, an internal state preventing seed germination even under favorable environmental conditions. Thus, seeds can germinate only if the sum of germination-promoting signals overcomes the depth of seed dormancy and positively outweighs the sum of germination-inhibiting signals [1, 2].

Because dormant seeds do not germinate even under favorable environmental conditions, seed dormancy status must be evaluated before performing germination assays. Seed dormancy is established during seed maturation. Thus dormancy is strong in freshly harvested seeds and gradually weakens after harvesting through the process of after-ripening [3]. Currently, the depth of seed dormancy can be determined only by a posteriori tests of seed

Andreas Hiltbrunner (ed.), *Phytochromes: Methods and Protocols*, Methods in Molecular Biology, vol. 2026,
https://doi.org/10.1007/978-1-4939-9612-4_12,

germination. Assays that test the effects of environmental conditions on seed germination should therefore be performed using seeds with similar depths of dormancy. Because it is difficult to determine the depth of seed dormancy prior to germination, seed germination assays that test the effects of environmental conditions are usually performed using fully after-ripened seeds due to their low seed dormancy levels.

Phytochromes are red and far-red light photoreceptors that mediate light conditions to promote seed germination [4–6]. Arabidopsis possesses five phytochromes: one phyA type, three phyB-types (phyB, phyD, and phyE), and one phyC type, with phyA and phyB being the major phytochromes that promote seed germination in response to different light conditions [7]. PhyA mediates the very low fluence response (VLFR) and far-red high irradiance response (FR-HIR), whereas phyB mediates red light low fluence response (LFR), promoting seed germination. The VLFR is irreversibly induced by a wide spectrum of very low fluence light (10^{-4} to 10^{-1} μmol m^{-2}); the LFR is reversibly induced by low fluence red light (1 to 10^{3} μmol m^{-2}) but nullified by far-red light; and the FR-HIR is irreversibly induced by prolonged far-red light [4–6, 8]. Thus, specific light regimens should be used to determine the effects of different phytochromes on seed germination. This chapter describes step-by-step protocols for phyA- and phyB-dependent seed germination assays.

2 Materials

2.1 Light Chambers

1. The LED light chambers used have either red (660~670 nm) or far-red (730~740 nm) LEDs: Controllers of light intensity, irradiation time, and temperature should be present. Light chambers should be inspected carefully with a photoradiometer to ensure that no light enters the chamber through crevices. For more sophisticated light treatments, LED light chambers should contain both red and far-red LEDs. LED chambers should be placed in a dark room. When the door is closed, the dark room should be completely dark.
2. Other light chambers should have white LEDs or fluorescent lights, as well as controllers of temperature, irradiation time, and preferably light intensity.

2.2 Seeds

Because seeds of good uniform quality are critical for germination assays, Arabidopsis plants should be carefully grown and monitored until harvested. The following steps describe our procedures for growing and harvesting Arabidopsis seeds for germination assays.

1. Pots (11 cm, 0.75 L) with moistened autoclaved soil mix are prepared, with the number of pots equal to the number of genotypes being compared. The soil should be watered until dripping using a watering spray gun and drain.
2. Two seeds are placed at four equidistant positions within a pot using a toothpick. The pots are covered with plastic wrap and labeled clearly.
3. The pots are placed onto a single plant tray (*see* **Note 1**), which is placed on a shelf in a growth room, maintained at 22 °C with a 16 h/8 h day/night cycle of light using a white LED (100 μmol/m^{-2} s^{-1}). One week later, the plastic wrap is removed, with only one seedling left in each position (four plants per pot).
4. The plant tray is filled with Hyponex (N-P-K:6-10-5) liquid fertilizer, diluted 1:2500. The tray is refilled whenever the top soil becomes slightly dry (*see* **Note 2**).
5. A sterilized vinyl covered steel rod is placed in each pot. Inflorescence stems in each pot are loosely tied to the steel rod using wire, preventing entanglement of stems from neighboring pots.
6. When more than half the siliques turn brown, seeds are harvested by gently tapping on a sheet of paper (*see* **Note 3**). The paper is folded into the shape of an envelope containing seeds. Each envelope is labeled with the genotype and harvesting date.
7. For after-ripening, the seed envelopes are placed in a chamber, maintained at 22 °C, for 2 months. The seeds are then transferred into microfuge tubes labeled with the genotype and harvesting date for storage. These seeds are ready for use in germination assays.

2.3 Media, Solutions, and Consumables

1. Diluted MS-agar plates for germination assay (*see* **Note 4**): To make 1 L of medium, 0.044 g MS salt, 0.5 g MES, adjusted to pH 5.7 with 5 M KOH, and 8 g phytoagar are added to a flask containing 1 L sterile distilled water. The flask is autoclaved for 15 min and allowed to cool to 60 °C. If necessary, filter-sterilized chemicals, such as antibiotics and hormones, are added. The plates are poured on a clean bench, covered with lids, and allowed to be dried overnight. The plates are labeled with type of medium, date, and initials, and placed in a vinyl bag. Each bag is sealed with tape, labeled, and stored in a refrigerator. The plates are allowed to warm to room temperature before use.
2. Seed sterilization solution: To make a sterilization solution containing 0.6% NaClO and 0.0175% Triton X-100, 7.5 mL of 4% (v/v) NaClO (Chlorox) and 17.5 mL of 0.05% (v/v) Triton X-100 are added to a 50 mL disposable sterile tube, with the volume made up to 50 mL with sterile distilled water.

3. Pots (11 cm, 0.75 L).
4. Soil (zeolite 7%, perlite 7%, vermiculite 3%, cocopeat 68%, peat moss 14.729%, fertilizer 0.243%, humectant 0.004%, pH regulator 0.024%)–vermiculite–perlite = 4:1:1 and sterilized with autoclave.
5. Hyponex (N-P-K:6-10-5) liquid fertilizer.
6. Petri dishes.
7. 3 M paper tape.
8. Sterile 1.7 mL microfuge tubes.

3 Methods

3.1 Seed Sterilization

1. Seed sterilization and sowing must be performed on a clean bench with laminar flow to avoid contamination. These steps should follow general guidelines for axenic culture protocols.
2. Sterile 1.7 mL microfuge tubes, each labeled with the appropriate genotype, are placed in a rack, and up to 50 μL of that genotype is added to each tube.
3. Following the addition of 1 mL sterilization solution to each microfuge tube, the tubes are mixed by vortexing for a few minutes. After the seeds sink to the bottom of the tubes, the sterilization solution is removed using a micropipette.
4. After adding 1 mL sterile distilled water, the tubes are mixed by vortexing. The seeds are allowed to sink to the bottom of each tube, and water is removed from each with a micropipette.
5. **Step 4** is repeated at least five more times.
6. After adding 1 mL sterile distilled water to each tube, the seeds are ready for sowing.

3.2 Sowing

1. Diluted MS-agar plates are taken out from the refrigerator and allowed to warm up on a clean bench.
2. Lines are drawn on the bottom of each plate with a marker, dividing each plate into sections. Each section is labeled with a genotype (*see* **Note 5**).
3. About 60 seeds are individually sown using a Pasteur pipette (*see* **Note 6**).
4. The plates are dried until no water remains around the seeds.
5. The plates are sealed with 3 M paper tape and left on the clean bench, followed by irradiation with red or far-red light (*see* Subheadings 3.3 and 3.4) 1 h after the start of sterilization (*see* **Note 7**).

3.3 PhyA-Dependent Seed Germination Assay

1. Three or more sets of seeded plates are prepared.
2. A far-red LED chamber placed in a dark room is set at a fluence rate of 3.2 $\mu mol \cdot m^{-2} \cdot s^{-1}$. Care should be taken that no light enters the room when the door is closed. In addition, the safety green light should be turned off and the indicator light on the LED chamber masked.
3. The paper-taped seeded plates are placed in the far-red LED chamber for 5 min (total fluence: 0.96 $mmol \cdot m^{-2}$) (*see* **Note 8**). While maintaining dark conditions, the plates are taken out and wrapped immediately with a few layers of aluminum foil, ensuring that the seeds are not exposed to any light after FR irradiation. This step inactivates phyB in seeds (*see* **Note 9**).
4. The foil-covered plates are incubated at 22 °C for 48 h to increase phyA protein levels in seeds. This increase enables phyA to induce seed germination in response to light.
5. One set of foil-covered plates is incubated at 22 °C for 5 more days (*see* **Note 10**). This set is used to determine germination frequency in phyAoff (and phyBoff) conditions. Wild-type and *phyA* and *phyB* mutant seeds should not germinate under these conditions.
6. Aluminum foil is removed from a second set of plates and the plates are incubated at 22 °C for 5 days in the presence of continuous white light (50~100 $\mu mol \cdot m^{-2} \cdot s^{-1}$). This set tests the viability of seeds; seeds that do not germinate under these conditions are either dead or dormant. Wild-type and both *phyA* and *phyB* mutant seeds should germinate.
7. Aluminum foil is removed from a third set of plates in the dark and the plates are incubated at 22 °C in the far-red LED chamber for 12 h (*see* **Note 11**).
8. Immediately after far-red light irradiation, the plates are wrapped with a few layers of aluminum foil and incubated at 22 °C for 5 more days, ensuring that these seeds are not exposed to any other light. This set is used to measure germination frequency under phyAon conditions. Both wild type and *phyB* mutant seeds should germinate, whereas *phyA* mutant seeds should not (Fig. 1).

3.4 PhyB-Dependent Seed Germination Assay

1. Three or more sets of seeded plates are prepared.
2. Far-red and red LED chambers set at fluence rates of 3.2 $\mu mol \cdot m^{-2} \cdot s^{-1}$ and 20 $\mu mol \cdot m^{-2} \cdot s^{-1}$, respectively, are placed in a dark room. Care should be taken that no light enters the room when the door is closed. In addition, the safety green light should be turned off and the indicator lights on the LED chambers masked.

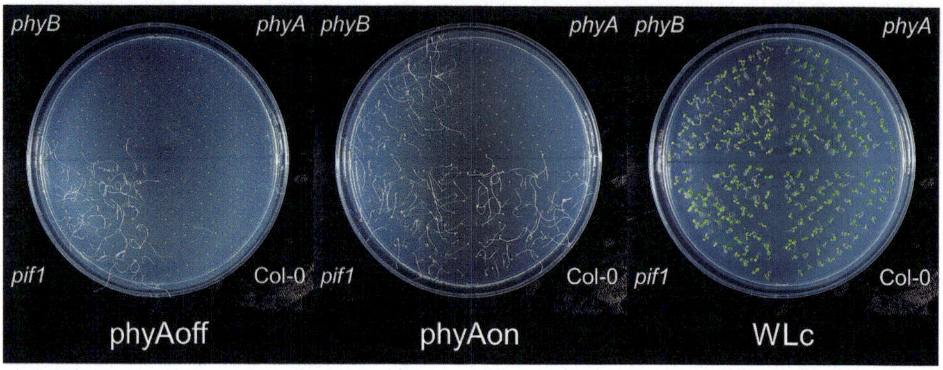

Fig. 1 PhyA-dependent seed germination assay. Sterilized and plated seeds are irradiated with a far-red light pulse and incubated at 22 °C for 48 h in the dark. After 48 h of dark incubation, seeds are irradiated with 0 h of far-red light (phyAoff) or 12 h of far-red light (phyAon). Irradiated seeds are further incubated at 22 °C for 5 days either in the dark or continuous white light (WLc). Wild type (Col-0), *phyA*, *phyB*, and *pif1* mutants are tested. PIF1 is a bHLH transcription factor inhibiting seed germination in the dark [8]

3. One set of seeded plates is placed in the far-red LED chamber for 5 min and immediately wrapped with a few layers of aluminum foil in the dark. The plates are incubated at 22 °C for five more days, making sure that these seeds are not exposed to any other light. This set is used to determine germination frequency in phyBoff (and phyAoff) conditions. Wild type and *phyA*, and *phyB* mutant seeds should not germinate under these conditions.
4. Another set of seeded plates is incubated at 22 °C for 5 more days in the presence of continuous white light (50~100 $\mu mol \cdot m^{-2} \cdot s^{-1}$) after removing aluminum foil. This set tests the viability of seeds; seeds that do not germinate under these conditions are either dead or dormant. Wild type and both *phyA* and *phyB* mutant seeds should germinate.
5. A third set of seeded plates is placed in the far-red LED chamber for 5 min to inactivate phyB in seeds. The plates are then immediately placed in the red LED chamber for 5 min at a fluence of 6 $mmol \cdot m^{-2}$ (*see* **Note 12**).
6. Immediately after red light irradiation, the plates are wrapped with a few layers of aluminum foil and incubated at 22 °C for 5 more days, ensuring that these seeds are not exposed to any other light. This set is used to measure germination frequency under phyBon conditions. Both wild type and *phyA* mutant seeds should germinate, whereas *phyB* mutant seeds should not (Fig. 2).

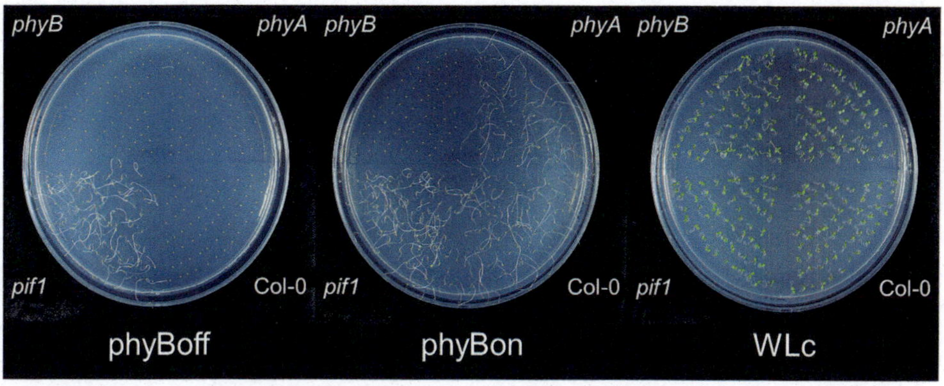

Fig. 2 PhyB-dependent seed germination assay. Sterilized and plated seeds are irradiated with either a far-red light pulse (phyBoff) or a far-red light pulse followed by a red light pulse (phyBon). Irradiated seeds are further incubated at 22 °C for 5 days either in the dark or continuous white light (WLc). Wild type (Col-0), *phyA*, *phyB*, and *pif1* mutants are tested

4 Notes

1. We have found that germination frequency depends on light, temperature, and watering conditions during growth. Therefore, we grow all genotypes in the same tray to ensure that all plants grow under the same conditions. We usually grow each set in triplicate in three different trays and seed them periodically to obtain different seed batches.
2. Plants and soil dryness should be checked every morning and care optimized. Trays should be filled with water on Fridays. Over long holidays or vacations, watering should be arranged with labmates.
3. We have found that germination frequency is reduced if we stop watering and dry plants before harvesting, presumably due to dried immature seeds.
4. We have observed that germination frequency is more reproducible when we use diluted MS-agar plates than when we use aqueous agar plates or moistened filter paper.
5. All genotypes being compared should be sown on the same plate to minimize unforeseen variations.
6. Practice and patience are required to spot seeds individually using a Pasteur pipette. Once mastered, it is more efficient than a micropipette or a toothpick.
7. To equalize the timing of irradiation among experiments, we irradiate 1 h after the start of sterilization. However, similar results were obtained when we irradiated 1 or 2 h after sterilization. Timing of all experiments should be standardized at 1 or 2 h.

8. The fluence we used is sufficient to inhibit phyB-induced germination of Arabidopsis, lettuce, and Nicotiana seeds, but seeds of other species may require a higher far-red fluence to inactivate phyB. To change the total fluence, we recommend to change a fluence rate rather than an irradiation time.
9. If seeded plates are incubated at 22 °C in the dark without preirradiation with far-red light, all seeds germinate due to active phyB in seeds. Thus, phyB must be inactivated by a short far-red pulse before activating phyA.
10. Because seed germination is dependent on temperature, temperature should be carefully monitored and kept constant throughout the germination assay.
11. We usually prepare additional sets of plates and incubate them in far-red LED chambers for 4, 8, and 12 h to determine the dependence of germination frequency on far-red fluence. Under our assay conditions, about 60% of wild type Arabidopsis (Col-0 ecotype) seeds germinate after irradiation for 8 h with far-red light.
12. The fluence we used is sufficient to promote phyB-induced germination of Arabidopsis, lettuce, and Nicotiana seeds, but seeds of other species may require a higher red fluence to activate phyB. To change the total fluence, we recommend to change a fluence rate rather than an irradiation time.

Acknowledgments

We thank the members of our laboratory for many technical tips. This work was supported in part by grants from the National Research Foundation of Korea (2018R1A3B1052617) to G.C.

References

1. Finch-Savage WE, Leubner-Metzger G (2006) Seed dormancy and the control of germination. New Phytol 171:501–523
2. Holdsworth MJ, Bentsink L, Soppe WJJ (2008) Molecular networks regulating Arabidopsis seed maturation, after-ripening, dormancy and germination. New Phytol 179:33–54
3. Hilhorst HWM (2007) Definitions and hypotheses of seed dormancy. In: Bradford KJ, Nonogaki H (eds) Seed development, dormancy and germination. Blackwell Publishing Ltd, Oxford, pp 50–71
4. Shinomura T, Nagatani A, Chory J, Furuya M (1994) The induction of seed germination in *Arabidopsis thaliana* is regulated principally by phytochrome B and secondarily by phytochrome A. Plant Physiol 104: 363–371
5. Shinomura T, Nagatani A, Hanzawa H, Kubota M, Watanabe M, Furuya M (1996) Action spectra for phytochrome A- and B-specific photoinduction of seed germination in *Arabidopsis thaliana*. Proc Natl Acad Sci U S A 93:8129–8133
6. Chory J, Chatterjee M, Cook RK et al (1996) From seed germination to flowering, light controls plant development via the pigment phytochrome. Proc Natl Acad Sci U S A 93:12066–12071
7. Mathews S (2006) Phytochrome-mediated development in land plants: red light sensing evolves to meet the challenges of changing light environments. Mol Ecol 15:3483–3503
8. Oh E, Kim J, Park E, Kim J-I, Kang C, Choi G (2004) PIL5, a phytochrome-interacting basic helix-loop-helix protein, is a key negative regulator of seed germination in *Arabidopsis thaliana*. Plant Cell 16:3045–3058

Chapter 13

Shade Avoidance and Neighbor Detection

Irma Roig-Villanova, Sandi Paulišić, and Jaime F. Martinez-Garcia

Abstract

Plants detect neighboring vegetation as potential competitors for resources. Vegetation proximity is perceived by changes in the red (R) to far-red (FR) ratio (R:FR) through the phytochrome photoreceptors. To face this challenge, many plants have evolved the strategy to avoid shade, displaying a series of responses known as the shade avoidance syndrome (SAS). The SAS responses have been mostly studied at the seedling stage, and cover hypocotyl elongation as well as cotyledon and primary leaf expansion. In adult stages, SAS responses include an increase in petiole elongation and a decrease in leaf expansion, and an increase in plant height. Thus, the analysis of these responses provides a valuable and simple way to study how vegetation proximity affects plant development in both seedlings and adult plants. Here we describe a simple protocol to simulate shade in the laboratory and to evaluate these responses. Overall, our protocol can be easily used to expand the set of SAS responses of plants at different stages of development.

Key words Simulated shade, Seedlings, Adult plants, Hypocotyls, Cotyledons, Leaves, Petioles, Elongation

1 Introduction

Vegetation proximity or canopy shade, a condition that can be associated with low light intensity, is undesirable for many species due to its negative effect on productivity. To deal with vegetation proximity or shade, two alternative strategies have emerged in plants: shade avoidance and shade tolerance [1, 2].

Shade-avoiding species display a variety of adaptive responses collectively referred to as the shade avoidance syndrome (SAS), many of which involve some form of organ elongation to escape from shade and better capture light. *Arabidopsis thaliana* seedlings, an excellent model to study SAS, strongly elongate their hypocotyls in shade, as well as expand cotyledons and primary leaves [3], a set of traits that form a typical SAS response in seedlings. Similarly, adult Arabidopsis plants elongate the petioles of the rosette leaves and place them in more upward position to

Andreas Hiltbrunner (ed.), *Phytochromes: Methods and Protocols*, Methods in Molecular Biology, vol. 2026,
https://doi.org/10.1007/978-1-4939-9612-4_13,

optimize light capture while inhibiting expansion of leaf blades [4]. Shade also strongly promotes flowering [5, 6] and reduces branching in favor of a main stem apical dominance [7].

Before actual shading occurs, it is crucial for a plant to sense the proximity of its neighbors which could outgrow it. Leaves and green tissues of plants selectively absorb light from the photosynthetically active radiation (PAR, 400–700 nm) region and reflect or transmit far-red light (FR, around 730 nm), overall lowering the red (R, around 665 nm)-to-FR ratio (R:FR) of propagated light [8]. These changes in the R:FR are perceived by the phytochrome photoreceptors. Five members of this family are described in Arabidopsis (phyA to phyE), of which phyA (photolabile) and phyB (photostable) have the major roles in SAS regulation [9–12]. Two photoconvertible forms of phytochromes exist: the active FR-absorbing Pfr form and the inactive R-absorbing Pr form [12]. When growing under light conditions, phytochromes naturally form a dynamic photoequilibrium between active (Pfr) and inactive (Pr) states, which depends on the R:FR of the incoming light. In low vegetation density (where R:FR is high, >1.0–1.5), active Pfr form prevails in the photoequilibrium, while the inactive Pr form dominates in environments of high vegetation density (such as proximity or canopy shade with low R:FR <1.0).

The importance of phytochromes for SAS is best observed in the Arabidopsis mutants deficient in phyA or phyB. The SAS-like aspect of *phyB* mutants in high R:FR, with long hypocotyls and early flowering phenotype, demonstrates that phyB is required for repressing SAS in open sunlight/low vegetation density. On the contrary, *phyA* mutant seedlings display exaggerated elongation in low and very low R:FR (<0.05, canopy shade), but in high and moderate R:FR (0.7–0.1, proximity shade) they are almost undistinguishable from wild type [13–15]. This antagonistic effect supports the idea that phyB is deactivated in all shade conditions to allow elongation, whereas phyA is activated in low and very low R:FR to prevent excessive elongation. These mutants can be successfully used as controls to differentiate proximity and canopy shade conditions (Fig. 1).

In SAS research it has become common to modify the R:FR to simulate shade/plant proximity. Many laboratories use FR to supplement white light (W) emitted from fluorescent tubes, a treatment named as simulated shade that lowers the R:FR of the incident light without reducing PAR. One of the difficulties in setting simulated shade conditions arises from how the R and FR are estimated [16, 17]. Usually this is done by measuring the fluence rate (i.e., the irradiance measured in μmol m^{-2} s^{-1}) of a bandwidth from 10 to 30 nm in the R and FR peaks, although this aspect is rarely detailed in the papers [11, 17–19]. This variability in the conditions employed among laboratories can produce slightly different seedling responses in the same genotype, complicating the comparisons of results produced by different laboratories.

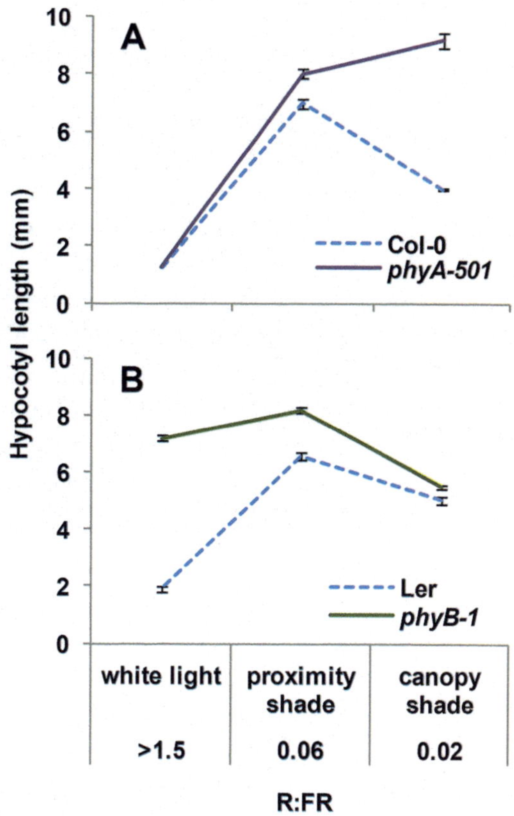

Fig. 1 Hypocotyl response of *phyA* and *phyB* mutants to simulated shade. Seedlings of wild-type, *phyA-501* (**a**) and *phyB-1* (**b**) lines were grown under W (R:FR >1.5), proximity (R:FR = 0.06) or canopy shade (R:FR = 0.02)

We have optimized our conditions to simulate two types of shade: moderate to low R:FR (~0.06), mimicking plant proximity, and very low R:FR (~0.02), mimicking canopy shade (*see* **Note 1**) [14]. In this chapter we will describe how to set up your shade treatment experiments for measuring physiological responses of seedlings and adult plants, including hypocotyl growth, cotyledon and primary leaf expansion as well as plant and petiole elongation.

2 Materials

2.1 Media and Solutions

1. Growth medium: half-strength Murashige and Skoog (0.5× MS–) mineral salts, 0.8% (w/v) agar and no sucrose (*see* **Note 2**). Mix 2.15 g MS salts and 0.5 g 2-(*N*-morpholino)ethanesulfonic acid (MES) and 900 mL water. Stir to dissolve and adjust the pH to 5.7 with KOH. Adjust the volume with water to 1 L, add 8 g Plant agar, and sterilize by autoclaving (*see* **Note 3**).

2. Imbibition solution: 0.1% (v/v) polyoxyethylene sorbitan monolaureate (Tween 20).
3. Sterilization solution: 0.1% (v/v) Tween 20, 10% (v/v) sodium hypochlorite.
4. Sterile milliQ water.

2.2 Soil

Mix in equal parts peat, perlite, and vermiculite.

2.3 Equipment and Software

1. Microcentrifuge tubes (1.5 mL).
2. Standard pipette tips (200–1000 μL).
3. Petri dishes (90 mm diameter and 16 mm height) to contain about 25 mL 0.5× MS medium.
4. Sterile nylon mesh (100 μM Nitex mesh, reference 03-100/44, Sefar, http://www.sefar.com/).
5. Porous surgical tape (e.g., tape 1.25 cm × 9.1 m).
6. Aluminum foil for stratification of seeds before any light treatment.
7. Light sources for treatment of seedlings: White light (W) is provided by cool-white fluorescent bulbs (e.g., Philips TL-D 36w/840). Deliver a range of photosynthetically active radiation (PAR) between 15 μmol m^{-2} s^{-1} and 20 μmol m^{-2} s^{-1} and an R:FR of about 1.5–3.0. W is supplemented with different amounts of monochromatic FR LEDs (e.g., GreenPower LED module from Philips, http://www.lighting.philips.com/main/products/horticulture/products.html) to simulate shade. Proximity shade is simulated by adding FR to intermediate and/or low R:FR (~0.6–0.05), whereas canopy shade is simulated by adding FR to very low R:FR (≤0.02) (*see* **Notes 1** and **4**, Fig. 2).
8. Light sources for treatment of adult plants: Grow plants under short day (SD, 8 h light/16 h darkness) or long day (LD, 16 h light/8 h darkness) photoperiods. W is provided by cool-white fluorescent bulbs, as in the treatments on seedlings, that deliver a higher light intensity (i.e., range of PAR between 90 μmol m^{-2} s^{-1} and 100 μmol m^{-2} s^{-1}) and an R:FR between 1.5 and 3.0. W is supplemented with monochromatic FR LEDs (as in the seedling treatments) only during the light period (*see* **Note 5**). This treatment results in a PAR of 90–100 μmol m^{-2} s^{-1} and an R:FR of ~0.06.
9. Spectroradiometer or equivalent to measure fluence rates (e.g., Spectrosense2 meter associated with a four-channel sensor; Skye Instruments Ltd., www.skyeinstruments.com), in the PAR (400–700 nm), B, R and FR regions (*see* **Note 6**).
10. Use ImageJ software (http://rsb.info.nih.gov/) or equivalent to measure the length of hypocotyls, cotyledons and primary leaves of the seedlings, as well as plant and petiole length and leaf area of the adult plants.

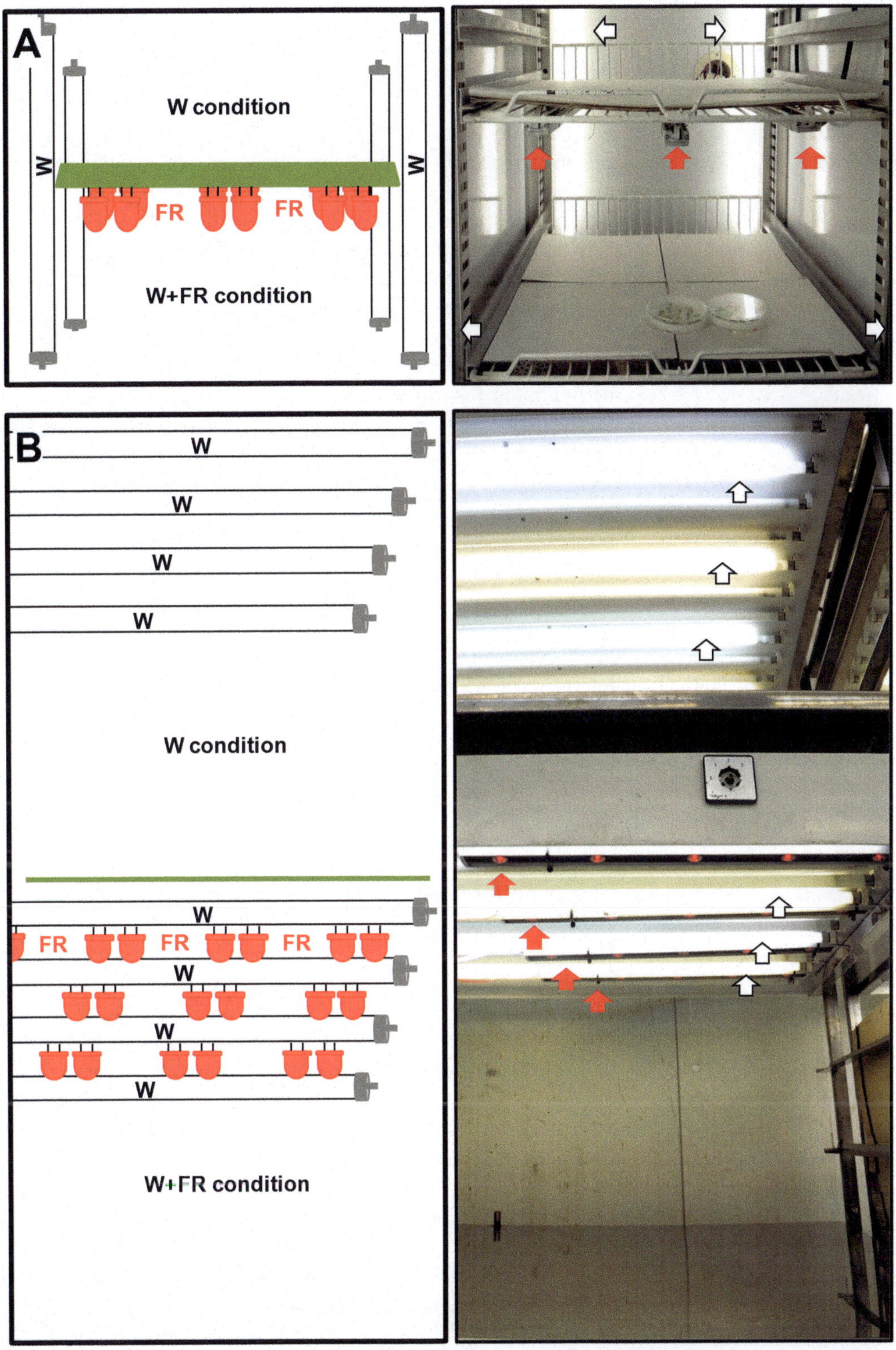

Fig. 2 Basic experimental setting for the simulated shade treatments. Schematic representation (left) and pictures (right) of light source setting. W and simulated shade (W + FR) conditions are arranged in the same chamber. W is obtained using vertical (**a**) or horizontal (**b**) W fluorescent tubes. Vegetation proximity or shade is simulated by supplementing W with FR. White arrows indicate the position of W tubes and red arrows the position of the FR LEDs

3 Methods

3.1 Seed Surface Sterilization

1. Aliquot the amount of seeds needed for the experiment in 1.5 mL microcentrifuge tubes for sterilization. The approximate weight of 100 Arabidopsis (accession Columbia-0) wild-type dry seeds is 0.002 g (*see* **Notes 7–9**).
2. Add 1 mL of imbibition solution for at least 20 min.
3. After removing the imbibition solution, add 1 mL of sterilization solution for 10 min. Do not incubate for longer than 10 min!
4. Remove the sterilization solution and wash the seeds five times with 1 mL of sterile milliQ water before sowing. The sowing methods will depend on the experimental treatment you plan to perform (*see* Subheading 3.2).

3.2 Preparation of Plates and Sowing of Seeds

1. After autoclaving, cool the growth medium to 55 °C in a water bath (*see* **Note 10**).
2. Pour approximately 25 mL of sterile growth medium in regular petri dishes (90 mm diameter and 16 mm height) (*see* **Note 11**).
3. For long term treatments, hypocotyl experiments and experiments with adult plants, sow the seeds directly on the solid growth medium one by one using a micropipette or a Pasteur pipette (*see* **Notes 12** and **13**).
4. For short treatments (e.g., analysis of gene expression) on seedlings, sow the seeds on a sterile nylon mesh (*see* **Note 14**).
5. After sowing, seal the plates using a porous tape (*see* **Note 15**).

3.3 Design of the Experiment

1. Always perform replicas of your experiments; at least triplicates (*see* **Note 9**).
2. Use growth medium stocks prepared at the same moment (*see* **Note 3**).
3. Always include a wild-type genotype (*see* **Note 8**).
4. For *Arabidopsis thaliana* perform the experiments as indicated in Subheadings 3.4 and 3.5. For other species *see* **Note 16**.

3.4 Light Treatments of Seedlings for Hypocotyl Elongation, Cotyledons and Primary Leaves Expansion Analysis

1. Stratification: Keep the plates containing the sown seeds in the dark at 4 °C for 2–5 days. This stratification treatment breaks dormancy and synchronizes the germination of the seeds.
2. For the simulated shade treatments, after stratification put plates to germinate in continuous W at 22 °C. After 2 days in W, maintain half the plates under W and transfer the other half to W supplemented with FR (day 2 in Fig. 3a).
3. Measure the hypocotyl length and the cotyledon and primary leaves expansion on day 7 from germination (after 5 days of differential light/shade treatment) (Fig. 3a) (*see* **Note 17**).

3.4.1 Measurement of Hypocotyls

1. Put the seedlings laying out flat on the same agar plates where they grew. If seedlings are too crowded, transfer them to new petri dishes containing a thin layer of agar growth medium before laying them out flat.
2. Take digital pictures of the plates. Include a reference scale (*see* **Note 18**).
3. Measure hypocotyl length by analyzing the digital images of the seedlings with Image J or similar software (Fig. 3b).

3.4.2 Measurement of Cotyledon and Primary Leaf Expansion

1. Remove the hypocotyls and put the seedling laying out flat on a transparent autoadhesive film.
2. Take digital pictures or scan the seedlings. Include a reference scale (*see* **Note 18**).
3. Measure cotyledons and primary leaves expansion by analyzing the digital images of the seedlings with Image J or similar software (Fig. 3c).

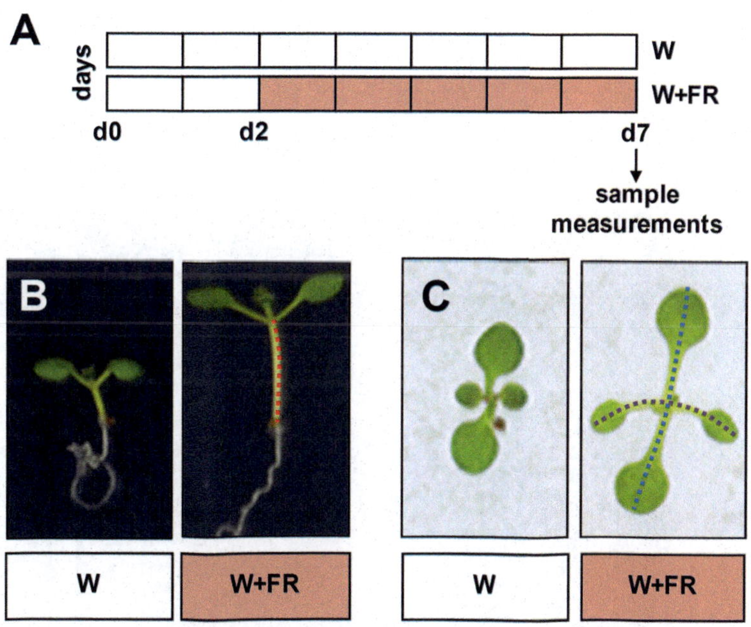

Fig. 3 Responses of Arabidopsis seedlings to simulated shade treatments. (**a**) Arabidopsis seedlings are grown for 7 days (d) in continuous W or 2 d in W and 5 d in W + FR (**b**, **c**) Aspect of seedlings grown as indicated in (**a**). Typically, seedlings grown in W + FR (right panels) show elongated hypocotyls (**b**), and expanded cotyledons and primary leaves (**c**), compared to W-grown seedlings (left panels). For easy visualization, dotted lines indicate the measurements of hypocotyl length (red), and cotyledon and primary leaf expansion (blue and purple, respectively) only on the W + FR treated samples

3.5 Light Treatments of Adult Plants for Plant and Petiole Length and Leaf Area Analysis

1. For the simulated shade treatments of adult plants, after stratification put plates to germinate in continuous W at 22 °C. On day 7, transfer the individual seedlings into pots (*see* **Note 7**, Fig. 4a). Cover them with a plastic film or a transparent lid to facilitate rooting of the seedlings during 2–3 days and keep them under the chosen photoperiod conditions (W). When analyzing petiole length, initiate simulated shade treatments once the rosette has about four leaves (around day 14 from germination; Fig. 4a). When analyzing plant height, initiate simulated shade treatment once plants have started to bolt (plants start to bolt around day 21–28 from germination; Fig. 4a) (*see* **Note 19**).

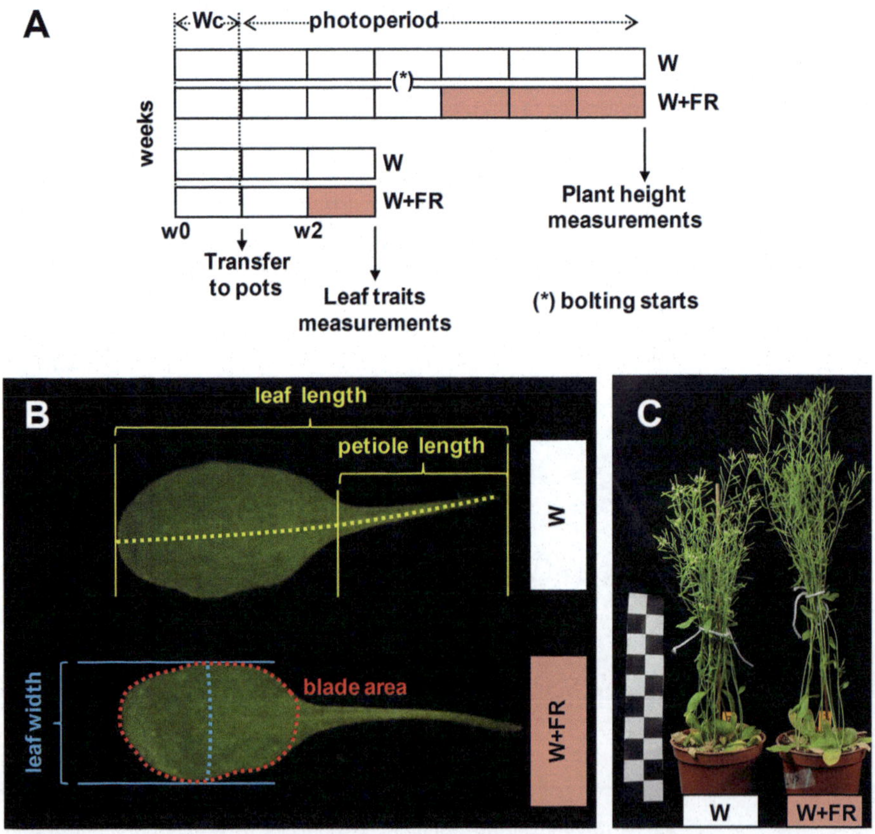

Fig. 4 Responses of Arabidopsis adult plants to simulated shade treatments. (**a**) Arabidopsis seedlings are grown on agar plates containing 0.5× MS– for 7 days (1 week) in continuous W. Selected seedlings are then transferred to pots containing soil to be grown under the chosen photoperiod. Later, once rosettes have around 4–5 leaves or flowering is induced, pots are either transferred to W + FR (e.g., on week 2 or 4–5 after germination) or kept in W. (**b**, **c**) Aspect of plants differentially grown in W or W + FR, as indicated in (**a**). Typically, plants grown in W + FR show expanded leaves with reduced blade area and elongated petioles (**b**) and stems (**c**). For easy visualization, dotted lines indicate the measurements of petiole/leaf length (yellow), leaf blade width (blue) and area (surface covered by the red line)

3.5.1 Measurement of Petiole Length and Leaf Area

1. Take the leaf of the rosette that you want to measure (*see* **Note 20**) and put it flat in between two glass plates.
2. Take digital pictures or scan the leaves. Include a reference scale (*see* **Note 18**).
3. Measure petiole length and leaf area by analyzing the digital images of the leaves with Image J or similar software.

3.5.2 Measurement of Plant Height

1. Take pictures of the individual plants standing up. Use a homogeneous and contrasting background. Alternatively, cut the plant from the base of the main stem and put it flat on a homogeneous background. Include a reference scale (*see* **Note 18**).
2. Measure plant length by analyzing the images tracing the main stems with Image J or similar software (*see* Subheading 2.3).

4 Notes

1. In nature, canopy shade involves a reduction in both the PAR and the R:FR. In the laboratory, however, W providing a PAR of around 15–60 μmol m^{-2} s^{-1} and supplemented with low or medium amounts of FR simulates vegetation proximity, while when supplemented with high or very high amounts of FR simulates canopy shade (even without lowering the PAR) [11]. The different characteristics or the W + FR conditions generated can be easily visualized by including *phyA* and *phyB* mutants in your experiment (*see* **Note 4**).
2. Addition of sucrose to the media interferes with the hypocotyl phenotypes in seedlings; thus avoid using sucrose.
3. Big amounts of growth medium can be prepared and stocked in 500 mL bottles for a few months. Write the preparation date on the bottle. When need it, melt the medium by heating the bottle with the lid slightly unscrewed in the microwave at low-medium power; avoid boiling for a long time which might evaporate part of the liquid and change the media concentration. Cool it at the appropriate temperature before pouring it (*see* Subheading 3.2).
4. Light intensity can have an impact on the phenotype of the plants. Also, comparison of light conditions between laboratories can be difficult due to differences in the measurements modes. You can include *phyA* and *phyB* mutants in your experiments to find out or confirm if you are providing proximity or canopy shade treatments (Fig. 1).
5. When working with adult plants, make sure that the FR LEDs are synchronized with the photoperiod, in order to avoid FR exposure during the night.

6. It is difficult to get comparable data in the light quantity and quality measurements between laboratories because the measured R and FR windows vary depending on the spectroradiometer used. Be aware of this and use the proper controls when needed (*see* **Note 4**).
7. When performing experiments with adult plants, sow and germinate the seeds in plates and grow them for 7 days. Avoid sowing the seeds directly in pots. This step will help to prevent contaminations; and importantly, by selecting seedlings properly germinated, it will result in healthy and comparable adult plants.
8. Always include the wild type of reference genotype together with the mutants of interest. When working with seedling phenotypes, it is particularly important to use seeds that came from contemporaneous mother plants (i.e., grown at the same time and in the same conditions), as any changes in the age or state of the seeds could affect germination and thus influence the hypocotyl phenotypes.
9. For the hypocotyl measurements and cotyledon and primary leaf expansion, use at least 20 seedlings for each data point and/or genotype. Repeat the experiments at least three times. This will help you to carry out statistical analyses of elongation data.
10. It is important to cool the growth medium before pouring it into the dishes. If medium is too hot, it can deform the plastic of petri dishes. Moreover, hormones, antibiotics or any other component added to the medium may be degraded above certain temperatures.
11. We normally use round petri dishes of 90 mm diameter and 16 mm height (the same ones used to grow bacteria). Squared, smaller or bigger plates can also be used, just make sure that medium fills half of the height of the plate.
12. After bleach sterilization and water washes, add 1 mL of sterile water. Seeds can be sown individually on top of the media using an automatic pipette and blue tips. Some seed batches are too big to get inside the blue tips. In those cases, sterile Pasteur pipettes coupled with a rubber sucker can be used. In any case, pull a small volume of water containing about ten seeds. They will easily fall one by one just by touching the medium surface.
13. It is important to distribute seeds separately to avoid negative effects of elongation because of plant proximity or touching. We normally sow about 100 seeds per regular petri dish, one by one. A paper with a grid placed at the bottom of the dish will help you to locate the seeds regularly spaced. A plate can be divided in a maximum of four sections; a genotype is sown

in each section (about 25 seeds per genotype). Make sure any remaining water dries out before sealing the plates to start the stratification.

14. If you want to harvest material to analyze gene expression or to extract proteins, sow the seeds on top of nylon mesh circles (previously cut and sterilized) placed on top of the media. It will be easier and faster to harvest the plant material for the further processing.
15. Usage of nonporous sealer (such as Parafilm or plastic wrap) is not recommended; we have observed that hypocotyl elongation is clearly affected in such closed atmosphere.
16. These protocols have been designed to analyze *Arabidopsis thaliana* plants. For other species, such as *Cardamine hirsuta* [20], that might be slightly bigger and have different developmental timings, consider using bigger plates and adjusting light conditions and timing of shade treatments to ensure correct germination and growth.
17. Be strict with the timing, especially when doing short light treatments. Start and perform the experiments always at the same time of the day, as circadian rhythms can influence your experiments. For instance, if you take the plates out from the stratification to induce germination at 11 am, transfer half of the plates into W + FR at day 2 at 11 am and end the experiment to take pictures on day 7 also at 11 am. This is particularly important when doing experiments with adult plants grown under photoperiods (i.e., under conditions in which the clock is entrained).
18. Remember to add a scale when taking the digital pictures for measurements. It will be needed as reference.
19. Timing of plant manipulation might need to be adjusted to the specific growth conditions of the laboratory, since minor changes in temperature, light and even watering affect plant growth.
20. Make sure that you always measure leaves that have the same age. For a good comparison, count the number of leaves from the oldest to the newest, and take the same one (e.g., leaf number 4).

Acknowledgments

Our research is supported by grants from the MINECO-FEDER (BIO2014-59895-P) and AGAUR (2014-SGR447 and Xarba) to JFM-G.

References

1. Gommers CM, Visser EJ, St Onge KR, Voesenek LA, Pierik R (2013) Shade tolerance: when growing tall is not an option. Trends Plant Sci 18:65–71
2. Roig-Villanova I, Martinez-Garcia JF (2016) Plant responses to vegetation proximity: a whole life avoiding shade. Front Plant Sci 7:236
3. Roig-Villanova I, Bou-Torrent J, Galstyan A, Carretero-Paulet L, Portoles S, Rodriguez-Concepcion M, Martinez-Garcia JF (2007) Interaction of shade avoidance and auxin responses: a role for two novel atypical bHLH proteins. EMBO J 26:4756–4767
4. de Wit M, Ljung K, Fankhauser C (2015) Contrasting growth responses in lamina and petiole during neighbor detection depend on differential auxin responsiveness rather than different auxin levels. New Phytol 208:198–209
5. Cerdan PD, Chory J (2003) Regulation of flowering time by light quality. Nature 423:881–885
6. Halliday KJ, Salter MG, Thingnaes E, Whitelam GC (2003) Phytochrome control of flowering is temperature sensitive and correlates with expression of the floral integrator FT. Plant J 33:875–885
7. Gonzalez-Grandio E, Poza-Carrion C, Sorzano CO, Cubas P (2013) BRANCHED1 promotes axillary bud dormancy in response to shade in Arabidopsis. Plant Cell 25:834–850
8. Ballare CL, Scopel AL, Sanchez RA (1990) Far-red radiation reflected from adjacent leaves: an early signal of competition in plant canopies. Science 247:329–332
9. Franklin KA, Quail PH (2010) Phytochrome functions in Arabidopsis development. J Exp Bot 61:11–24
10. Schepens I, Duek P, Fankhauser C (2004) Phytochrome-mediated light signalling in Arabidopsis. Curr Opin Plant Biol 7:564–569
11. Martinez-Garcia JF, Galstyan A, Salla-Martret M, Cifuentes-Esquivel N, Gallemí M, Bou-Torrent J (2010) Regulatory components of shade avoidance syndrome. Adv Bot Res 53:65–116
12. Casal JJ (2012) Shade avoidance. The Arabidopsis book 10:e0157
13. Johnson E, Bradley M, Harberd NP, Whitelam GC (1994) Photoresponses of light-grown *phyA* mutants of Arabidopsis. Plant Physiol 105:141–149
14. Martinez-Garcia JF, Gallemi M, Molina-Contreras MJ, Llorente B, Bevilaqua MR, Quail PH (2014) The shade avoidance syndrome in Arabidopsis: the antagonistic role of phytochrome A and B differentiates vegetation proximity and canopy shade. PLoS One 9:e109275
15. Yanovsky MJ, Casal JJ, Whitelam GC (1995) Phytochrome A, phytochrome B and HY4 are involved in hypocotyl growth responses to natural radiation in Arabidopsis: weak de-etiolation of the *phyA* mutant under dense canopies. Plant Cell Environ 18:788–794
16. Fankhauser C, Casal JJ (2004) Phenotypic characterization of a photomorphogenic mutant. Plant J 39:747–760
17. Franklin KA, Whitelam GC (2007) Light-quality regulation of freezing tolerance in *Arabidopsis thaliana*. Nat Genet 39:1410–1413
18. Leivar P, Tepperman JM, Cohn MM, Monte E, Al-Sady B, Erickson E, Quail PH (2012) Dynamic antagonism between phytochromes and PIF family basic helix-loop-helix factors induces selective reciprocal responses to light and shade in a rapidly responsive transcriptional network in Arabidopsis. Plant Cell 24:1398–1419
19. Tao Y, Ferrer JL, Ljung K, Pojer F, Hong F, Long JA, Li L, Moreno JE, Bowman ME, Ivans LJ, Cheng Y, Lim J, Zhao Y, Ballare CL, Sandberg G, Noel JP, Chory J (2008) Rapid synthesis of auxin via a new tryptophan-dependent pathway is required for shade avoidance in plants. Cell 133:164–176
20. Hay AS, Pieper B, Cooke E, Mandakova T, Cartolano M, Tattersall AD, Ioio RD, McGowan SJ, Barkoulas M, Galinha C, Rast MI, Hofhuis H, Then C, Plieske J, Ganal M, Mott R, Martinez-Garcia JF, Carine MA, Scotland RW, Gan X, Filatov DA, Lysak MA, Tsiantis M (2014) *Cardamine hirsuta*: a versatile genetic system for comparative studies. Plant J 78:1–15

Chapter 14

A Simple Method for Quantification of Protochlorophyllide in Etiolated *Arabidopsis* Seedlings

Matthew J. Terry and Sylwia M. Kacprzak

Abstract

Etiolated seedlings accumulate the chlorophyll biosynthesis intermediate protochlorophyllide (Pchlide) and measuring Pchlide can be important for characterizing photomorphogenic mutants that may be affected in chloroplast development. In this chapter we outline a simple and sensitive method for quantifying Pchlide in extracts of Arabidopsis seedlings using fluorescence spectroscopy. This method can be easily adapted to study chloroplast development in a wide range of plant species.

Key words Tetrapyrroles, Chlorophyll synthesis, Chloroplast development, Etioplasts, Fluorescence spectroscopy

1 Introduction

1.1 The Importance of Measuring Pchlide in Photomorphogenic Research

This chapter describes a simple and sensitive method to estimate protochlorophyllide (Pchlide) levels in Arabidopsis seedlings that can usefully be applied to a range of different photomorphogenic mutants. It is also suitable for use with most plant species with modification to the solvent volumes used. In dark-grown or etiolated seedlings of angiosperms the synthesis of the light-harvesting tetrapyrrole, chlorophyll is blocked at the biosynthetic intermediate Pchlide, which then accumulates [1]. Figure 1 shows the Arabidopsis tetrapyrrole pathway. The block occurs because the enzyme Pchlide oxidoreductase (POR) in angiosperms requires light for activity [2]. In lower plants and algae, a light-independent enzyme exists and Pchlide does not accumulate in darkness [1]. In angiosperms the synthesized Pchlide is normally held bound to POR and NADPH in a complex that makes up a major component of the prolamellar body of etioplasts. Once a seedling is exposed to light, Pchlide is rapidly converted to chlorophyllide (Chlide) and then esterified to provide chlorophyll for the new photosystems, and POR is degraded resulting in the breakdown of the prolamellar body [3].

Andreas Hiltbrunner (ed.), *Phytochromes: Methods and Protocols*, Methods in Molecular Biology, vol. 2026,
https://doi.org/10.1007/978-1-4939-9612-4_14,

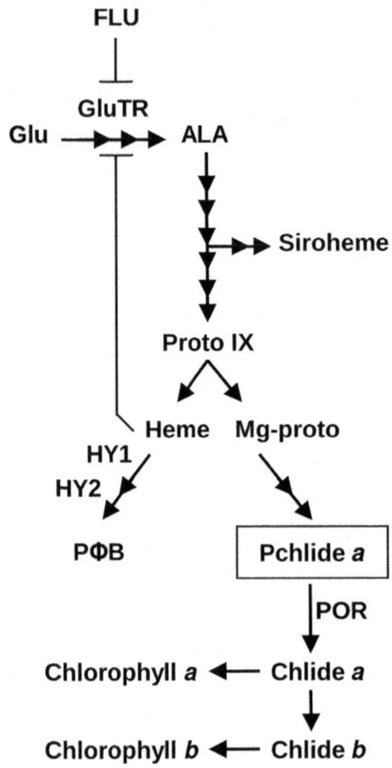

Fig. 1 The tetrapyrrole pathway in Arabidopsis. ALA, 5-aminolevulinic acid, Chlide, chlorophyllide; FLU, repressor of GluTR; Glu, glutamate; GluTR, glutamyl tRNA reductase encoded by *HEMA* genes; HY1, heme oxygenase 1; HY2, phytochromobilin synthase; Mg-proto, magnesium protoporphyrin IX; POR, protochlorophyllide oxidoreductase; Pchlide, protochorophyllide; Proto IX, protoporphyrin IX

Photoreceptor regulation of chloroplast development is an important part of de-etiolation and photomorphogenesis [4] and measurement of Pchlide levels in dark-grown seedlings can be an extremely useful in the characterization of photomorphogenic mutants. The understanding of two classes of photomorphogenic mutants in particular has benefited from the ability to measure Pchlide levels in etiolated seedlings. Firstly, mutants that have a constitutively photomorphogenic phenotype, that is respond in the dark as if they are in the light [5], frequently show elevated levels of Pchlide if the signaling components have a role in chloroplast development. Examples include the *cop1* mutant [6], the *pif1* [7] and *pif3* [8] mutants, and the *pifQ* mutant lacking PIF1, PIF3, PIF4 and PIF5 [9]. *See* Fig. 2 for an illustration of the higher Pchlide levels observed in the *pifQ* mutant. Various hormone response pathways also regulate Pchlide levels via these components, including, for example, gibberellins [10] and ethylene [11]. In all cases, the signaling pathways result in an increase in Pchlide via induction of *HEMA1* expression and increased activity of

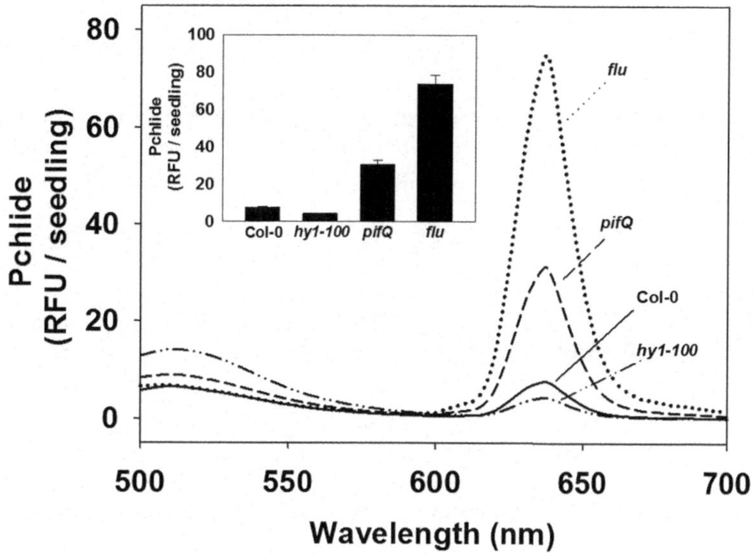

Fig. 2 Measurement of protochlorophyllide (Pchlide) by fluorescence spectroscopy. Fluorescence spectra taken following excitation at 440 nm in wild-type (Col-0) seedlings and three mutants, *flu* (the *fluorescent* mutant lacking a repressor of GluTR), *pifQ* (comprised of *pif1, pif3, pif4* and *pif5*) and *hy1* (deficient in heme oxygenase 1). See text for further information on these mutants. Inset, quantification of relative fluorescence (RFU) at the ~636 nm peak from three independent biological repeats. Data shown are mean ± standard error

glutamyl tRNA reductase (GluTR), the rate limiting step for all tetrapyrrole synthesis [8, 12]. In this type of analysis the measurement of Pchlide acts as a proxy for a role in chloroplast development without the added complications of interpreting the effect of white light on chlorophyll accumulation. Under conditions that result in the accumulation of Pchlide, seedlings photobleach on transfer to white light. This can be seen for example in the *flu* mutant, which lacks a repressor of GluTR activity, resulting in accumulation of Pchlide in the dark (*see* Fig. 2) and photobleaching in the light [13]. The effect of elevated Pchlide is exacerbated under conditions in which POR levels are reduced, for example by far-red light that inhibits expression of *POR* genes, but does not permit Pchlide photoconversion [14]. Photobleaching on transfer to white light is the result of the production of singlet oxygen gen erated by the reaction of excited Pchlide with molecular oxygen. This results in inhibition of nuclear-encoded photosynthesis gene expression [15] and can ultimately lead to cell death [16]. The production of singlet oxygen from elevated Pchlide levels has also been demonstrated in *pif* mutants [17].

The second class of photomorphogenic mutants for which Pchlide measurements have proved useful is the phytochrome chromophore-deficient mutants [18]. These mutants lack heme oxygenase and phytochromobilin synthase activities (*see* Fig. 1), resulting in long hypocotyl phenotypes due to phytochrome deficiency. This class of mutants includes *elongated hypocotyl 1* (*hy1*) and *hy2* [19–21]. Analysis of Pchlide levels in the equivalent tomato mutants *yg-2* and *au* showed that they also contained reduced levels of Pchlide [22] supporting a model in which heme feeds back to inhibit the plant tetrapyrrole pathway [23]. Reduced Pchlide has also been demonstrated for chromophore-deficient mutants of pea and Arabidopsis [24] as shown in Fig. 2 for *hy1*.

1.2 Approaches to Measuring Pchlide

There are a range of different methods that can be used to measure Pchlide. Most of these require extraction from seedlings by solvents, but it is also possible to measure Pchlide in situ. Low-temperature (77 K) fluorescence spectroscopy has been used to identify four different fluorescent Pchlide forms corresponding to different molecular organizations of the POR proteins binding the Pchlide [25]. This technique has been applied to analysis of both constitutively photomorphogenic mutants [26] and phytochrome chromophore-deficient mutants [27]. More frequently, Pchlide is measured following solvent extraction, where again a wide range of methods can be used for analysis. Some laboratories use high-performance liquid chromatography (HPLC) where the highly absorbent and fluorescent properties of chlorophyll precursors make detection relatively easy [28–30]. Pchlide has also been successfully analyzed by mass spectrometry following separation by HPLC [31, 32]. This method makes it easier to distinguish the different forms of Pchlide (*see* later).

For routine measurement of Pchlide, spectroscopic methods are far more straightforward. For reliable quantification, absorption spectroscopy is the best method as the Pchlide concentration can be calculated using its molar absorption coefficient in acetone of 31,100 M^{-1} cm^{-1} at 626 nm [33]. This method was used for the analysis of tomato seedlings [22], but the amount of tissue needed is not ideal for routine analysis in Arabidopsis seedlings. Instead, it is possible to take advantage of the highly fluorescent properties of chlorophyll precursors and estimate Pchlide amounts using fluorescence spectroscopy. It should be noted that fluorescence does not have units and all measurements are of relative fluorescence enabling only a comparison with other genotypes or treatments in the experiment. Fluorescence can also be an unreliable measure of the amount of a compound because it can be subject to fluorescence quenching in which photons emitted from the fluorescent molecules are absorbed by other molecules in the sample and not detected by the spectrophotometer. This will result in lower values at higher concentrations of Pchlide. Nevertheless, the small amounts of Arabidopsis tissue required and the ease and reliability of the measurements makes this a very useful, routine assay for characterizing the effect of light signaling mutants on chloroplast development.

2 Materials

1. Growth media: 1% (w/v) agar with 0.5× MS salts.
2. Extraction solvent: Acetone–0.1 M NH_4OH (9:1; v/v) should be prepared in advance and kept at 4 °C prior to use (*see* **Note 1**).
3. 1.5 mL microcentrifuge tubes and polypropylene pestles.
4. Chilled benchtop centrifuge (*see* **Note 2**).
5. 1 mL fluorescence glass cuvettes.
6. Fluorescence spectrophotometer.

3 Methods

3.1 Pchlide Extraction

Arabidopsis seedlings should be grown in complete darkness to ensure there is no conversion of Pchlide to Chlide. There are a variety of standard methods for doing this, but typically seedlings are grown on petri dishes containing 1% (w/v) agar with 0.5× MS media [34]. Germination is promoted by 2–3 days at 4 °C followed by 2 h white light (~100 μmol.m^{-2}.s^{-1}) before transfer to the dark. Five days of dark growth at 22 °C should be sufficient for maximum Pchlide accumulation. Even when grown in a dark growth cabinet it is recommended that petri dishes are wrapped twice in foil (or placed in a tin). Harvesting should take place in a dark room under very low levels of green safelight as the POR enzyme is not activated at these wavelengths. The extraction method used is based on the method of Rebeiz et al. [35].

1. Excise 20–30 seedlings (count exactly how many) by cutting near the base of the hypocotyl and transfer to a labeled microcentrifuge tube kept on ice (*see* **Note 3**).
2. Add 0.4 mL ice-cold acetone–0.1 M NH_4OH (9:1; v/v).
3. Homogenize tissue using a polypropylene pestle.
4. Centrifuge for 5 min in a chilled benchtop centrifuge (*see* **Note 2**) at top speed (> 12,000 × *g* is recommended).
5. Transfer the supernatant into a new, labeled microcentrifuge tube.
6. Add a further 0.4 mL ice-cold acetone–0.1 M NH_4OH (9:1; v/v) to the first tube with the pellet and repeat the extraction.
7. Repeat centrifugation step.
8. Combine supernatants from the two extraction steps (**steps 5–7**). This is your Pchlide sample for analysis (*see* **Note 4**).
9. Put samples on ice, cover in foil, and bring out of the dark room (*see* **Note 5**).

3.2 Pchlide Measurement

1. Set the baseline of the fluorescence spectrophotometer to zero using 80% (v/v) acetone.
2. Measure samples at the peak value of ~636 nm (*see* **Notes 6–8**) following excitation at 440 nm. As there is some variation in background fluorescence between samples this can be normalized by also recording emission at 750 nm. Pchlide does not emit at this wavelength.
3. Calculate Pchlide as relative fluorescence on a per seedling basis, or if seedling morphology is different, on a per mg fresh weight basis.

4 Notes

1. The solvent solution is stable over a long period but, as with all solvents, do not pipette directly from the bottle to avoid plastic contamination.
2. If a chilled benchtop centrifuge is unavailable then a room temperature one can be used. In this case, keep samples chilled before and after centrifugation.
3. A lot more Pchlide is present in the cotyledons than in the hypocotyls and a measurement per seedling will mostly reflect cotyledon Pchlide. If the morphology of the seedlings being tested is quite different then it can be best practice to also weigh seedlings so that can be presented on a per seedling and per gram fresh weight basis.
4. The Pchlide extracted using acetone–0.1 M NH_4OH (9:1; v/v) is a mixture of Pchlide and esterified Pchlide, or protochlorophyll (Pchl). The Pchl, which is generally in lower amounts, can be removed by washing with an equal volume of hexane [35], but where examined there was little observed difference between relative amounts of Pchlide and Pchl [22]. For this reason, a hexane wash is not recommended for Pchlide estimation in Arabidopsis given the small volumes involved.
5. Once Pchlide is in solution it does not have to be kept in safe-light conditions. As with all pigments keep cold and covered where possible to reduce light-mediated degradation, which will happen over time.
6. When measuring fluorescence always record the peak height not the height at a single wavelength (e.g., 636 nm). Contamination of a sample can shift the position of the peak, but has less effect on peak height. In addition, as discussed earlier, sample readings can be inaccurate due to fluorescence quenching. It is important to be measuring Pchlide when fluorescence increases linearly with sample concentration. The solvent volumes and tissue amounts suggested here should ensure this

is the case, but if in doubt this can be determined empirically using dilutions of the most concentrated samples.

7. If the seedlings have been exposed to light some of the Pchlide will have been converted to Chlide. This will be seen as a peak (or shoulder) at ~670 nm. If this is absent then your samples are of good quality.
8. There are actually numerous forms of Pchlide in plants, and the fluorescent measurements recorded will be the sum of these. In etiolated seedlings photoactive Pchlide is generally in the form of 3,8-divinyl Pchlide *a* [36]. Monovinyl Pchlide *a* is also commonly detected [28], but Pchlide *b,* which differs in having an aldehyde group instead of a methyl group at the C7 position, has only been detected in green plants and not in etiolated seedlings (e.g., [30]).

Acknowledgments

Work on tetrapyrroles in the MJT laboratory has been supported by the UK Biotechnology and Biological Sciences Research Council. SMK was supported by the Gatsby Charitable Foundation. This chapter is dedicated to the memory of Margareta Ryberg.

References

1. Masuda T, Takamiya K (2004) Novel insights into the enzymology, regulation and physiological functions of light-dependent protochlorophyllide oxidoreductase in angiosperms. Photosynth Res 81:1–29
2. Lebedev N, Timko MP (1998) Protochlorophyllide photoreduction. Photosynth Res 58:5–23
3. Solymosi K, Aronsson H (2013) Etioplasts and their significance in chloroplast biogenesis. In: Biswal B, Krupinska K, Biswal UC (eds) Plastid development in leaves during growth. Springer, Dordrecht, pp 39–71
4. Waters MT, Langdale JA (2009) The making of a chloroplast. EMBO J 28:2861–2873
5. Xu X, Paik I, Zhu L, Huq E (2015) Illuminating progress in phytochrome-mediated light signaling pathways. Trends Plant Sci 20:641–650
6. Sperling U, Franck F, van Cleve B, Frick G, Apel K, Armstrong GA (1998) Etioplast differentiation in Arabidopsis: both PORA and PORB restore the prolamellar body and photoactive protochlorophyllide-F655 to the *cop1* photomorphogenic mutant. Plant Cell 10:283–296
7. Huq E, Al-Sady B, Hudson M, Kim C, Apel K, Quail PH (2004) Phytochrome interacting factor 1 is a critical bHLH regulator of chlorophyll biosynthesis. Science 305:1937–1941
8. Stephenson PG, Fankhauser C, Terry MJ (2009) PIF3 is a repressor of chloroplast development. Proc Natl Acad Sci U S A 106:7654–7659
9. Shin J, Kim K, Kang H, Zulfugarov IS, Bae G, Lee C-H, Lee D, Choi G (2009) Phytochromes promote seedling light responses by inhibiting four negatively acting phytochrome-interacting factors. Proc Natl Acad Sci U S A 106:7660–7665
10. Cheminant S, Wild M, Bouvier F, Pelletier S, Renou JP, Erhardt M, Hayes S, Terry MJ, Genschik P, Achard P (2011) DELLAs regulate chlorophyll and carotenoid biosynthesis to prevent photooxidative damage during seedling deetiolation in *Arabidopsis.* Plant Cell 23:1849–1860
11. Zhong S, Shi H, Xue C, We N, Guo H, Deng XW (2014) Ethylene-orchestrated circuitry coordinates a seedling's response to soil cover

and etiolated growth. Proc Natl Acad Sci U S A 111:3913–3920

12. McCormac AC, Fischer A, Kumar AM, Söll D, Terry MJ (2001) Regulation of *HEMA1* expression by phytochrome and a plastid signal during de-etiolation in *Arabidopsis thaliana*. Plant J 25:549–561
13. Meskauskiene R, Nater M, Goslings D, Kessler F, op den Camp R, Apel K (2001) FLU: a negative regulator of chlorophyll biosynthesis in *Arabidopsis thaliana*. Proc Natl Acad Sci U S A 98:12826–12831
14. Sperling U, van Cleve B, Frick G, Apel K, Armstrong GA (1997) Overexpression of light dependent PORA or PORB in plants depleted of endogenous POR by far-red light enhances seedlings survival in white light and protects against photooxidative damage. Plant J 12:649–658
15. Page MT, McCormac AC, Smith AG, Terry MJ (2017) Singlet oxygen initiates a plastid signal controlling photosynthetic gene expression. New Phytol 213:1168–1180
16. Danon A, Miersch O, Felix G, op den Camp RG, Apel K (2005) Concurrent activation of cell death-regulating signaling pathways by singlet oxygen in *Arabidopsis thaliana*. Plant J 41:68–80
17. Chen D, Xu G, Tang W, Jing Y, Ji Q, Fei Z, Lin R (2013) Antagonistic basic helix-loop-helix/ bZIP transcription factors form transcriptional modules that integrate light and reactive oxygen species signaling in *Arabidopsis*. Plant Cell 25:1657–1673
18. Terry MJ (1997) Phytochrome chromophore-deficient mutants. Plant Cell Environ 20:740–745
19. Koornneef M, Rolff E, Spruitt CJP (1980) Genetic control of light-inhibited hypocotyl elongation in *Arabidopsis thaliana* L. Heynh. Z Pflanzenphysiol 100:147–160
20. Muramoto T, Kohchi T, Yokota A, Hwang I, Goodman HM (1999) The *Arabidopsis* photomorphogenic mutant *hy1* is deficient in phytochrome chromophore biosynthesis as a result of a mutation in a plastid heme oxygenase. Plant Cell 11:335–347
21. Kohchi T, Mukougawa K, Frankenberg N, Masuda M, Yokota A, Lagarias JC (2001) The *Arabidopsis HY2* gene encodes phytochromobilin synthase, a ferredoxin-dependent biliverdin reductase. Plant Cell 13:425–436
22. Terry MJ, Kendrick RE (1999) Feedback inhibition of chlorophyll synthesis in the phytochrome chromophore-deficient *aurea* and *yellow-green-2* mutants of tomato. Plant Physiol 119:143–152
23. Cornah JE, Terry MJ, Smith AG (2003) Green or red: what stops the traffic in the tetrapyrrole pathway? Trends Plant Sci 8:224–230
24. Terry MJ, Ryberg M, Raitt CE, Page AM (2001) Altered etioplast development in phytochrome chromophore-deficient mutants. Planta 214:314–325
25. Böddi B, Ryberg M, Sundqvist C (1992) Identification of four universal protochlorophyllide forms in dark-grown leaves by analyses of the 77 K fluorescence emission spectra. J Photochem Photobiol B Biol 12:389–401
26. Lebedev N, van Cleve B, Armstrong G, Apel K (1995) Chlorophyll synthesis in a deetiolated (det340) mutant of Arabidopsis without NADPH–protochlorophyllide (PChlide) oxidoreductase (POR) A and photoactive PChlide-F655. Plant Cell 7:2081–2090
27. Ryberg M, Terry MJ (2002) Analysis of protochlorophyllide reaccumulation in the phytochrome chromophore-deficient *aurea* and *yg-2* mutants of tomato by in vivo fluorescence spectroscopy. Photosynth Res 74:195–203
28. Shioi Y, Takamiya KI (1992) Monovinyl and divinyl protochlorophyllide pools in etiolated tissues of higher plants. Plant Physiol 100:1291–1295
29. Schoefs B, Bertrand M, Lemoine Y (1995) Separation of photosynthetic pigments and their precursors by high performance liquid chromatography using a photodiode array detector. J Chromatogr A 692:239–245
30. Scheumann V, Klement H, Helfrich M, Oster U, Schoch S, Rüdiger W (1999) Protochlorophyllide *b* does not occur in barley etioplasts. FEBS Lett 445:445–448
31. Moulin M, Smith AG (2008) A robust method for determination of chlorophyll intermediates by tandem mass spectrometry. In: Allen J, Gantt E, Golbeck J, Osmond B (eds) Photosynthesis: energy from the sun. 14th international congress on photosynthesis. Springer, Heidelberg, pp 1221–11228
32. Moulin M, McCormac AC, Terry MJ, Smith AG (2008) Tetrapyrrole profiling in Arabidopsis seedlings reveals that retrograde plastid nuclear signaling is not due to mg-protoporphyrin IX accumulation. Proc Natl Acad Sci U S A 105:15178–15183
33. Kahn A (1983) Spectrophotometric quantitation of protochlorophyll(ide): specific absorption and molar extinction coefficients reconsidered. Physiol Plant 59:99–102

34. Murashige T, Skoog F (1962) A revised medium for rapid growth and bioassays with tobacco tissue cultures. Physiol Plant 15:473–497
35. Rebeiz CA, Mattheis JR, Smith BB, Rebeiz CC, Dayton DF (1975) Chloroplast biogenesis. Biosynthesis and accumulation of protochlorophyll by isolated etioplasts and developing chloroplasts. Arch Biochem Biophys 171:549–567
36. Tanaka R, Tanaka A (2007) Tetrapyrrole biosynthesis in higher plants. Annu Rev Plant Biol 58:321–346

Chapter 15

Measuring Phytochrome-Dependent Light Input to the Plant Circadian Clock

Rachael J. Oakenfull, James Ronald, and Seth J. Davis

Abstract

The circadian clock allows plants to synchronize their internal processes with the external environment. This synchronization occurs through daily cues, one of which is light. Phytochromes are well established as light-sensing proteins and have been identified in forming multiple signaling networks with the central circadian oscillator. However, the precise details of how these networks are formed are yet to be established. Using established promoter-luciferase lines for clock genes crossed into mutant lines, it is possible to use luciferase-based imaging technologies to determine whether specific proteins are involved in phytochrome signaling to the circadian oscillator. The methods presented here use two automated methods of luciferase imaging in Arabidopsis to allow for high-throughput measurement of circadian clock components under a range of different light conditions.

Key words Circadian clock, Phytochrome, Arabidopsis, Luciferase, TopCount-luminescence measurement

1 Introduction

The circadian clock allows plants as sessile organisms to synchronize with their external environment [1]. The ability to efficiently synchronize with the external environment is an indicator of fitness in plants [2]. Entrainment is the daily process of synchronization with these external cues called "*zeitgebers*" [3]. Entrainment allows plants to anticipate, prepare for, and better survive diurnal stresses, such as drought [4] or cold [5]. This occurs through a host of interactions between the central oscillator and pathways controlling intracellular processes [6].

Daily *zeitgebers* allow the circadian clock to reset each day at dawn [7]. The major *zeitgeber* in this process is light [8]. There are ten known circadian-associated photoreceptors in Arabidopsis [9]; one subcategory of which are phytochromes [10]. There are five phytochromes present in Arabidopsis, named phyA-E, respectively [11]. Each has at least a partially specific role and function [12].

Andreas Hiltbrunner (ed.), *Phytochromes: Methods and Protocols*, Methods in Molecular Biology, vol. 2026,
https://doi.org/10.1007/978-1-4939-9612-4_15,

Phytochromes primarily detect red and far-red light [13], as well as being involved in blue and white-light detection, also potentially detecting temperature [14]. Phytochromes have rhythmic expression patterns [15, 16] and have been shown to interact with the central circadian oscillator to form negative feedback loops [17]. However, some of these interactions have yet to be formalized.

Luciferase is a commonly used biological tool across a range of organisms, frequently used as a marker to measure changes in gene expression in vivo [18]. Luciferase constructs are used for measuring diurnal expression of a particular clock gene [19]. Luciferase breaks down faster than fluorescent tags. This avoids over accumulation, which would mask the fluctuations in gene expression across a 24-h measurement window [18, 20]. Circadian experiments are time-intensive experiments and take a number of days with measurements taken every hour or two [18]. Luminescence based examinations of clock rhythms have to date provided the best resolution and depth in clock studies of plants.

Here we present two possible automated methods of measuring clock-gene expression in Arabidopsis using luciferase expression, with a particular consideration of tests in a phytochrome null background. We explore specific considerations for growth and entrainment conditions in such phytochrome clock-input experiments. Luciferase measurements are examined for both a TopCount scintillation counter and a CCD camera-imaging platform, then circadian data analysis with a focus on period estimation.

2 Materials

2.1 Equipment, Solutions, and Software

1. Clear 96-well microtiter plates (PerkinElmer Life Sciences).
2. Ball bearings.
3. Plate seals (PerkinElmer Life Sciences).
4. TopCount NXT scintillation counter (PerkinElmer Life Sciences).
5. TopCount plate barcodes (PerkinElmer Life Sciences).
6. Luciferin solution: Prepare a stock solution of 50 mM potassium luciferin in pH 8.0 1 M tris-phosphate; store the stock solution in the dark at −80 °C. To make a working solution, dilute the stock solution to 1 mM in 0.001% (w/v) Triton X-100 solution and filter-sterilize before use.
7. Amber glass spray bottles with spray atomizer.
8. Packard results analysis software (PerkinElmer Life Sciences).
9. TopTemp and BioDare2 (*see* **Note 1**).

10. Rumed temperature controlled chamber (Rumed Rubarth Apparate GmbH).
11. LED light panels (custom assembled).
12. Pixis 512 CCD camera (Princeton Instruments, USA).
13. Computer and controller units (Vistron Systems GmbH).
14. MetaVue and Metamorph research imaging system (Molecular Devices, USA).

2.2 Plants

To look at interactions between a gene expression of interest and phytochrome, a genotype is constructed harboring luciferase driven by the promoter region of the test gene in a phytochrome loss of function background. A number of well-established phytochrome null lines are currently used by the circadian community [13]. Many such lines are available from the Arabidopsis Biological Resource Centre (ABRC) and other seed stores. The promoter luciferase (LUC) line of interest is most ideal when crossed into the phytochrome null background, such that expression-level variation can be additionally explored (*see* **Note 2**). For many clock genes [21], promoter luciferase lines are also available from stock centres. Similarly a great number of clock mutants can also be obtained if phytochrome clock double mutants are to be studied (*see* **Note 3**).

For example, to establish control methods of phyB on *CCR2* (*COLD AND CIRCADIAN REGULATED 2*) expression, the *PHYB* loss of function line would be crossed with the *CCR2* luciferase lines. In the segregating population, a doubly homozygous line for the *PHYB* mutation and the *CCR2* transgene could be isolated. Such lines could then be used in the following experiments.

2.3 Growth Conditions and Entrainment

In circadian experiments, growth conditions from germination are often referred to as entraining conditions. Entrainment is used to simulate a particular environment to test for a response, for example, for shade response light conditions with supplemental far-red wavelengths of light should be used. One of the most important considerations in entrainment is to ensure that the seeds are placed in the required entrainment conditions from the very first stages of seed germination [22]. Circadian experiments, particularly those looking at interactions between phytochromes and the clock, have a large number of possible growth and entrainment conditions, discussed in more detail below.

2.3.1 Day Length

It is possible to entrain plants under a series of different light regimes. These could include cycles of 12 h light–dark (12:12), short days (8:12), or long days (16:8). Different photoperiods have been shown to induce different circadian responses in

Arabidopsis. Short-day entrainment causes a phase shortening leading to earlier metabolic timing; conversely, long-day entrainment causes a phase delay [22]. For luminescence experiments, 12:12 entrainment is usually chosen. As entrainment conditions determine the length of circadian period, and have an effect on the future behavior of the circadian oscillator [23], it is recommended to follow this regime such that resultant data are consistent with other published work.

2.3.2 Wavelengths of Light

Phytochromes (phy) regulate different aspects of the response to red and far-red light. In Arabidopsis, phyA is a dominant regulator of the response to far-red light, while phyB and the related phyC-E, regulates the response to red light (Fig. 1) [24]. phyA is also involved in the perception of low-intensity red light and blue light and regulates subsequent downstream signaling events [25]. Therefore, a range of different light conditions can be tested to determine an interaction between phytochromes and a target gene expression, as assayed by perturbations of luminescence. For example, plants could be grown under constant red light, or constant white light with supplemental amounts of far-red light to stimulate the shade avoidance response [26]. It is possible to measure phase in diurnal conditions using the TopCount method described in Subheading 3.1 with lights switching on-and-off as required to simulate light–dark cycles [22]. However, phytochrome experiments are typically carried out under constant light, rather than the diurnal (light–dark) cycles of entrainment [27]. Under constant light, often termed “LL,” wild-type plants will retain robust rhyth-

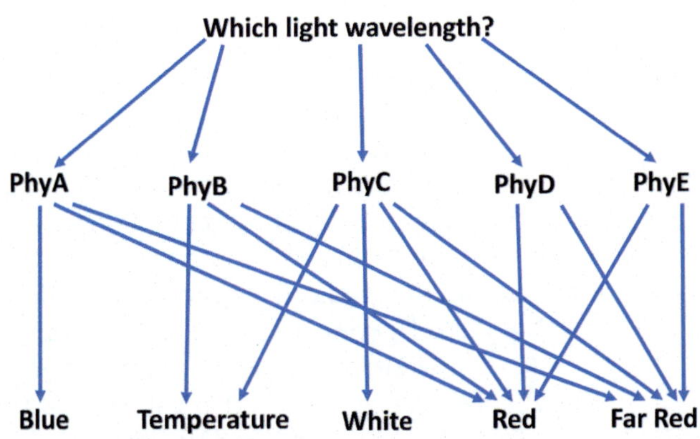

Fig. 1 Suggested light wavelengths to start testing the effects of each of the five Arabidopsis phytochromes. It is possible to test multiple wavelengths simultaneously or in combination with temperature. It is also possible to test the effects of darkness on each phytochrome (not shown). Green and UV wavelengths have been omitted as it is not possible to measure their input using the equipment described in this chapter

micity and circadian driven expression of LUC [21]. Phytochrome assays can also be carried out under constant darkness, referred to as "DD," and pulses of light can also be applied to examine cross talk between the clock and light-regulated gene expression. The expression LUC fused to a light-inducible promoter serves well and the *CAB2* (also called *LHCB1.1*) promoter has been a "model system" for such work. Such experiments can generate phase response curves; these have been previously carried out to establish the relationship of phytochromes and clock genes [28–30].

2.3.3 Temperature

Temperature is an important factor in entrainment conditions [31]. In plants it is possible to entrain solely for light in which case a constant optimal temperature is chosen [2]. For Arabidopsis, this is usually a constant temperature of between 12 °C and 27 °C [32]. A cycle of warm, cold temperatures can be used as part of entrainment with the duration of the warm temperature matching the duration of light and the cold temperature matching the dark phase. For example, a day with 12 h-light/12-h dark with warm/cold would have 12 h at 22 °C followed by 12 h of 15 °C [33]. It is also possible to entrain using temperature fluctuations in constant light or dark conditions [34] (*see* **Note 4**).

2.3.4 Growth Media

Murashige and Skoog media (MS) is a commonly used plant growth media [35]. Circadian experiments particularly those with a prolonged dark phase should use MS media containing supplemental sucrose. Typically 1% (w/v) is sufficient, although historically up to 3% (w/v) has been used [36]. For experiments requiring exact concentrations of macro elements and micro elements, for example experiments removing particular elements, Hoagland's media [37] should be used instead of MS. However, the chosen growth media and sucrose concentration should be used throughout the experiment, for both seedling plates and microtiter plates for experimental consistency.

(a) Murashige and Skoog media (MS) [35] plus sucrose plates: MS Basal salts (Duchefa) at 4.4 g per liter with 1–3% (w/v) sucrose, 0.5 g/litre MES (2-*N*-Morpholinoethane sulfonic acid) and 1.5% (w/v) phyto-agar (Duchefa). Adjust the pH to 5.7 with KOH (do not use NaOH) and autoclave before use (*see* **Note 5**).

(b) Hoagland's media [37] plus sucrose plates: Stock solutions of the macro and micro elements described in Hoagland et al. [37] diluted to the specified concentration, with 1–3% (w/v) sucrose, 0.5 g/L MES. Adjust the pH to 5.7 with KOH before adding the iron chelate Fe-HBED (*N,N'*-bis(2-hydroxyphenyl) ethylenediamine-*N,N'*-diacetic acid). Add 1.5% (w/v) phyto-agar (Duchefa) before autoclaving (*see* **Note 5**).

3 Methods

3.1 Luciferase Measurements Using a TopCount

A TopCount scintillation counter is a high throughput method of measuring seedling luminescence in 96-well microtiter plates via photomultiplier detectors. Each experimental run can contain up to 20 96-well plates, which can be read approximately once per hour for up to 10 days (time measurements of a given seedling greatly depends on the number of plates assayed). If LED light panels are attached, this method also allows for luciferase expression to be measured under a range of different light conditions. The luminescence per well (per plant) is quantified when the results are downloaded. The advantage of this method is that it allows in vivo quantification of production of a specific protein in a non-invasive way. Seedlings of interest could therefore be transplanted and grown on to mature plants for further experiments or seed production [19] (*see* **Note 6**).

3.1.1 Experimental Set Up

1. Surface-sterilize seeds as described in Hanano et al. [38] and plate onto MS/sucrose plates (*see* Subheading 2.3.4) with the seeds sufficiently spaced to avoid subsequent seedlings crowding after germination. A sterile 0.1% agar solution can aid in plating seeds that are not self-clinging. Stratify for 2 days at 4 °C in the dark and transfer to growth chambers with the required entrainment conditions (e.g., 12:12 constant 20 °C).
2. In a laminar flow hood, surface-sterilize 96-well microtiter plates with an ethanol solution and allow to fully air-dry while maintaining sterility. Then transfer 200 μl of molten MS/sucrose agar media into each well and allow to solidify. These plates can be made in advance and stored by wrapping in commercial cling film, where we have not detected contamination even though these are not sold as sterile (*see* **Note** 7).
3. Transfer 6-day-old seedlings with tweezers taking care not to crush the plant (Fig. 2). Gently holding the hypocotyl, transplant one seedling per well to the microtiter plate ensuring that the roots are placed into the MS media with as little disruption to the media as possible. Where more than one genotype is used, set up the plate in rows rather than columns so that the light is equal across the replicate plants for both the control and test plants (*see* **Note 8**).
4. Add 15 μl of 1 mM luciferin solution to each well of the microtiter plate (Fig. 2).
5. Cover the plate with an adhesive plate top seal (PerkinElmer) and use a small needle to pierce the film above each well to allow gas exchange, taking care not to touch the cotyledon with the needle.

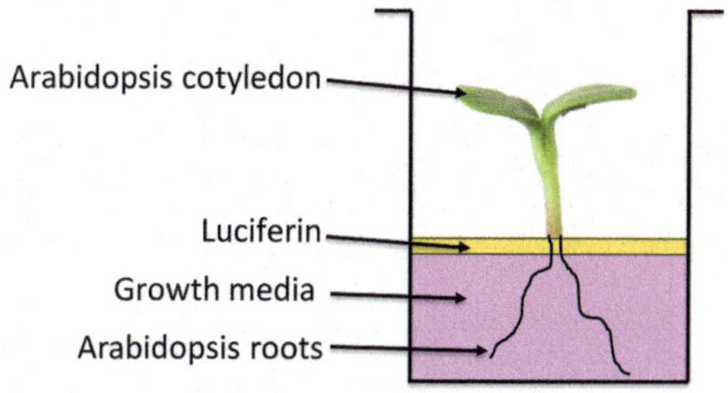

Fig. 2 Assembly of microtiter plate wells

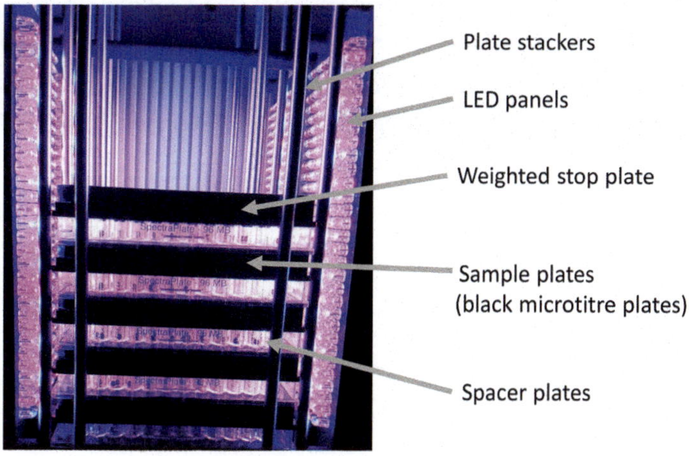

Fig. 3 Plate loading for a TopCount assay with LED panels

6. Using a razor blade, carefully trim away any of the sections of the plate seal that are overlapping the edges.
7. Stick a barcode to the middle of the edge of the right-hand side of the plate (*see* **Note 9**).
8. Transfer the plates back to the original growth chamber for 24 h or until subjective dusk on day 7 of growth.
9. At subjective dusk, transfer the plates to the TopCount, if using LED lighting panels (*see* **Notes 10** and **11**) start by loading a reflector plate followed by a black microtiter plate and continue to alternate reflector and microtiter plates until the last plate which should be a black "stop" microtiter plate (shown by the two barcodes on the side) (Fig. 3).
10. Create and run a TopCount assay that measures luminescence indefinitely taking readings after a 1-min delay for each plate (*see* **Note 12**).
11. Often the plants entrain for a further light–dark cycle in the TopCount before changing to free-run conditions, where the timer is removed from the lights or the timing is altered to create the desired lighting regime (*see* **Note 13**).

12. Experiments are run from 5 to 10 days in the TopCount, 10 days being the maximum due to gradual desiccation of the media.

3.1.2 TopCount Data Analysis

1. Download the data from the TopCount machine using the inbuilt results analysis software. Usually “read time” and “sample cps” (counts per second) are downloaded as an ASC2 file. This produces a text file for each individual read of a plate, these individual reads can be merged into one data set per plate using command prompt to merge all files referring to an individual plate.
2. The resulting merged file can be loaded into an online analysis program called BioDare2 [39–41] (*see* **Note 1**).
3. BioDare2 can then be used to run analyses such as FFT-NLS to produce period, phase and amplitude estimations for the selected data set [39] (Fig. 4).

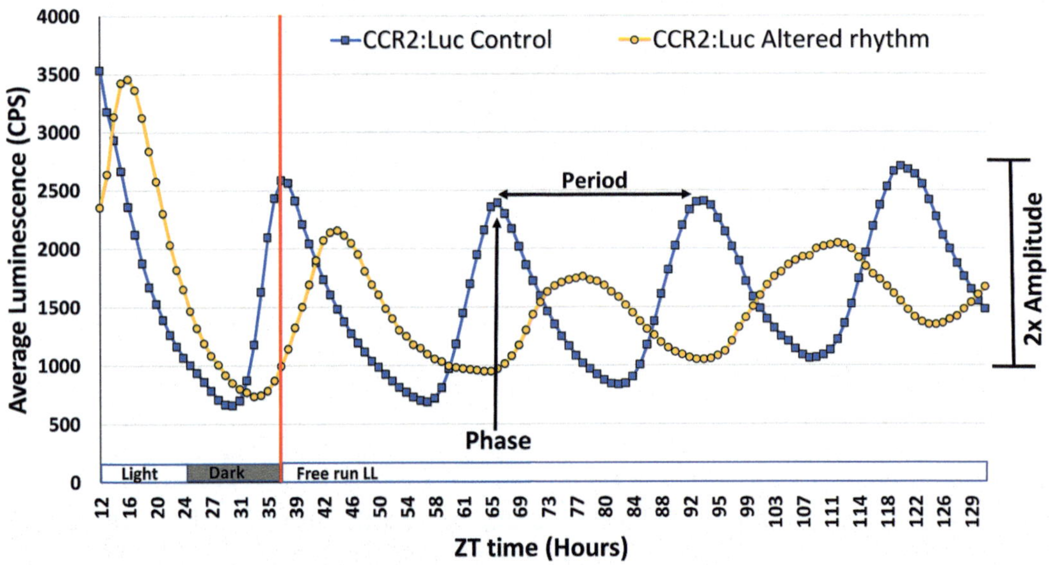

Fig. 4 Example Arabidopsis circadian data. Seedlings were entrained in 12:12 light–dark cycles at 20 °C constant temperature and then transferred to a top count to further entrain at 12:12 light–dark. From ZT 36 (indicated by a red line) the seedlings were subjected to a free run under constant light and temperature for the rest of the experiment. Blue (squares) represents a control circadian rhythm from *CCR2:Luciferase* (*CCR2:LUC*) with commonly analyzed features annotated. Period; the distance between two peaks of expression (or troughs), Phase; the highest point in the expression peak (or troughs), Amplitude; one half the distance between the peaks and troughs in expression. Yellow (circles) illustrates a mutant harboring *CCR2:LUC* with altered period, phase, and amplitude, as the main rhythm phenotypes detected in phytochrome or cryptochrome mutant lines

3.2 Luciferase Measurements Using a CCD Camera

As TopCount plate readers are a large expensive piece of equipment, a more frequently available piece of equipment is a CCD camera (charge-coupled device). Experiments can be carried out in a similar way to top count experiments and run in an automated way for a number of days. CCD cameras are internally cooled to, for example −80 °C, to improve image quality and stability [42]. An advantage of a CCD camera is that it is possible to export individual images for publication. In an ideal equipment setup, the camera is mounted in a temperature-controlled chamber, such as a RUMED chamber (*see* **Note 14**) with mounted computer controllable LED lights. The distance between the samples and camera should be adjusted to account for the number and size of plants for imaging. It is possible to image seedlings in a similar manner to the TopCount or to image mature plants in soil by adjusting the sample stage for the best focus on each sample (*see* **Note 15**). Again in an ideal set up the camera and LEDs are then controlled by an external computer running a program, such as MetaVue (Molecular Devices). Pictorial representation of equipment set up can be found in Hall and Brown [20] (*see* **Note 16**).

3.2.1 Experimental Setup

1. Sterilize seed and plate onto MS sucrose media (*see* Subheading 2.3.4) in ordered rows leaving space in between each seed to allow the resulting plants to grow without overlapping (*see* **Note 17**), stratify at 4 °C for 2 days, and transfer to a growth chamber to entrain for 6 days.
2. At subjective dusk on day 7, spray the seedlings with 1 mM luciferin and transfer to the growth chamber to begin imaging at the beginning of the dark phase (*see* **Note 18**).
3. Imaging is controlled by the MetaVue software. The experiment is run from a preprogrammed journal controlling both the lights and camera. The journals can be set to match the desired light/dark conditions, such as long day, or to free run with imaging. For imaging in the light phase, the lights are switched off for 1 min before imaging in the dark (*see* **Note 12**), and then the light are turned on after the image is acquired and saved. The image exposure time can be adjusted accordingly to the strength of signal from the seedlings usually between a 5 and 20-min exposure. Images can be taken at intervals of 1 h with the lights switching on after each image is taken and off again before the start of the next. Each image is automatically stored as part of an image library. This will make an image stack, which can be played out as a movie. Other automated software for CCD image acquisition and storage also exists.

3.2.2 CCD Camera Data Analysis

1. MetaMorph software is the ideal software to pair with the MetaVue system described in Subheading 3.2.1. However, there are other image acquisition software packages available,

the results may also be analyzed with other software capable of this type of analysis, such as Image J (*see* **Note 19**) [43].

2. Load the image library into MetaMorph. On the first image draw around the regions of interest (each seedling) and one control region to account for background noise [20]. The shape for region of interest can then be copied onto each subsequent image (*see* **Note 17**).
3. The luminescence for each region is quantified using the data exchange function to export the values to an Excel workbook. The graph intensities function should then be used to choose the time points and associated images to analyze.
4. To format the data export correctly for BioDare2 analysis, the data log should only save image name, image plane, image time and date, region name, and integrated. Then start the export using initiate measurement.
5. The finished measurements can then be used for FFT-NLS analysis in BioDare2 (*see* Subheading 3.1.2) or in Excel to draw graphs such as in Fig. 4.

4 Notes

1. BioDare2 can be accessed at the following web address: https://biodare2.ed.ac.uk/welcome
2. When selecting phytochrome loss of function lines, be careful of the ecotype used. Many promoter luciferase lines are in the WS-2 ecotype. WS-2 is lacking a phytochromeD so all phytochrome mutants in the WS-2 ecotype are essentially double mutants of your chosen phytochrome as well as *PHYD*. Alternatively, it would also be possible to complete the same experiments with the overexpression lines rather than loss of function.
3. The example used is phytochrome mutants; however, the same experiment could be carried out using mutants of potential signaling intermediates of phytochrome and the circadian clock to establish whether there are interactions.
4. To maintain temperature fluctuations throughout the measurement phase of the experiment, the CCD-camera method would prove easiest as the chamber could be programmed to specific temperature and time points. To carry out temperature entrainment while measuring with a top count system the temperature would need to be manually altered in both the room and the machine, temperatures within the top count machine are limited to a range of 15 °C–30 °C.
5. When working with media containing sucrose, extra care should be taken to maintain good sterile technique, as the

plates will become contaminated more easily than plates without sucrose.

6. It would be possible to carry out a similar style experiment without a TopCount system. Instead a single plate reader could be used inside a plant growth chamber, so long as the plate is ejected from the reader in between each reading [44].
7. Plates can be surface-sterilized by submerging in 100% ethanol and air-drying in a laminar flow hood before adding media. Plates containing media can be stored for up to 24 h before use.
8. For experiments using multiple genotypes appropriate controls need to be included on each plate. For example if using *phyB* mutants with *CCR2:LUC* in the WS-2 background, controls of *CCR2:LUC* in WS-2 wild type would need to be included. Replicates of 12 plants as a minimum need to be used for each.
9. The barcodes serve as unique identifiers for each plate; it is essential that a barcode is added to each black plate and that the plates are loaded in the specific orientation that allows the barcode to be read, so that the results will be recorded properly. Clear plates do not require barcodes as they are not to be measured.
10. Often circadian experiments involve a free-running period under constant light. LED banks are attached to the sides of the plate stack to allow the plants to remain in constant light conditions (Fig. 3). The plates are arranged alternately with clear microtiter plates to allow light to pass through as shown in Fig. 3. Clear plates used alternately in the microtiter plate stack are identical to the black microtiter plates, but contain three small metal ball bearings per well to reflect light and add weight to the plate. Clear plates need to be stacked in the same orientation as the black plates but do not need barcodes; instead, orientate them by the cut-out corner on each plate. For experiments running in darkness, the clear reflector plates are not needed.
11. Where LEDs are used, red and blue are used in equal proportions to be considered "white" light, in that major photoreceptors are all activated, assumingly as other wavelengths of light have not yet been determined as essential.
12. A 1-min delay is added to the assay to allow any delayed fluorescence to dissipate before the photomultiplier tubes start measurements. This eliminates the possibility of any false readings.
13. Although TopCount experiments are often used to record free-running period, it is also possible to record circadian rhythms under entrainment conditions by changing the light regieme in

the TopCount stackers (e.g., long days to short days or to intervals of 6 h light–6 h dark repeatedly). This can easily be set up by adding a timer switch to the lighting controller. Analog timers that the LED panels connect through the electrical outlet work well for this.

14. If a RUMED or another temperature controlled chamber is not available, an alternative can be the CCD camera in a dark box placed in a temperature-controlled room, as described in [20].
15. The whole plant could be imaged using this method. Particular organs or parts of the plant could then be quantified separately. This would be done by highlighting the individual sections of interest at the data analysis stage and quantifying them separately.
16. The camera, chambers, and program suggested in this method are one example of a large range of available CCD cameras, chambers, and software systems that are available and would be capable of carrying out the experiment described.
17. If multiple rows and plates are to be made, the seeds need to be sown in an identical pattern on each plate.
18. Imaging could also begin at the beginning of the light phase starting day 8 depending on entrainment conditions chosen.
19. If using software other than MetaMorph, care needs to be taken to ensure that the format of results produced can be input into BioDare if needed.

Acknowledgments

This work was supported by the Biotechnology and Biological Sciences Research Council (BBSRC) award number BB/N018540/1.

References

1. Dodd AN, Salathia N, Hall A, Kévei E, Tóth R, Nagy F, Hibberd JM, Millar AJ, Webb AAR (2005) Plant circadian clocks increase photosynthesis, growth, survival, and competitive advantage. Science 309:630–633
2. Michael TP, Salomé PA, McClung CR (2003) Two *Arabidopsis* circadian oscillators can be distinguished by differential temperature sensitivity. Proc Natl Acad Sci U S A 100:6878–6883
3. Harmer SL, Hogenesch JB, Straume M, Chang HS, Han B, Zhu T, Wang X, Kreps JA, Kay SA (2000) Orchestrated transcription of key pathways in *Arabidopsis* by the circadian clock. Science 290:2110–2113
4. Habte E, Müller LM, Shtaya M, Davis SJ, Von Korff M (2014) Osmotic stress at the barley root affects expression of circadian clock genes in the shoot. Plant Cell Environ 37:1321–1337
5. Nakamichi N, Kusano M, Fukushima A, Kita M, Ito S, Yamashino T, Saito K, Sakakibara H, Mizuno T (2009) Transcript profiling of an Arabidopsis PSEUDO RESPONSE REGULATOR arrhythmic triple mutant reveals a role for the circadian clock in cold stress response. Plant Cell Physiol 50:447–462

6. Bujdoso N, Davis SJ (2013) Mathematical modeling of an oscillating gene circuit to unravel the circadian clock network of *Arabidopsis thaliana*. Front Plant Sci 4:3
7. Michael TP, Salomé PA, Yu HJ, Spencer TR, Sharp EL, McPeek MA, Alonso JM, Ecker JR, McClung CR (2003) Enhanced fitness conferred by naturally occurring variation in the circadian clock. Science 302:1049–1053
8. Wenden B, Kozma-Bognár L, Edwards KD et al (2011) Light inputs shape the Arabidopsis circadian system. Plant J 66:480–491
9. Edwards KD, Guerineau F, Devlin PF, Millar AJ (2015) Low-temperature-specific effects of PHYTOCHROME C on the circadian clock in Arabidopsis suggest that PHYC underlies natural variation in biological timing. bioRxiv
10. Devlin PF, Kay SA (2000) Cryptochromes are required for phytochrome signaling to the circadian clock but not for rhythmicity. Plant Cell 12:2499–2510
11. Sharrock RA, Quail PH (1989) Novel phytochrome sequences in *Arabidopsis thaliana*: structure, evolution, and differential expression of a plant regulatory photoreceptor family. Genes Dev 3:1745–1757
12. Mathews S, Sharrock RA (1997) Phytochrome gene diversity. Plant Cell Environ 20:666–671
13. Somers DE, Devlin PF, Kay SA (1998) Phytochromes and cryptochromes in the entrainment of the *Arabidopsis* circadian clock. Science 282:1488–1490
14. Franklin KA, Davis SJ, Stoddart WM, Vierstra RD, Whitelam GC (2003) Mutant analyses define multiple roles for phytochrome C in *Arabidopsis* photomorphogenesis. Plant Cell 15:1981–1989
15. Bognár LK, Hall A, Adám E, Thain SC, Nagy F, Millar AJ (1999) The circadian clock controls the expression pattern of the circadian input photoreceptor, phytochrome B. Proc Natl Acad Sci U S A 96:14652–14657
16. Toth R, Kevei E, Hall A, Millar AJ, Nagy F, Kozma-Bognar L (2001) Circadian clock-regulated expression of phytochrome and cryptochrome genes in Arabidopsis. Plant Physiol 127:1607–1616
17. Más P, Devlin PF, Panda S, Kay SA (2000) Functional interaction of phytochrome B and cryptochrome 2. Nature 408:207–211
18. Millar AJ, Short SR, Chua NH, Kay SA (1992) A novel circadian phenotype based on firefly luciferase expression in transgenic plants. Plant Cell 4:1075–1087
19. Southern MM, Millar AJ (2005) Circadian genetics in the model higher plant, *Arabidopsis thaliana*. In: Young MW (ed) Methods in enzymology. Academic Press, Cambridge, MA, pp 23–35
20. Hall A, Brown P (2007) Monitoring circadian rhythms in *Arabidopsis thaliana* using luciferase reporter genes. In: Circadian rhythms. Humana Press, Totowa, NJ, pp 143–152
21. Millar AJ, Carré IA, Strayer CA, Chua NH, Kay SA (1995) Circadian clock mutants in *Arabidopsis* identified by luciferase imaging. Science 267:1161–1163
22. Darrah C, Taylor BL, Edwards KD, Brown PE, Hall A, McWatters HG (2006) Analysis of phase of *LUCIFERASE* expression reveals novel circadian quantitative trait loci in Arabidopsis. Plant Physiol 140:1464–1474
23. Boikoglou E, Ma Z, von Korff M, Davis AM, Nagy F, Davis SJ (2011) Environmental memory from a circadian oscillator: the *Arabidopsis thaliana* clock differentially integrates perception of photic vs. thermal entrainment. Genetics 189:655–664
24. Reed JW, Nagatani A, Elich TD, Fagan M, Chory J (1994) Phytochrome A and phytochrome B have overlapping but distinct functions in *Arabidopsis* development. Plant Physiol 104:1139–1149
25. Chun L, Kawakami A, Christopher DA (2001) Phytochrome A mediates blue light and UV-A-dependent chloroplast gene transcription in green leaves. Plant Physiol 125:1957–1966
26. Salter MG, Franklin KA, Whitelam GC (2003) Gating of the rapid shade-avoidance response by the circadian clock in plants. Nature 426:680–683
27. Genoud T, Millar AJ, Nishizawa N, Kay SA, Schäfer E, Nagatani A, Chua NH (1998) An Arabidopsis mutant hypersensitive to red and far-red light signals. Plant Cell 10:889–904
28. Covington MF, Panda S, Liu XL, Strayer CA, Wagner DR, Kay SA (2001) ELF3 modulates resetting of the circadian clock in Arabidopsis. Plant Cell 13:1305–1315
29. Kolmos E, Davis SJ (2007) *ELF4* as a central gene in the circadian clock. Plant Signal Behav 2:370–372
30. McWatters HG, Bastow RM, Hall A, Millar AJ (2000) The *ELF3 zeitnehmer* regulates light signalling to the circadian clock. Nature 408:716–720
31. Thines B, Harmon FG (2010) Ambient temperature response establishes ELF3 as a required component of the core *Arabidopsis* circadian clock. Proc Natl Acad Sci U S A 107:3257–3262
32. McClung CR, Davis SJ (2010) Ambient thermometers in plants: from physiological outputs towards mechanisms of thermal sensing. Curr Biol 20:R1086–R1092

33. Anwer MU, Boikoglou E, Herrero E, Hallstein M, Davis AM, James GV, Nagy F, Davis SJ (2014) Natural variation reveals that intracellular distribution of ELF3 protein is associated with function in the circadian clock. elife 2014:e02206
34. Rensing L, Ruoff P (2002) Temperature effect on entrainment, phase shifting, and amplitude of circadian clocks and its molecular bases. Chronobiol Int 19:807–864
35. Murashige T, Skoog F (1962) A revised medium for rapid growth and bio assays with tobacco tissue cultures. Physiol Plant 15:473–497
36. Bläsing OE, Gibon Y, Günther M, Höhne M, Morcuende R, Osuna D, Thimm O, Usadel B, Scheible W-R, Stitt M (2005) Sugars and circadian regulation make major contributions to the global regulation of diurnal gene expression in *Arabidopsis*. Plant Cell 17:3257–3281
37. Hoagland D, Arnon D (1950) The water-culture method for growing plants without soil. Circular California Agricultural Experiment Station 347:1–32
38. Hanano S, Stracke R, Jakoby M, Merkle T, Domagalska MA, Weisshaar B, Davis SJ (2008) A systematic survey in *Arabidopsis thaliana* of transcription factors that modulate circadian parameters. BMC Genomics 9:182
39. Plautz JD, Straume M, Stanewsky R, Jamison CF, Brandes C, Dowse HB, Hall JC, Kay SA (1997) Quantitative analysis of *Drosophila period* gene transcription in living animals. J Biol Rhythm 12:204–217
40. Dixon LE, Knox K, Kozma-Bognar L, Southern MM, Pokhilko A, Millar AJ (2011) Temporal repression of core circadian genes is mediated through EARLY FLOWERING 3 in *Arabidopsis*. Curr Biol 21:120–125
41. Zielinski T, Moore AM, Troup E, Halliday KJ, Millar AJ (2014) Strengths and limitations of period estimation methods for circadian data. PLoS One 9:e96462
42. Welsh DK, Imaizumi T, Kay SA (2005) Real-time reporting of circadian-regulated gene expression by luciferase imaging in plants and mammalian cells. In: Young MW (ed) Methods in enzymology. Academic Press, Cambridge, MA, pp 269–288
43. Schindelin J, Rueden CT, Hiner MC, Eliceiri KW (2015) The ImageJ ecosystem: an open platform for biomedical image analysis. Mol Reprod Dev 82:518–529
44. Perales M, Portolés S, Más P (2006) The proteasome-dependent degradation of CKB4 is regulated by the Arabidopsis biological clock. Plant J 46:849–860

Chapter 16

Characterization of Flowering Time Mutants

Xu Wang, Qing Liu, Wenjin He, Chentao Lin, and Qin Wang

Abstract

Timing of flowering, which is adapted to the ambient environment, is one of the key traits to ensure the reproductive success of plants. Our current understanding of the complex genetic control network of this trait is mostly derived from the studies in the model plant species *Arabidopsis thaliana*. *Arabidopsis thaliana* is an annual facultative long-day plant, whose flowering time is controlled by numerous environmental and endogenous factors. Here we briefly summarize the genetic pathways that promote flowering of *Arabidopsis* and describe standard protocols to characterize the flowering time phenotype of *Arabidopsis* mutants under laboratory conditions.

Key words Flowering time, Flowering pathways, *Arabidopsis thaliana*

1 Introduction

Flowering, also known as floral initiation, is a physiological event that plants transit from vegetative growth to reproductive growth. During this transition, the shoot apical meristem of plants starts to produce flower primordial instead of leaves. Plants have evolved the abilities to adapt their growth to the ever-changing environment and various endogenous cues, to ensure the correct timing of flowering for successful reproduction and species survival [1–3]. The genetic control of flowering time has been well studied in the model plant species *Arabidopsis thaliana*. Numerous genes involved in this process have been isolated through both forward and reverse genetic approaches. Genes that promote or repress the flowering could be identified either as mutations that delay or accelerate flowering time, or as overexpressions that accelerate or delay flowering time. The flowering time of *Arabidopsis* can be measured by both the number of days to flowering and the number of leaves. Alleles causing late flowering extend the duration of vegetative growth and therefore increase the number of leaves formed before flowering; whereas early flowering causes precease of vegetative growth, thereby producing less leaves.

Andreas Hiltbrunner (ed.), *Phytochromes: Methods and Protocols*, Methods in Molecular Biology, vol. 2026,
https://doi.org/10.1007/978-1-4939-9612-4_16,

Arabidopsis can sense and respond to not only various environmental changes, like day length and ambient temperature, but also many endogenous cues, like hormone levels and age. These factors profoundly affect the flowering time of plants. Yet, six major genetic pathways that promote flowering have been characterized in *Arabidopsis* [4–6] (Fig. 1). The photoperiod, vernalization, and thermosensory pathways act in response to day length, winter temperature, and ambient temperature change, respectively, while the autonomous gibberellin (GA) and age pathways respond more to endogenous stimuli rather than environmental cues [5].

Photoperiod has long been known to regulate flowering in many plant species. According the photoperiod type that plants respond to initiate flowering, plants could fall into three different groups: plants that only flower as day length increases are categorized as long-day plants; plants that flower only as day length wanes are known as short-day plants; while plants whose flower-

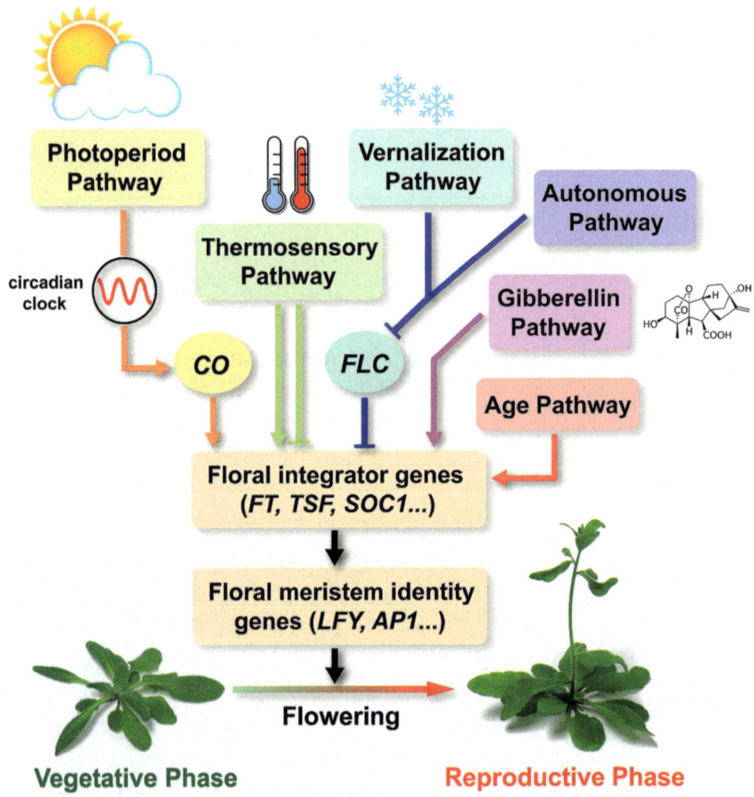

Fig. 1 Simplified genetic pathways that control flowering time in *Arabidopsis*. Six major genetic pathways that control flowering time have been characterized in *Arabidopsis*. The photoperiod, vernalization, and thermosensory pathways act in response to day length, winter temperature, and ambient temperature change, respectively, while the autonomous gibberellin (GA) and age pathways respond more to endogenous stimuli rather than environmental cues

ing is insensitive to day length change are called day-neutral plants. *Arabidopsis thaliana* is a facultative long-day plant, its flowering could be strongly induced by long-day photoperiod (16 h light; 8 h dark as the laboratory condition); albeit short-day photoperiod (8 h light; 16 h dark as the laboratory condition) is noninductive, *Arabidopsis* grown under this condition would eventually flower as well. The genetic control of the day-length measurement in *Arabidopsis* is not fully understood yet, the known mechanisms mainly involve the circadian regulated-expression and light-regulated degradation of *CONSTANS* (*CO*) gene and its protein [7–9].

Arabidopsis natural accessions have evolved two different flowering types, summer-annual and winter-annual habits. Summer-annual accessions, also known as rapid-cycling accessions, include many commonly used lab strains, like Columbia. They finish their life cycle rapidly (~6 weeks), flower and produce seeds within one spring and summer; whereas, winter-annual accessions normally germinate and grow in fall, overwinter as rosettes and flower in the next spring. To prevent flowering in the nonoptimal time as autumn, those winter-annual plants have evolved this special strategy using a prolonged cold period, known as vernalization, to induce flowering in spring [10]. The principal effect of vernalization at molecular level is thought to shut off the expression of a flowering repressor gene, *FLOWERING LOCUS C* (*FLC*) [11, 12]. The maintenance of the silence state of *FLC* after vernalization implicates intricate epigenetic mechanisms [13, 14], which are not completely clear yet.

Gibberellins (GA) are plant hormones that play important roles in determining the flowering time in *Arabidopsis* [15]. Studies on the effect of application of exogenous GA and the mutations impaired of GA biogenesis and signaling have revealed that GA has a minor influence on flowering in long days, but in non-inductive short days, GA pathway functions more pronounced [15–17].

The signals passing down from diverse flowering pathways can be eventually converged to activate the expressions of floral integrator genes, such as *FLOWERING LOCUS T* (*FT*) and *SUPPRESSOR OF OVEREXPRESSION OF CONSTANS 1* (*SOC1*) [5, 18, 19]. The activated floral integrators ultimately stimulate the expressions of floral meristem identity genes *APETALA1* (*AP1*) and *LEAFY* (*LFY*) [20–22], to initiate the formation of floral primordium.

In this chapter, we describe a set of standard protocols to characterize the flowering time phenotype of *Arabidopsis* mutants under laboratory conditions and discuss how to assess the gene functions in specific flowering pathways.

2 Materials

1. Plant growth chambers or walk-in rooms with controlled temperature, light, humidity, and other environmental conditions.
2. Soil.
3. Plastic pots and trays.
4. Water source.
5. GA_3 solution: Prepare 10 mM GA_3 stock solution in ethanol; dilute stock solution in water to 100 μM GA_3 and add 0.02% (v/v) Silwet 77.
6. Mock solution: 1% (v/v) ethanol, 0.02% (v/v) Silwet.
7. Marker pen (for labeling leaves).

3 Methods

In this section, different protocols are designed for assessing if the flowering time mutant falls into a specific pathway.

3.1 Preparation of Plants

1. Autoclave the soil before use, not only to kill eggs of common pests and any stray seeds but also to make the soil moistened.
2. Fill the pots with autoclaved soil and put the pots into a plastic tray.
3. Pour enough water into the secondary container to let the soil absorb until the soil become uniformly moistened.
4. Sow both wild-type and mutant seeds of *Arabidopsis* on the top of soil in different pots (*see* **Notes 1** and **2**). Keep the soil in the dark at 4 °C for 3 days to synchronize seed germination, then seedlings can be grown under different conditions.

3.2 Measure Flowering Time Under Photoperiodic Conditions

1. Move the seeds from 4 °C to both long-day (16 h light; 8 h dark; 22 °C) and short-day (8 h light; 16 h dark; 22 °C) conditions with cool-white fluorescent lights at 100 μmol m^{-2} s^{-1}. Record the date as the first day for counting the days to flower.
2. Let the plants grow until bolting (*see* **Note 3**). Count the number of primary rosette leaves (*see* **Note 4**) and calculate the number of days to flowering under both long-day and short-day conditions.

Arabidopsis thaliana is a facultative long-day plant, whose flowering is induced by long-day (16 h light; 8 h dark) photoperiod and delayed by short-day (8 h light; 16 h dark) photoperiod. Therefore, *Arabidopsis* flowers much earlier under long days than that under short days. If the long-day grown mutants flower just as late as they do under short days, it suggests that the floral initiation in the mutant plants is insensitive to the inductive photoperiod. Then we may assign this mutant to the photoperiod pathway.

3.3 Measure Flowering Time with the Vernalization Treatment

1. Move the seeds from 4 °C to the growth chamber in short-day (8 h light; 16 h dark; 22 °C) conditions under cool-white fluorescent lights at 100 μmol m^{-2} s^{-1}. Keep the plants grow there for 7 days (*see* **Note 5**).
2. Transfer plants needed for vernalization treatment to 4 °C in short days (8 h light; 16 h dark) for 6 weeks. Leave plants that do not need vernalization treatment at 22 °C in short days until bolting.
3. After vernalization treatment, move plants to 22 °C in short days until bolting. Count the number of primary rosette leaves for both vernalized and nonvernalized plants.

Vernalization strongly promotes the flowering for the winter-annual *Arabidopsis* accessions, like San Feliu-2 (SF2). Therefore, after vernalization, the wild-type plants should have many fewer rosette leaves at flowering than nonvernalized plants. The mutant plants which flower later than wild-type plants after vernalization treatment may be assigned to the vernalization pathway.

3.4 Measure Flowering Time with the Application of GA

1. Transfer the seeds from 4 °C to the growth chamber in short-day (8 h light; 16 h dark; 22 °C) conditions under cool-white fluorescent lights at 100 μmol m^{-2} s^{-1}. Record the date as the first day for counting the days to flower. Keep the plants grow there for 7 days (*see* **Note 6**).
2. Spray the plants either with the GA_3 solution or the Mock solution twice per week until bolting (*see* **Note 7**).
3. Count the number of primary rosette leaves and calculate the number of days to flowering for both GA-treated and mock-treated plants.

GA is a promoter of flowering in *Arabidopsis*. If the mutant is impaired in the GA response, the flowering time of GA-treated mutants would not be earlier than that of mock-treated mutants.

4 Notes

1. Because flowering time can be delicately influenced by multiple uncontrollable environmental factors, such as nutrition and biotic and abiotic stress, it is imperative that only plants grown at the same time and in the same place are comparable. Also, a conspicuous variation of flowering time exists among different ecotypes, one should keep in mind that only those plants in the same genetic background can be compared. Spacing between plants and even positions of pots will potentially affect the flowering time, it is advisable to keep at least 4–5 cm between plants and distribute the pots randomly in the tray.
2. Consider both the growth space and adequate sample size for statistical analyses; a reasonable number of plants should be

used in each set of experiment. Usually, ~20 wild-type control plants and ~20 mutant plants to be tested are used.

3. The time to flowering can be measured by either the number of days to flowering or the number of rosette leaves, but it is a better way to use both when characterizing the flowering phenotype of the mutants to be tested. The day of flowering can be practically determined as either when the inflorescence stem is 1 cm height or when the first time seeing the opened flower. However, it is more often to count rosette leaves when the inflorescence stem reaches 1 cm height, at this moment, the last rosette leaf and the first cauline leaf can be easily distinguished.
4. Only primary rosette leaves should be considered when counting the leaf numbers, since sometimes secondary leaves which look like rosette leaves but with smaller sizes would be appeared at the base of the inflorescence. Those leaves are more commonly seen under short-day conditions or in some late-flowering mutants. Because the secondary leaves tend to emerge at or after flowering [23], it is advisable to start counting the number of primary rosette leaves before flowering, especially for those plants grown under short-days or have a delayed flowering time. One may use the marker pen to label those leaves already counted.
5. Vernalization treatment can be also done with the seeds sown on MS-agar medium. After sowing the seeds, keep the MS plate at 4 °C for 3 days to synchronize seeds germination. Then move the plate to short days (8 h light; 16 h dark) at 22 °C and grow for 7 days. Transplant the seedlings that do not need vernalization to soil and grow at 22 °C in short days. Meanwhile, transfer the seedlings needed for vernalization to 4 °C in short days for 6 weeks. After vernalization, move the seedlings to 22 °C in short days to grow 10 days before transplant to soil. Count the number of primary rosette leaves at bolting for both vernalized and nonvernalized plants.
6. Since GA signaling has quite a minor influence on the flowering in long days but has more pronounced functions in short days, it should be easier to observe the effect of exogenous GA on flowering in short days than that in long days.
7. To prevent any chance that GA spread onto the mock-treated plants, it is advisable to grow the plants for different treatments in separate trays and spray the plants with different solutions in separate places.

Acknowledgements

The work in authors' laboratory is supported by grants of the National Institute of Health (GM56265 to CL), National Science Foundation of China (31500991 to QW) and National Science

Foundation of Fujian Province in China (2017J01604 to XW). The authors thank the UCLA-FAFU (Fujian Agriculture and Forestry University) Joint Research Center, Haixia Institute of Science and Technology, and Fujian Provincial Key Laboratory of Haixia Applied Plant Systems Biology for institutional supports.

References

1. de Montaigu A, Tóth R, Coupland G (2010) Plant development goes like clockwork. Trends Genet 26:296–306
2. Jarillo JA, Piñeiro M (2011) Timing is everything in plant development. The central role of floral repressors. Plant Sci 181:364–378
3. Michaels SD (2009) Flowering time regulation produces much fruit. Curr Opin Plant Biol 12:75–80
4. Andrés F, Coupland G (2012) The genetic basis of flowering responses to seasonal cues. Nat Rev Genet 13:627–639
5. Fornara F, de Montaigu A, Coupland G (2010) SnapShot: control of flowering in Arabidopsis. Cell 141:550, 550.e1–2
6. Wellmer F, Riechmann JL (2010) Gene networks controlling the initiation of flower development. Trends Genet 26:519–527
7. Hayama R, Coupland G (2004) The molecular basis of diversity in the photoperiodic flowering responses of Arabidopsis and rice. Plant Physiol 135:677–684
8. Imaizumi T, Kay SA (2006) Photoperiodic control of flowering: not only by coincidence. Trends Plant Sci 11:550–558
9. Song YH, Shim JS, Kinmonth-Schultz HA, Imaizumi T (2015) Photoperiodic flowering: time measurement mechanisms in leaves. Annu Rev Plant Biol 66:441–464
10. Johanson U, West J, Lister C, Michaels S, Amasino R, Dean C (2000) Molecular analysis of *FRIGIDA*, a major determinant of natural variation in *Arabidopsis* flowering time. Science 290:344–347
11. Kim D-H, Doyle MR, Sung S, Amasino RM (2009) Vernalization: winter and the timing of flowering in plants. Annu Rev Cell Dev Biol 25:277–299
12. Michaels SD, He Y, Scortecci KC, Amasino RM (2003) Attenuation of FLOWERING LOCUS C activity as a mechanism for the evolution of summer-annual flowering behavior in *Arabidopsis*. Proc Natl Acad Sci U S A 100:10102–10107
13. Song J, Angel A, Howard M, Dean C (2012) Vernalization - a cold-induced epigenetic switch. J Cell Sci 125:3723–3731
14. Zografos BR, Sung S (2012) Vernalization-mediated chromatin changes. J Exp Bot 63:4343–4348
15. Wilson RN, Heckman JW, Somerville CR (1992) Gibberellin is required for flowering in *Arabidopsis thaliana* under short days. Plant Physiol 100:403–408
16. Harberd NP (2003) Relieving DELLA restraint. Science 299:1853–1854
17. Schwechheimer C, Willige BC (2009) Shedding light on gibberellic acid signalling. Curr Opin Plant Biol 12:57–62
18. Kardailsky I, Shukla VK, Ahn JH, Dagenais N, Christensen SK, Nguyen JT, Chory J, Harrison MJ, Weigel D (1999) Activation tagging of the floral inducer FT. Science 286:1962–1965
19. Lee H, Suh SS, Park E, Cho E, Ahn JH, Kim SG, Lee JS, Kwon YM, Lee I (2000) The AGAMOUS-LIKE 20 MADS domain protein integrates floral inductive pathways in *Arabidopsis*. Genes Dev 14:2366–2376
20. Abe M, Kobayashi Y, Yamamoto S, Daimon Y, Yamaguchi A, Ikeda Y, Ichinoki H, Notaguchi M, Goto K, Araki T (2005) FD, a bZIP protein mediating signals from the floral pathway integrator FT at the shoot apex. Science 309:1052–1056
21. Weigel D, Alvarez J, Smyth DR, Yanofsky MF, Meyerowitz EM (1992) *LEAFY* controls floral meristem identity in Arabidopsis. Cell 69:843–859
22. Wigge PA, Kim MC, Jaeger KE, Busch W, Schmid M, Lohmann JU, Weigel D (2005) Integration of spatial and temporal information during floral induction in *Arabidopsis*. Science 309:1056–1059
23. Möller-Steinbach Y, Alexandre C, Hennig L (2010) Flowering time control. Methods Mol Biol 655:229–237

Chapter 17

Rapid Detection of Hormonal Involvement in Light Responses

Filip Vandenbussche

Abstract

Many aspects of light-controlled metabolism and development of plants depend on hormonal pathways. Here, a method is described to identify such hormonal dependence in light-regulated processes. A number of compounds—hormones and chemicals which interfere with hormonal pathways—are listed because of their usefulness in pharmacological treatment experiments. As an example for practical use of such compounds, elongation growth is discussed. An experimental setup is described in which plants are grown so that their structures develop predominantly in a two-dimensional plane. Time-lapse imaging is used to follow the plants in time, and image analysis reveals changes in plant morphology.

Key words Elongation, Arabidopsis, Auxin, Gibberellin, Brassinosteroid, Imaging

1 Introduction

Light effects on plant physiology and morphogenesis are very often depending on or interacting with hormonal responses. These include, among others, pigment accumulation, stomatal opening, germination, stem elongation, and phototropism. Interestingly, different species have different hormone dependencies of their light responses [1], warranting studies of hormonal control of photobiological processes in many species.

In order to know whether a light controlled response is hormone dependent, a number of tools exist. For some species marker genes for the output of hormonal pathways have been described. For Arabidopsis, currently used markers exist for auxin (*DR5::GFP*, *DR5::Venus*, *DII-Venus*), gibberellin (*pRGA::RGA-GFP*), cytokinin (*TCS::GFP*), brassinosteroids (*BZR1-GFP*), jasmonates (*JAZ1::GFP*), ethylene (*EIN3-GFP*, *EBS::GUS*), and abscisic acid (*RD29A*) [2–10]. These constructs can sometimes be used heterologously in other species, but implementation of markers is hampered because of the lack of knowledge on true functional orthologs.

Andreas Hiltbrunner (ed.), *Phytochromes: Methods and Protocols*, Methods in Molecular Biology, vol. 2026,
https://doi.org/10.1007/978-1-4939-9612-4_17,

However, morphological assays are often the most easily implemented and accessible experiments. The effect on elongation of stem like organs is among the most studied light responses in terms of photomorphogenesis. In sun plants, light with increasing intensity or fluence rate inhibits elongation. A notable exception is far red light, which in shade avoiding plants inactivates stable phytochromes which, when phyA levels are low, leads to elongation. When photoreceptors are mutated, elongation becomes less inhibited, yielding elongated plants. This is the case for mutants of cryptochromes in blue light, phytochrome A in far red light and phytochrome B in red light. The mutants thus partly mimic skotomorphogenic growth or the shade avoidance syndrome in the wild type.

A number of hormones are involved in elongation and participate in or interfere with light regulated elongation. Auxins [11], gibberellins [12], and brassinosteroids [13] all stimulate elongation, while ethylene can either inhibit or stimulate the same, depending on the species or the conditions [14]. As light signaling may share components with hormone signaling [15], chemicals that interfere with hormone signaling may be of use to unravel the necessity of a particular hormonal pathway. For many hormone biosynthesis pathways, rather specific chemical inhibitors exists (Table 1). Pharmacological treatments based on hormones or inhibitors have the advantage that they are applicable across species, if the chemical is taken up by the plant. The use of such compounds to dissect hormonal pathways involved in photomorphogenesis will be discussed in the methodological example below. In this example, the focus will be on elongation as output, yet for other processes, such chemicals may also be of use.

Morphological traits, including hypocotyl elongation, can be quantified by measuring structures on the plants. This is very often achieved by photographing the sample plant and subsequent image analysis. A setup for automated time-lapse imaging suitable for following elongation in seedlings will be described.

2 Materials

2.1 Plant Growth, Hormones, and Light Sources

1. Plant Medium: Murashige and Skoog (MS) salts at half strength, 1% sucrose, 0.8% plant tissue culture agar, pH 5.7 (adjusted with 1 M KOH).
2. Square petri dishes.
3. Sterilization solution: 5% NaOCl (1/3 commercial NaOCl 15%, 2/3 water), 2 drops of Tween 20 per 100 ml.
4. Sterile tweezers or forceps.
5. Sterile demineralized water.

Table 1
List of hormones and commonly used chemicals that can interfere with their function

Compound	Concentration range	Stock solution	Function	Target	Assay	Suggested supplier	Reference
N-1-naphtylphtalamic acid	0.1–10 μM	10 mM, DMSO	Auxin efflux inhibitor	ABCB chaperone TWD1	Arabidopsis (differential) elongation	Sigma-Aldrich, Olchemim	[18]
Picloram	1–10 μM	10 mM, DMSO	Synthetic Auxin	Auxin receptors AFB4, AFB5	Arabidopsis hypocotyl elongation	Duchefa	[19]
Abamine SG	100 μM	10 mM, DMSO	ABA biosynthesis inhibitor	9-cis-Epoxycarotenoid dioxygenase NCED	Stomatal closure		[20]
Brassinazole	0.1–10 μM	1 mM, DMSO	BR biosynthesis inhibitor	DWF4, Steroid 22-alpha hydroxylase	Arabidopsis hypocotyl elongation	Sigma-Aldrich, TCI	[21, 22]
Bikinin	5–30 μM	30 mM, DMSO	BR signaling stimulator	GSK kinases	Arabidopsis hypocotyl elongation	Chembridge, ApexBio	[23]
Brassinolide	0.1–100 nM	100 μM, ethanol	Plant hormone	BR receptor BRI1	Arabidopsis hypocotyl elongation	Sigma-Aldrich	[24]
Paclobutrazol	0.1–1 μM	1 mM, ethanol or DMSO	GA biosynthesis inhibitor	GA3, Kaurene oxidase	Arabidopsis hypocotyl elongation	Sigma-Aldrich, Duchefa	[25]
GA_3	0.1–10 μM	10 mM, ethanol	Plant hormone	GA receptors GID	Arabidopsis hypocotyl elongation; germination	Duchefa	[26]
TIS108	1–3 μM		Strigolactone biosynthesis inhibitor	cytP450?	Shoot bushiness		[27]

(continued)

Table 1 (continued)

Compound	Concentration range	Stock solution	Function	Target	Assay	Suggested supplier	Reference
GR24	10 μM	10 mM, acetone	Synthetic strigolactone	SL receptor DWARF14/D14	Arabidopsis hypocotyl elongation	Strigolab, Olchemim	[28]
ACC	0.1–50 μM	100 mM, in water	Ethylene precursor	ACC oxidase	Arabidopsis hypocotyl elongation; apical hook opening	Sigma	[29]
1-MCP	250 ppm	gaseous	Ethylene antagonist	Ethylene receptor	Arabidopsis hypocotyl elongation	AgroFresh	[16]

6. Seeds.
7. One 100 ml and two 1000 ml beakers.
8. Miracloth (Merck-Millipore) and stapler.
9. Micropore tape (3 M) and Parafilm M (Bemis Company, Oshkosh, WI, USA).
10. Laminar flow.
11. Fridge or cold room.
12. Growth chamber with suitable conditions. For Arabidopsis this is 21–22 °C and 50–70% humidity.
13. A holder to maintain petri dishes in a vertical position by supporting the edges.
14. Light sources are according to desire and scientific question. The following are regularly used: Philips Greenpower LEDs of 450 nm, 660 nm and 730 nm, cool white T8 TL-D fluorescent light (Philips).
15. For working in "darkness" or simulating dark conditions, respectively dim green safety light and 940 nm LED light are used (*see* **Note 1**).
16. Light sources are installed overhead. Fluence rates are measured at plant height by a radiometer.
17. Hormones and hormone inhibitors: Dissolve the hormones or chemicals to influence plant physiology in DMSO (dimethylsulfoxide) to 1000× to 10,000× their effective concentration range (Table 1) in the medium (*see* **Note 2**). Filter-sterilize and use immediately or keep at −20 °C in aliquots. Avoid repetitive freeze–thaw cycles.

2.2 Image Analysis

1. Camera: This can be any DSLR, network or compact camera, or smartphone, with built-in or compatible suitable computer control software for time-lapse imaging, and over 2 Mpixel resolution. For imaging in darkness they should be infrared enabled (e.g., LifePixel.com). However, a low threshold system based on a USB webcam with at least 640 × 480 resolution, where the infrared filter can be removed, such as Hercules Optical Glass (Hercules), works well (*see* **Note 3**).
2. Computer, able to run and containing the Active Webcam software program (PY software; http://www.pysoft.com/ActiveWebCamMainpage.htm) and ImageJ (NIH).

3 Methods

3.1 Plant Growth

Depending on the experimental setup, sterile conditions can be used to grow plants from seed. The following method works well for sterilizing small seeds.

1. Switch on the laminar flow cabinet and clean the table surface with denatured ethanol. Wait for 15 min.
2. Enter all materials (*see* Subheading 2.1), after cleaning the glassware with denatured ethanol.
3. Let the autoclaved medium cool down to 50 °C (liquid but touchable with bare hands).
4. Thaw compound stock solution at room temperature if frozen. Add the compound in the desired concentration to the medium. Add solvent in volume equal to the solvent (e.g., DMSO) volume used for the highest concentration of compound (*see* **Note 4**).
5. Mix well by swirling.
6. Pour the medium in the square petri dishes and let solidify and cool down to room temperature.
7. Per seed line, combine seeds into a bag made of Miracloth and sealed with a staple.
8. Surface-sterilize seeds by submerging the Miracloth bag with seeds for 15 min at room temperature in sterilization solution (*see* **Note 5**).
9. Rinse the seeds with water for 3 min by replacing the sterilization solution. Rinse an additional four times for 3 min with sterile water by replacing the previous solution (*see* **Note 6**).
10. Open the Miracloth bag in a sterile empty petri dish.
11. Transfer the seeds to the medium using sterile tweezers. For Arabidopsis seedlings on vertical petri dishes, leave a distance of 1–2 mm between the seeds.
12. Seal the petri dish at three sides with Parafilm (the future bottom and lateral sides), followed by an optional complete sealing with Micropore (3 M) tape (*see* **Note 7**).
13. Stratify seeds by keeping petri dishes horizontal for 36–72 h at 4 °C in the dark.
14. Give a 6 h light pulse of 100 μmol m^{-2} s^{-1} cool white light at 22 °C.
15. Put petri dishes upright in the respective light conditions at 22 °C (*see* **Notes 8** and **9**).

3.2 Image Acquisition and Analysis

For meaningful physiological data, a single image after a number of days of compound treatment may be enough. However, much more informative is a setup in which at regular intervals a picture of the growing biological material is taken, time-lapse acquisition. This may yield additional data on the time window when a hormone is most active.

1. Set up the cameras at a distance from the petri dishes suitable for monitoring one or various plants at the same time.
2. Have the computer recognize the cameras and install the drivers (usually plug and play).
3. Open the Active Webcam software and use the search camera function (*see* **Note 10**).
4. Adjust the focal plane on the camera to obtain sharp images.
5. In the tab "Recording," change "Not active" to "Create snapshots all the time"; specify "snapshot settings" including frame rate (= number of pictures per time unit), and number of total pictures to be taken.
6. Stop the experiment when desired (e.g., after 10 days for studying the growth of Arabidopsis seedlings).
7. Put a ruler instead of the petri dish, and have a picture taken for scaling purposes.
8. Using ImageJ, analyze the length of the tissue with the segmented line tool.
9. Move values from ImageJ to a spreadsheet software for further analysis and graph generation. Generic typical output graphs for time-lapse experiments are shown in Fig. 1.

3.3 Experimental Steps to Determine Hormonal Dependence of a Continuous Light Response

1. Select a compound for pharmacological treatment (*see* **Note 11**).
2. Determine when saturation occurs in the response to light: Analyze the plant response at end point (e.g., after 7–10 days of elongation growth in *A. thaliana* seedlings reaches a plateau) to an increasing range of fluence rates until saturation is observed. The resulting graph for elongation will be of the form shown in Fig. 2. This will set the fluence rate to be used in the following steps.
3. If the compound acts independently of light treatments, it will have a relatively similar effect in darkness and in light. For example, the effect on elongation can be either negative (Fig. 1a) or positive (Fig. 1b).
4. Using saturating light conditions, it is also possible to assess whether a hormone or chemical has a contrasting effect to that of the light. In case the effect is caused by a hormone, and the effect is not visible in darkness, then this hormone or its signal-

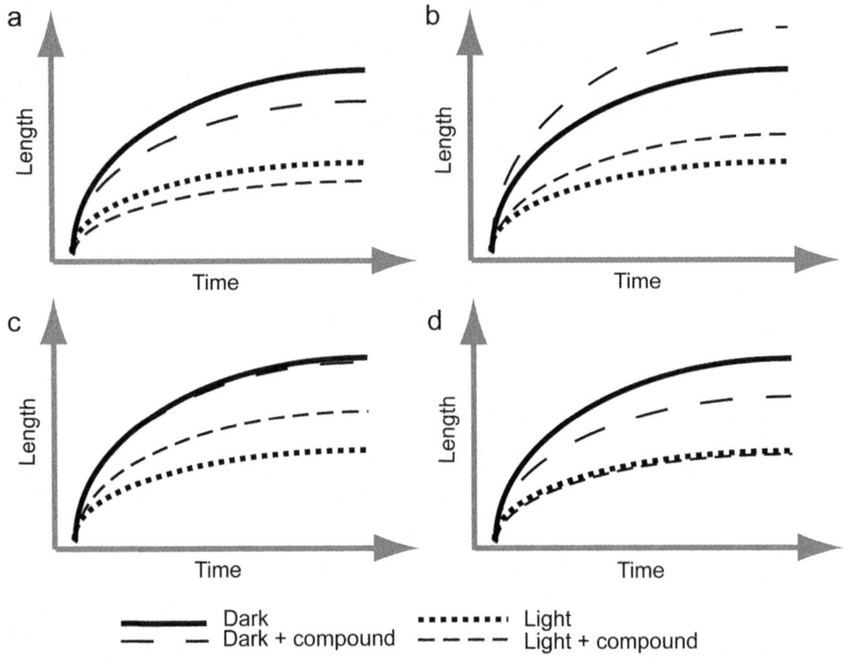

Fig. 1 Generic graphs derived from time-lapse experiments in which the length of a plant hypocotyl is followed in the presence or absence of light saturating for the light response, and the presence or the absence of a hormone or interfering chemical (compound). (**a**, **b**) Light and compound act independently. (**a**) Situation where the compound has the same relative negative effect on elongation irrespective of the light conditions. (**b**) Situation where the compound has the same relative positive effect on elongation irrespective of the light conditions. (**c**) Situation where the compound counteracts the light effect. (**d**) Situation where the compound (partially) mimics the light effect

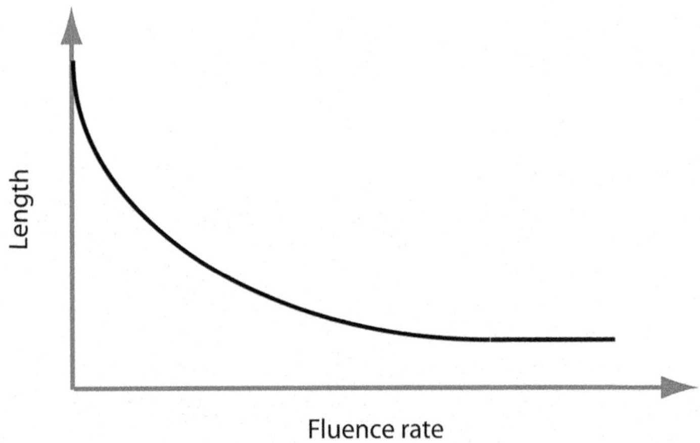

Fig. 2 Generic graph showing saturation of the light response (inhibition of elongation) with increasing fluence rate

ing pathway is a possible target for light signaling (Fig. 1c) (e.g., gibberellin) (*see* **Note 12**).

5. Corroborating the above, a compound interfering with the function of the hormone will not have an additional effect at saturating light conditions (e.g., paclobutrazol), provided the light response also confers saturation of the inhibition of hormone response. By contrast, its effect will be clearly visible in dark-grown tissues (Fig. 1d), or in nonsaturating light (*see* **Note 13**). In case the hormone response is not completely inhibited, the response should be relatively less pronounced than in nonsaturating light conditions (*see* **Note 14**).

3.4 Determining Additivity, Synergism, or Antagonism

1. To determine whether the effects of light and hormones are additive or synergistic, nonsaturating levels of light should be used (*see* **Note 15**).
2. Determine values at a certain point in time (e.g., 7 days after germination).
3. Make a graph displaying the response (length) versus compound concentration at different fluence rates (Fig. 3a–c). Transformation to relative values (i.e., hypocotyl length in the presence of the compound divided by the hypocotyl length in the absence of the compound) will show immediate details on the nature of the response (Fig. 3d–f).

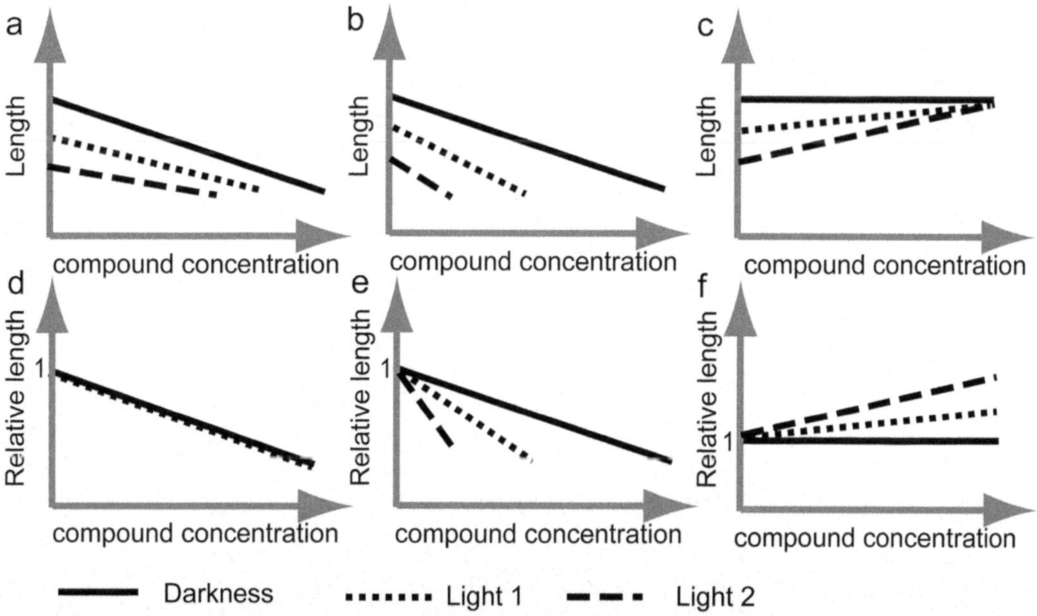

Fig. 3 Generic graphs showing examples of additivity (**a**, **d**), synergism (**b**, **e**), and antagonism (**c**, **f**) of the effect of a compound and light. The range of light intensities and compound concentration are chosen in a window of linear decrease of length for both. Fluence rates increase in the order darkness < Light 1 < Light 2. Lower graphs are relative values of the upper graphs

4. Additive effects are characterized by a relatively equal response to the compound at each light intensity (Fig. 3d).
5. Synergistic effects are characterized by a relatively stronger response with increasing compound concentration than to be expected from an additive effect (Fig. 3e) (*see* **Note 15**).
6. Antagonistic effects are characterized by a relatively weaker, or even a contrasting response with increasing compound concentration than expected from an additive effect (Fig. 3f).

4 Notes

1. The use of 940 nm IR LEDs for manipulation of materials in simulated darkness is the best with respect to compatibility with plant physiology. For this, a system to monitor in IR is needed and can consist of cameras or, better still, goggles which allow stereoscopic viewing. A more accessible alternative to this is the use of ambient dim green safety light for short periods of manipulation. This will work well in most of the cases without affecting the physiological responses under research. A test for morphological effects versus true darkness should be considered before using dim green safety light.
2. DMSO appears the most universal solvent for (bio) chemicals. It is advised not to surpass 1% of DMSO in the media, to avoid side effects due to DMSO. In some cases (e.g., gibberellin, paclobutrazol), ethanol is equally suitable as solvent but is less universal, which is a drawback when comparing multiple compounds in a factorial experiment.
3. When connecting more than one webcam to the same computer, for hardware recognition by the computer, it is best to use different camera models.
4. In case of using 1-MCP, treatment should be done in a gas-tight container to allow the 1-MCP to accumulate. Alternating with a couple of hours of air refreshment per day, the 1-MCP treatment should be repeated daily [16].
5. This protocol is suitable for small seeds. Larger seeds are better sterilized using chlorine gas in a desiccator, a protocol that can also be used for small seeds (http://www.plantpath.wisc.edu/fac/afb/vapster.html).
6. The last rinsing solution should not smell of chlorine. If so, rinse more with sterile water.
7. If Micropore tape would interfere with the light conditions, lighting from above is best preserved omitting this optional step.

8. Vertical medium allows plants to grow with their hypocotyl (and root) in a two-dimensional plane corresponding to the focal plane of the camera. This setup avoids loss of information due to three-dimensional effects caused by loss of gravity perception, as is the case with seedlings growing in red light or pulsed far red light.
9. In the special case where the compound inhibits seed germination (e.g., paclobutrazol), let seeds germinate on 1/2× MS medium without additives, or in water. Once a radicle is visible, transfer to compound containing medium (in a laminar flow cabinet, using tweezers). Do this in green safety light or infrared light.
10. The software continuously draws image frames from the cameras and may overload the internal memory. To avoid this, once the cameras are found, click on the image from the camera of interest. Under tab "devices," adjust the preview frame rate if the processor speed or graphical card has trouble with generating images.
11. To study hormonal involvement, the ideal compounds (hormone or chemical that interferes with the hormone) have the following characteristics. Either the compound driven response mimics the dark situation, and the compound has little effect in darkness in the concentration range used, or the compound response mimics the light condition and can yield an effect similar to saturating light conditions.
12. Additional information to this fast indication on hormonal involvement should be gathered. It will be necessary to confirm that the morphological observation matches changes at the molecular level (e.g., with marker genes), as it is also possible that light does not directly control this hormone pathway, but controls another pathway that interacts with it. Mutants are needed to dissect light–hormone interaction pathways genetically.
13. Even though the described chemicals (Table 1) have been thoroughly tested in the past, not every physiological situation or tissue has been evaluated, and hence they may have side effects. Therefore, when using biosynthesis inhibitors, add a "rescue" experiment. In this experiment, additional to the inhibitor, the hormone is added in a concentration normally yielding saturating response. Reversal of the inhibitor effect indicates that it is really hormone specific. In addition, where possible, different inhibitor classes for one pathway can be used to confirm the results.
14. In all cases, for clarity a next step would be quantification of the hormone, usually by LC-MS, SIFT-MS, or photoacoustic

detection, depending on the hormone. Generally, such analysis is performed in a specialized laboratory.

15. Synergism can be an expression of regulation of closely related genes, or of the control of a node in a network controlled by interacting light and hormone conditions [17]. The latter may thus be an indication that light does not directly interfere with the hormonal pathway but with the output of it. Such a regulation may involve direct control of light on the process but can also be mediated by another hormone.

References

1. Symons GM, Reid JB (2003) Interactions between light and plant hormones during de-etiolation. J Plant Growth Regul 22:3–14
2. Brunoud G, Wells DM, Oliva M et al (2012) A novel sensor to map auxin response and distribution at high spatio-temporal resolution. Nature 482:103–106
3. Yang J, Yuan Z, Meng Q et al (2017) Dynamic regulation of auxin response during rice development revealed by newly established hormone biosensor markers. Front Plant Sci 8:256
4. Guo H, Ecker JR (2003) Plant responses to ethylene gas are mediated by $SCF^{EBF1/EBF2}$-dependent proteolysis of EIN3 transcription factor. Cell 115:667–677
5. Vandenbussche F, Petrášek J, Žádníková P et al (2010) The auxin influx carriers AUX1 and LAX3 are involved in auxin-ethylene interactions during apical hook development in *Arabidopsis thaliana* seedlings. Development 137:597–606
6. Pauwels L, Barbero GF, Geerinck J et al (2010) NINJA connects the co-repressor TOPLESS to jasmonate signalling. Nature 464:788–791
7. Zhao J, Peng P, Schmitz RJ, Decker AD, Tax FE, Li J (2002) Two putative BIN2 substrates are nuclear components of brassinosteroid signaling. Plant Physiol 130:1221–1229
8. Zürcher E, Tavor-Deslex D, Lituiev D, Enkerli K, Tarr PT, Müller B (2013) A robust and sensitive synthetic sensor to monitor the transcriptional output of the cytokinin signaling network in planta. Plant Physiol 161:1066–1075
9. Fu X, Harberd NP (2003) Auxin promotes *Arabidopsis* root growth by modulating gibberellin response. Nature 421:740–743
10. Yamaguchi-Shinozaki K, Shinozaki K (1993) Characterization of the expression of a desiccation-responsive *rd29* gene of *Arabidopsis thaliana* and analysis of its promoter in transgenic plants. Mol Gen Genet 236:331–340
11. Lavy M, Estelle M (2016) Mechanisms of auxin signaling. Development 143:3226–3229
12. Hedden P, Sponsel V (2015) A century of gibberellin research. J Plant Growth Regul 34:740–760
13. Singh AP, Savaldi-Goldstein S (2015) Growth control: brassinosteroid activity gets context. J Exp Bot 66:1123–1132
14. Pierik R, Tholen D, Poorter H, Visser EJW, Voesenek L a CJ (2006) The Janus face of ethylene: growth inhibition and stimulation. Trends Plant Sci 11:176–183
15. Lau OS, Deng XW (2010) Plant hormone signaling lightens up: integrators of light and hormones. Curr Opin Plant Biol 13:571–577
16. De Paepe A, De Grauwe L, Bertrand S, Smalle J, Van Der Straeten D (2005) The *Arabidopsis* mutant *eer2* has enhanced ethylene responses in the light. J Exp Bot 56:2409–2420
17. Pérez-Pérez JM, Candela H, Micol JL (2009) Understanding synergy in genetic interactions. Trends Genet 25:368–376
18. Zhu J, Bailly A, Zwiewka M et al (2016) TWISTED DWARF1 mediates the action of auxin transport inhibitors on actin cytoskeleton dynamics. Plant Cell 28:930–948
19. Prigge MJ, Greenham K, Zhang Y, Santner A, Castillejo C, Mutka AM, O'Malley RC, Ecker JR, Kunkel BN, Estelle M (2016) The *Arabidopsis* auxin receptor F-box proteins AFB4 and AFB5 are required for response to the synthetic auxin picloram. G3 (Bethesda) 6:1383–1390
20. Kitahata N, Han S-Y, Noji N et al (2006) A 9-cis-epoxycarotenoid dioxygenase inhibitor for use in the elucidation of abscisic acid action mechanisms. Bioorg Med Chem 14:5555–5561
21. Asami T, Min YK, Nagata N, Yamagishi K, Takatsuto S, Fujioka S, Murofushi N, Yamaguchi I, Yoshida S (2000) Characterization of brassinazole, a triazole-type brassinosteroid biosynthesis inhibitor. Plant Physiol 123:93–100
22. Nagata N, Min YK, Nakano T, Asami T, Yoshida S (2000) Treatment of dark-grown

Arabidopsis thaliana with a brassinosteroid-biosynthesis inhibitor, brassinazole, induces some characteristics of light-grown plants. Planta 211:781–790

23. De Rybel B, Audenaert D, Vert G et al (2009) Chemical inhibition of a subset of *Arabidopsis thaliana* GSK3-like kinases activates brassinosteroid signaling. Chem Biol 16:594–604
24. Watanabe T, Fujioka S, Yokota T, Takatsuto S (1998) Synthesis and biological activity of 2,3-diol stereoisomers of 28-homobrassinolide and brassinolide. J Chem Res 0:744–745
25. Dalziel J, Lawrence D (1984) Biochemical and biological effects of kaurene oxidase inhibitors, such as paclobutrazol. In: Menhenett R, Lawrence DK (eds) Monograph-British Plant Growth Regulation Group. British Plant Growth Regulation Group, Wantage, pp 43–47
26. Davière J-M, Achard P (2013) Gibberellin signaling in plants. Development 140:1147–1151
27. Ito S, Umehara M, Hanada A, Yamaguchi S, Asami T (2013) Effects of strigolactone-biosynthesis inhibitor TIS108 on *Arabidopsis*. Plant Signal Behav 8:e24193
28. Tsuchiya Y, Vidaurre D, Toh S, Hanada A, Nambara E, Kamiya Y, Yamaguchi S, McCourt P (2010) A small-molecule screen identifies new functions for the plant hormone strigolactone. Nat Chem Biol 6:741–749
29. Vandenbussche F, Vancompernolle B, Rieu I, Ahmad M, Phillips A, Moritz T, Hedden P, Van Der Straeten D (2007) Ethylene-induced *Arabidopsis* hypocotyl elongation is dependent on but not mediated by gibberellins. J Exp Bot 58:4269–4281

Chapter 18

Phytochrome and Light Signaling in *Marchantia*

Keisuke Inoue, Ryuichi Nishihama, and Takayuki Kohchi

Abstract

Bryophytes, which comprise liverworts, mosses, and hornworts, are one of the earliest diverging lineages of extant land plants and a key plant group for understanding evolutionary aspects of land plant adaptation. *Marchantia polymorpha*, a liverwort, has recently been established as a model plant species having molecular genetic tractability. In *M. polymorpha*, phytochrome is encoded by a single-copy gene, Mp*PHY*, with Mpphy regulating various physiological responses through PHYTOCHROME INTERACTING FACTOR (PIF)-mediated transcriptional regulation. The phytochrome signaling system of *M. polymorpha*, with its single Mpphy and single PIF (MpPIF), is relatively simple compared with other model plants carrying multiple phytochromes and PIFs. Consequently, investigation of phytochrome signaling using *M. polymorpha* may provide novel insights into fundamental mechanisms and roles of phytochrome during the course of land plant evolution. This chapter provides a number of basic procedures, along with some tips, for designing and performing experiments with *M. polymorpha* to study phytochrome signaling.

Key words Gemma germination, Liverwort, *Marchantia polymorpha*, Orthotropic growth, Phytochrome, Subcellular fractionation

1 Introduction

Liverworts represent an early diverging lineage of extant land plants [1, 2]. The liverwort *Marchantia polymorpha* is being increasingly used as a model plant species because it possesses characteristics that are advantageous for genetics analysis, such as a haploid-dominant life cycle and tractability to genetic manipulation (reviewed in [3]). In addition, the *M. polymorpha* genome has been sequenced, with the genome and transcriptome data now available ([4]; http://marchantia.info/). Compared with other sequenced land plants, *M. polymorpha* exhibits low genetic redundancy, especially in regards to regulatory pathways, which suggests that its genome resembles that of the common ancestor

Andreas Hiltbrunner (ed.), *Phytochromes: Methods and Protocols*, Methods in Molecular Biology, vol. 2026,
https://doi.org/10.1007/978-1-4939-9612-4_18,

of land plants [4]. Molecular genetics using *M. polymorpha* is therefore a straightforward way to understand evolution and adaptation of land plants.

Physiological studies of phytochrome-mediated responses of *M. polymorpha* date back more than 50 years. A pulse of far-red light (FR) at the end of the light period (the so-called end-of-day FR) leads to orthotropic growth of thalli and a decrease in the amount of chlorophyll. The effect of end-of-day FR is cancelled by a subsequent pulse of red light (R), which suggests the involvement of phytochrome in these responses [5, 6]. Senescence of thalli and germination of gemmae also seem to be regulated by phytochrome according to their R/FR photoreversibility [7–9]. Furthermore, recent molecular genetic studies have revealed that the single phytochrome in *M. polymorpha* (Mpphy) may possess biochemical and cell-biological characteristics of both type-I and type-II phytochromes of angiosperms [10]. Mpphy functions in the nucleus and regulates gene expression by interacting with a basic helix–loop–helix transcription factor, MpPIF, which suggests that a light-dependent transcriptional regulatory mechanism was already present in early land plants [11]. Here, we summarize the basic procedures to investigate phytochrome signaling using *M. polymorpha*.

2 Materials

2.1 Equipment

1. A red LED illuminator (MIL-R18, Sanyo Electric; peak emission at 657 nm, half-bandwidth of 12 nm).
2. A far-red LED illuminator (MIL-IF18, Sanyo Electric; peak emission at 734 nm, half-bandwidth of 13 nm).
3. A diurnal plant growth chamber equipped with a far-red LED illuminator.
4. An LI-250A light meter (LI-COR Biosciences).
5. An SKR 110 red/far-red sensor (Skye Instruments).
6. A viridian transparent cutting sheet (431C, Nakagawa Chemical Inc.)
7. Green safelight (A green fluorescent tube wrapped in a viridian transparent cutting sheet; peak wavelength at 520 nm, half-bandwidth of 30 nm).
8. Aluminum foil.
9. Real-Time PCR system.
10. Refrigerated benchtop centrifuge (for 1.5 and 2 mL tubes).
11. Mortar and pestle.
12. 70-μm cell strainer or equivalent.
13. Equipment for running SDS-PAGE and immunoblotting.

2.2 Media, Molecular Biology Reagents and Kits

1. Half-strength Gamborg's B5 medium [12]: half-strength Gamborg's B5 basal salt mixture, 0.5 g/L 2-(*N*-morpholino) ethanesulfonic acid (MES), and 1.0% (w/v) agar, adjusted to pH 5.5 with KOH (*see* **Note 1**).
2. TRIzol reagent (Thermo Fisher Scientific).
3. RNeasy Plant Mini kit (Qiagen).
4. Primer pairs for quantitative real-time PCR (qPCR): for *LHCB* (Mapoly0139s0012.1), 5′-GGGATTTCGTTCAGTGGA TGG-3′ and 5′-GCTCGCTCCAATTCCGTTC-3′; for *POR* (Mapoly0030s0118.1), 5′-GTCTGATGAAGCGAGCAA CC-3′ and 5′-GACGGAACTCTAAGCGAAACC-3′; for *EF* (Mapoly0024s0116.1), 5′-AAGCCGTCGAAAAGAAGG AG-3′ and 5′-TTCAGGATCGTCCGTTATCC-3′; for *APT*(Mapoly0100s0027.1), 5′-CGTCGTTGTTGGAATTGAA GC-3′ and 5′-TTCTTGGGCTTTCGTTGAGG-3′.
5. Lysis buffer: 20 mM Tris–HCl(pH 8.0), 25%(v/v) glycerol, 10 mM $MgCl_2$, 250 mM sucrose, 5 mM dithiothreitol (DTT), 1 mM phenylmethylsulfonyl fluoride (PMSF), 20 μM MG132, and 1× cOmplete EDTA-Free Protease Inhibitor (Roche).
6. Wash buffer: 20 mM Tris–HCl(pH 8.0), 25%(v/v) glycerol, 2.5 mM $MgCl_2$, 0.2%(v/v) Triton X-100, 5 mM DTT, 1 mM PMSF, 20 μM MG132, and 1× cOmplete EDTA-Free Protease Inhibitor (*see* **Note 2**).
7. Extraction buffer: 20 mM Tris–HCl(pH 8.0), 5 mM EDTA, 150 mM NaCl, 1%(v/v) Triton X-100, 0.5%(w/v) sodium deoxycholate, 0.1%(w/v) sodium dodecyl sulfate (SDS), 5 mM DTT, 1 mM PMSF, 20 μM MG132, and 1× cOmplete Protease Inhibitor (Roche).
8. 2× Laemmli sample buffer: 100 mM Tris–HCl(pH 6.8), 4%(w/v) SDS, 10%(v/v) 2-mercaptoethanol, and 20%(v/v) glycerol.
9. Liquid nitrogen.
10. Antibodies for Mpphy and MpPIF [11].

3 Methods

3.1 Phytochrome-Mediated Gemma Germination

Under standard growth conditions (half-strength B5 medium; 22 °C; 50–60 μmol photons m^{-2} s^{-1}; continuous white light), gemma cups are formed on the thallus ~2 weeks after plating, with gemmae suitable for analysis obtained after roughly another week (*see* **Note 3**). Gemma germination is photoreversibly regulated by R and FR: R induces germination, whereas R followed by FR can inhibit induction of germination (Fig. 1a). This pattern suggests that the mode of action of gemma germination in *M. polymorpha* can be classified as a typical LFR [11]. Here, we describe how to assess the phytochrome regulation of gemma germination.

3.1.1 Assessing R-Induced Gemma Germination

1. Prepare solid plates containing half-strength B5 medium with 1%(w/v) sucrose.
2. Plant ~50 gemmae per plate on three plates under a green safelight (*see* **Note 3**).
3. Illuminate all three plates with FR (30 μmol photons m^{-2} s^{-1}) for 15 min to completely inactivate phytochrome present in the gemmae.
4. Wrap the plates in aluminum foil and incubate in the dark for 24 h at 22 °C to allow water imbibition by gemmae.
5. Illuminate the first plate with R (10–30 μmol photons m^{-2} s^{-1}) for 5 min. As a control for R-induced germination, illuminate the second plate with FR (30 μmol photons m^{-2} s^{-1}) for 5 min. Keep the third plate wrapped in foil.
6. Wrap the illuminated plates in aluminum foil and incubate in the dark at 22 °C.
7. Unwrap the plates in 6–8 days to assess germination. Gemma germination is scored by recording the occurrence of rhizoid elongation and/or morphological changes in meristem (Figs. 1a and 2). Wild-type gemmae treated with FR do not exhibit these responses, whereas R-treated gemmae show high germination rates [11].

3.1.2 Assessing FR Inhibition of Gemma Germination

1. Prepare solid plates containing half-strength B5 medium with 1%(w/v) sucrose.
2. Plant ~50 gemmae on the plates under a green safelight (*see* **Note 3**).
3. Illuminate the plates with FR (30 μmol photons m^{-2} s^{-1}) for 15 min to completely inactivate phytochrome already present in the gemmae.

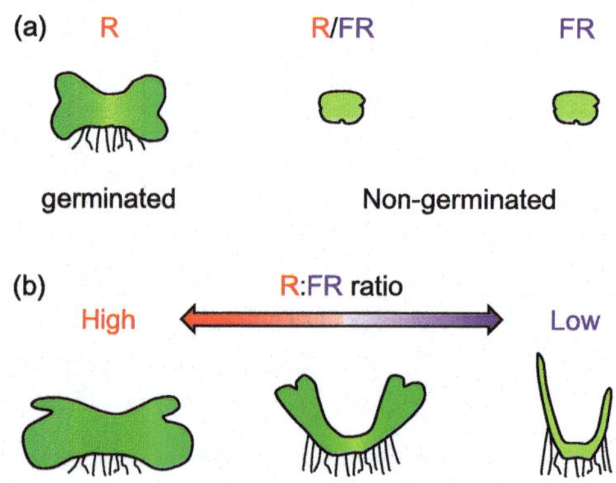

Fig. 1 Phytochrome in *Marchantia polymorpha* regulates gemma germination and thallus morphology. Phytochrome regulation of gemma germination (**a**) and thallus morphology (**b**) under various light conditions is illustrated. *R* red light, *FR* far-red light, *R/FR* R followed by FR

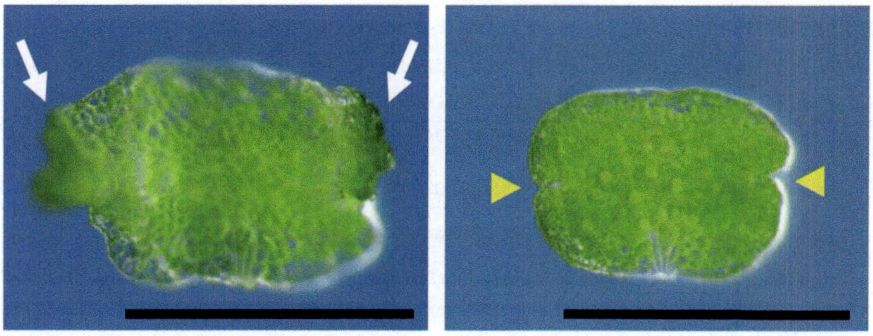

Fig. 2 Photographs of wild-type gemmae of *Marchantia polymorpha*. Germinated (left panel) and nongerminated (right panel) gemmae are shown. White arrows indicate developing tissues after germination. Yellow arrowheads indicate apical notches in nongerminated gemmae. Bars = 1 mm

4. Wrap the plates in aluminum foil and incubate in the dark for 1 day at 22 °C to allow water imbibition by gemmae.
5. Illuminate the plates with R (10–30 μmol photons m^{-2} s^{-1}) for 5 min and then immediately with FR (30 μmol photons m^{-2} s^{-1}) for 5 min.
6. Wrap the illuminated plates in aluminum foil and incubate in the dark at 22 °C.
7. Unwrap the plates in 6–8 days to assess germination.

3.2 Phytochrome-Mediated Inhibition of Orthotropic Growth

Thallus morphology is also controlled in a phytochrome-dependent manner. End-of-day FR treatment causes the notches of thalli to change their growth orientation from horizontal to nearly vertical with narrower lobes. Similar morphological changes occur in response to a low R to FR ratio, which suggests that this response is a shade-avoidance syndrome (Fig. 1b).

1. Prepare solid plates containing half-strength B5 medium.
2. Plant ~4 gemmae per plate on three plates.
3. Incubate the first plate under a short-day (8 h of white light and 16 h of darkness) light cycle supplemented with FR (30 μmol photons m^{-2} s^{-1}) for 5 min at the end of the light period. As controls for the end-of-day FR irradiation, incubate the second plate under the same short-day cycle without FR illumination, and incubate the third plate under the short-day cycle supplemented with FR for 5 min followed immediately by R for 5 min.
4. Observe thallus morphology in ~4 weeks.
5. Under end-of-day FR conditions, thallus growth is orthotropic with narrower, more elongated lobes than that under control conditions (Fig. 1b).

3.3 Phytochrome-Mediated Gene Expression

Most phytochrome responses in Arabidopsis are mediated by regulation of gene expression. Mpphy can also regulate gene expression in an R/FR reversible manner [11]. Here, we describe how to investigate the phytochrome regulation of gene expression.

1. Prepare solid plates containing half-strength B5 medium.
2. Plant ~9 gemmae per plate on three plates.
3. Incubate the plates under standard growth conditions (22 °C; 50–60 μmol photons m^{-2} s^{-1}; continuous light).
4. Ten days after plating, wrap the plates in aluminum foil and incubate in the dark at 22 °C for dark adaptation.
5. After 2–3 days, illuminate the first plate with R (10–30 μmol photons m^{-2} s^{-1}) for 5 min. As a control for phytochrome-mediated gene expression, illuminate the second plate with R (10 μmol photons m^{-2} s^{-1}) for 5 min followed immediately by FR (30 μmol photons m^{-2} s^{-1}) for 5 min. Collect ~100 mg of thalli from the third plate for the dark control, freeze in liquid nitrogen, and store at −80 °C until use.
6. Wrap the illuminated plates in aluminum foil and incubate in the dark for 1–6 h.
7. Collect ~100 mg of thalli, freeze in liquid nitrogen, and store at −80 °C until use.
8. Isolate RNA with TRIzol reagent (Thermo Fisher Scientific) or equivalent. Column purification methods, such as the RNeasy Plant Mini kit (Qiagen), are also useful for RNA extraction from *M. polymorpha* thalli.
9. Synthesize cDNA using 0.5–2 μg of total RNA.
10. Analyze gene expression by qPCR. The expressions of *LHCB* (Mapoly0139s0012.1) and *POR* (Mapoly0030s0118.1) are increased after R irradiation, whereas R followed by FR inhibits R-dependent induction of these genes [11]. Several reliable reference genes for normalization of qPCR have been reported [13].

3.4 Subcellular Fractionation for Immunodetection of Nuclear-Localized Mpphy and MpPIF

Phytochromes are translocated from the cytosol into the nucleus upon conversion to the Pfr form [14]. Although Mpphy localizes in the nucleus in response to both R and FR, Mpphy in the nucleus appears to undergo FR-dependent post-translational modifications that can be detected by immunoblot analysis [11]. In addition, MpPIF has been shown to degrade in an R-dependent manner [11]. Detection of these nuclear events requires isolation and enrichment of nuclei. Here, we describe a method to fractionate the nuclei of *Marchantia* for detection of nuclear-localized Mpphy and MpPIF.

1. Prepare solid plates containing half-strength B5 medium.
2. Plant ~9 gemmae per plate on the plates.

3. Incubate the plates under standard growth conditions (22 °C; 50–60 μmol photons m^{-2} s^{-1}; continuous light).
4. Ten days after plating, wrap the plates in aluminum foil and incubate in the dark at 22 °C for dark adaptation.
5. After 2–3 days, illuminate the plates for 10 min with R (e.g., 30 μmol photons m^{-2} s^{-1}) or FR (e.g., 30 μmol photons m^{-2} s^{-1}).
6. Collect ~300 mg of thalli, freeze in liquid nitrogen under R or FR illumination, and store at −80 °C until use. The following procedures should be performed under a green safelight.
7. Grind the thallus tissues into a fine powder with a mortar and pestle in liquid nitrogen.
8. Add 1 mL of lysis buffer to the mortar; using the pestle, mix well until the frozen lysate is completely melted.
9. Filtrate the lysate through a 70-μm cell strainer or equivalent.
10. Centrifuge at 1500 × *g* for 10 min at 4 °C.
11. Collect the supernatant as a nucleus-depleted fraction. Mix an aliquot with an equal volume of 2× Laemmli sample buffer and heat at 95 °C for 5 min.
12. Wash the pellet from **step 10** two times with 1 mL of wash buffer (*see* **Note 4**).
13. Discard the supernatant and add one pellet volume of extraction buffer. A pellet volume is typically 50–70 μL.
14. Incubate on ice for 10 min, add an equal volume of 2× Laemmli sample buffer, and boil at 95 °C for 5 min.
15. Centrifuge the boiled samples at 16,000 × *g* for 15 min at room temperature, and use the supernatants for immunoblot analysis (*see* **Note 5**).

4 Notes

1. Half-strength B5 medium is normally used for growing *M. polymorpha*. Compared with 1× B5 or other media, such as Murashige and Skoog (MS) and M51C, half-strength B5 medium allows more regular branching patterns and formation of flattened thalli. Growth and gemma production can be enhanced by the addition of 1% sucrose(w/v).
2. The specified concentration of Triton X-100 is very important for nuclear isolation. A lower concentration causes insufficient lysis of chloroplasts, whereas a higher concentration leads to low yields of nuclei.
3. The status of parental plants is important for preparing homogeneous gemmae. Gemmae sometimes germinate within gemma cups because of water absorption from dew condensa-

tion. For reproducible results, take care to avoid using germinated gemmae. Gemmae obtained from parental plants grown under short-day conditions (8 h of light and 16 h of darkness) tend to exhibit better dormancy than those obtained from plants under continuous light conditions.

4. Repeat the procedure until the pellet is no longer green. After washing, the nuclear pellet should be gray.
5. Samples should be immediately subjected to SDS-PAGE, as nuclear-localized Mpphy appears to be fragile even after heating in Laemmli sample buffer.

Acknowledgments

We thank Barbara Goodson, PhD, from Edanz Group (www.edanzediting.com/ac) for editing a draft of the manuscript.

References

1. Qiu YL, Li L, Wang B, Chen Z, Knoop V, Groth-Malonek M, Dombrovska O, Lee J, Kent L, Rest J, Estabrook GF, Hendry TA, Taylor DW, Testa CM, Ambros M, Crandall-Stotler B, Duff RJ, Stech M, Frey W, Quandt D, Davis CC (2006) The deepest divergences in land plants inferred from phylogenomic evidence. Proc Natl Acad Sci U S A 103:15511–15516
2. Bowman JL, Floyd SK, Sakakibara K (2007) Green genes—comparative genomics of the green branch of life. Cell 129:229–234
3. Ishizaki K, Nishihama R, Yamato KT, Kohchi T (2016) Molecular genetic tools and techniques for *Marchantia polymorpha* research. Plant Cell Physiol 57:262–270
4. Bowman JL, Kohchi T, Yamato KT, Jenkins J, Shu S, Ishizaki K, Yamaoka S, Nishihama R, Nakamura Y, Berger F, Adam C, Aki SS, Althoff F, Araki T, Arteaga-Vazquez MA, Balasubrmanian S, Barry K, Bauer D, Boehm CR, Briginshaw L, Caballero-Perez J, Catarino B, Chen F, Chiyoda S, Chovatia M, Davies KM, Delmans M, Demura T, Dierschke T, Dolan L, Dorantes-Acosta AE, Eklund DM, Florent SN, Flores-Sandoval E, Fujiyama A, Fukuzawa H, Galik B, Grimanelli D, Grimwood J, Grossniklaus U, Hamada T, Haseloff J, Hetherington AJ, Higo A, Hirakawa Y, Hundley HN, Ikeda Y, Inoue K, Inoue S, Ishida S, Jia Q, Kakita M, Kanazawa T, Kawai Y, Kawashima T, Kennedy M, Kinose K, Kinoshita T, Kohara Y, Koide E, Komatsu K, Kopischke S, Kubo M, Kyozuka J, Lagercrantz U, Lin SS, Lindquist E, Lipzen AM, Lu CW, Luna ED, Martienssen RA, Minamino N, Mizutani M, Mizutani M, Mochizuki N, Monte I, Mosher R, Nagasaki H, Nakagami H, Naramoto S, Nishitani K, Ohtani M, Okamoto T, Okumura M, Phillips J, Pollak B, Reinders A, Rövekamp M, Sano R, Sawa S, Schmid MW, Shirakawa M, Solano R, Spunde A, Suetsugu N, Sugano S, Sugiyama A, Sun R, Suzuki Y, Takenaka M, Takezawa D, Tomogane H, Tsuzuki M, Ueda T, Umeda M, Ward JM, Watanabe Y, Yazaki K, Yokoyama R, Yoshitake Y, Yotsui I, Zachgo S, Schmutz J (2017) Insights into land plant evolution garnered from the *Marchantia polymorpha* genome. Cell 171:287–304
5. Hartmann E, Jenkins GI (1984) Photomorphogenesis of mosses and liverworts. In: Dyer AF, Duckett JG (eds) The experimental biology of bryophytes. Academic Press, London
6. Fredericq H, De Greef J (1966) Red (R), far-red (FR) photoreversible control of growth and chlorophyll content in light-grown thalli of *Marchantia polymorpha* L. Naturwissenschaften 53:337
7. Otto K-R, Halbsguth W (1976) Die Förderung der Bildung von Primärrhizoiden an Brutkörpern von *Marchantia polymorpha* L. durch Licht und IES. Z Pflanzenphysiol 80:197–205

8. De Greef J, Butler WL, Roth TF (1971) Control of senescence in *Marchantia* by phytochrome. Plant Physiol 48:407–412
9. Ninnemann H, Halbsguth W (1965) Rolle des phytochroms beim etiolement von *Marchantia polymorpha*. Naturwissenschaften 52:110–111
10. Inoue K, Nishihama R, Kohchi T (2017) Evolutionary origin of phytochrome responses and signaling in land plants. Plant Cell Environ 40:2502–2508
11. Inoue K, Nishihama R, Kataoka H, Hosaka M, Manabe R, Nomoto M, Tada Y, Ishizaki K, Kohchi T (2016) Phytochrome signaling is mediated by PHYTOCHROME INTERACTING FACTOR in the liverwort *Marchantia polymorpha*. Plant Cell 28:1406–1421
12. Gamborg OL, Miller RA, Ojima K (1968) Nutrient requirements of suspension cultures of soybean root cells. Exp Cell Res 50:151–158
13. Saint-Marcoux D, Proust H, Dolan L, Langdale JA (2015) Identification of reference genes for real-time quantitative PCR experiments in the liverwort *Marchantia polymorpha*. PLoS One 10:e0118678
14. Klose C, Viczian A, Kircher S, Schafer E, Nagy F (2015) Molecular mechanisms for mediating light-dependent nucleo/cytoplasmic partitioning of phytochrome photoreceptors. New Phytol 206:965–971

Chapter 19

Analysis of *Physcomitrella* Phytochrome Mutants via Phototropism and Polarotropism

Anna Lena Ermert, Fabian Stahl, Tanja Gans, and Jon Hughes

Abstract

In mosses such as *Physcomitrella patens* phytochrome photoreceptors steer directional/vectorial responses to unilateral/polarized light. In this chapter, we describe procedures to assay phototropism and polarotropism quantitatively in wild type and mutant lines. Protonemata are placed on agar-based medium in square Petri dishes in darkness for 1 week, allowing caulonemata to develop and grow negatively gravitropically. For phototropism, the dishes are placed vertically in black boxes and unilaterally irradiated with continuous red light. For polarotropism, Petri dishes are placed horizontally and irradiated with linearly polarized red light from above. After irradiation, the filaments are photographed using a macroscope with CCD camera and the bending angles measured using image processing software. The data are transfered to a spreadsheet program, placed into 10° bending angle classes and illustrated using a circular histogram.

Key words *Physcomitrella patens*, Phytochrome mutants, Physiology, Phototropism, Polarotropism

1 Introduction

Phytochromes are red/far-red photoreceptors in higher and lower plants, algae, fungi, as well as in photosynthetic and nonphotosynthetic bacteria [1]. In higher plants, they have diverse morphogenic functions such as regulating germination, hypocotyl elongation, deetiolation, entrainment of the circadian clock, and the onset of flowering [2–8]. These functions are executed via the activation of the photoreceptor by light, its subsequent translocation to the nucleus and the hierarchical orchestration of gene expression via master transcription factors, such as PHYTOCHROME INTERACTING FACTORs (PIFs), ELONGATED HYPOCOTYL 5 (HY5), and CONSTANS (CO) [9–20]. In mosses, however, phytochromes not only regulate nuclear responses (e.g., spore germination) but also steer cytoplasmic responses to unilateral and polarized red light [21–23]. As a matter of principle, these responses cannot be transduced via transcription-translation because thereby all directional information

Andreas Hiltbrunner (ed.), *Phytochromes: Methods and Protocols*, Methods in Molecular Biology, vol. 2026,
https://doi.org/10.1007/978-1-4939-9612-4_19,

is lost [24]. A separate, more direct signaling mechanism has therefore been proposed, probably localized in the cytoplasm or at the plasma membrane [24–26]. The haploid protonemata (the first developmental stage following spore germination) respond to unilateral or polarized red light via fluence-rate-dependent negative or positive phototropism and perpendicular growth in relation to the *E*-vector of the incident light, respectively [21, 27]. How phytochromes mediate these responses is unknown, but knockout mutagenesis implies that the *PHY4* gene product, coding for one of the seven phytochromes in *Physcomitrella*, is involved [25]. Additionally, chloroplast movement is mediated by red and blue light by a cooperation between phytochromes and the blue-light-perceiving phototropins [25, 28]. This too seems to be mediated by *PHY4*.

We attempted to characterize the signal transduction mechanism involved by identifying cytoplasmic or plasma membrane-associated interacting phy4-partners [25]. In particular, we identified phototropins as phy4-partners [26]. As phototropins are attached to the plasma membrane they might immobilize the otherwise soluble phytochrome at the cell periphery, thereby providing a possibility to sense and transduce vectorial information (*see* [24]). Yeast 2-hybrid screens using full-length phy4 holophytochrome in the photoactivated Pfr state also revealed a diverse set of phy4-interacting proteins which could be verified as *in planta* interactors via BiFC [29]. Although some of them are promising candidate signal transducers from the phy4-phot complex, their physiological relevance still needs to be proven [29]. It seems that the other end of the signal transduction chain is able to reorganize the cytoskeleton such that actin microfilaments congregate at the point of prospective outgrowth, leading to spatial delivery of new membrane and cell wall material to the cell tip and thus, finally, resulting in bending [30, 31]. To study putative components of the system, we have developed methods to assay phototropism and polarotropism, expecting the corresponding mutants to show aberrant response phenotypes [25]. The following protocols thus describe the quantitative analysis of *Physcomitrella* responses to unidirectional and polarized red light, respectively.

2 Materials

1. *Physcomitrella patens* wild type and mutant protonemata grown for 1 week on BCE225 Petri dishes overlaid with cellophane.
2. Growth chamber (continuous white light, ca. 50–80 μmol m^{-2} s^{-1} photosynthetically active radiation (PAR), at 21 °C).
3. Sterile bench and equipment for sterile handling of cultures (sterile 50 ml plastic tubes, disposable 3 ml plastic pipettes, etc.)

4. Facility for aeration of liquid cultures:
   (a) Sterile 100 ml and 500 ml flasks with suitable air-permeable cellulose plugs, which are centrally perforated with a pointed metal stick and equipped with a 5 ml glass pipette so that—after assembly of flask and plug—the cylindrical end of the pipette points into the flask and nearly touches the bottom. The flask-plug-pipette-junction is wrapped in aluminum foil and the conical end of the pipette is connected to a 5–8 cm piece of a flexible hose. The covering of the latter is closed with aluminum foil for autoclaving.
   (b) Syringe sterile filter with cellulose mixed ester membrane (0.22 μm pore size, Roth) connected to each culture flask after inoculation of the moss culture.
   (c) Network of hoses connected by T-fittings linking the aeration system to the individual culture flasks. The hoses are connected to each flask via a sterile filter.
   (d) The aeration network is driven by an electrical air pump. In order to prevent evaporative losses from the cultures the air is first bubbled through a bottle containing autoclaved H_2O. Contamination problems can be eliminated by passing the input air through a glass olive filled with sterile cotton wool.
5. Mechanical shredder (e.g., T18 basic Ultra-Turrax, IKA, Staufen, Germany).
6. Round and square plastic Petri dishes.
7. Micropore surgical tape.
8. BCE225 liquid and solid medium [32]: 1× Solution B, 1× Solution C, 1× Solution E, 1× Hoagland's trace element solution, 0.2% (w/v) glucose, 0.8% (w/v) agar for plant cultivation (Applichem) where required. Autoclave, cool to ca. 50 °C, and add 1× Solution N and 1× Solution Ca through a sterile filter.
   (a) Solution B (100×): 0.1 M $MgSO_4$. Autoclave and store at 4 °C.
   (b) Solution C (100×): 180 mM KH_2PO_4 (adjust pH to 6.5 with 4 M KOH). Autoclave and store at 4 °C.
   (c) Solution E (100×): 1 M KNO_3, 1 mM Fe(III)$C_6H_5O_7$ (ferric citrate). Autoclave and store at 4 °C.
   (d) Hoagland's trace element solution (1000×): 10 mM H_3BO_3, 0.12 mM $Al_2(SO_4)_3 \bullet K_2SO_4$, 0.22 mM $CuSO_4$, 0.24 mM KBr or 0.24 mM $KBrO_3$, 0.66 mM LiCl, 0.10 mM Na_2MoO_4 (precipitates), 1.97 mM $MnCl_2$, 0.23 mM $CoCl_2$, 0.19 mM $ZnSO_4$, 0.17 mM KI, 0.15 mM $SnCl_2$.

(e) Solution Ca (100×): 200 mM $CaCl_2$. Autoclave separately from other BCE225 components.

(f) Solution N (100×): 500 mM $C_4H_{12}N_2O_6$ (diammonium tartrate). Autoclave separately from other BCE225 components.

9. Contamination control medium: 1 mM KNO_3, 0.1 mM $CaCl_2$, 1 mM KH_2HPO_4, 40 μM $MgSO_4$, 10 μM Fe(III) $C_6H_5O_7$ (ferric citrate), 1× Hoagland's trace element solution, 27 mM glucose, 0.3% (w/v) yeast extract, 1.2% (w/v) agar for plant cultivation (Applichem). Autoclave and pour small Petri dishes.
10. Hydrated sterile cellophane foils: squares/rectangles (either matching the dimensions of the square Petri dish or smaller) and circles (8 cm in diameter) matching standard circular Petri dishes. Place individual foils in an autoclavable container with H_2O, let them soak for ca. 30 min, refresh the water, and autoclave directly on the day of use (this best avoids contamination).
11. Black stands to hold the Petri dishes upright and to orient them accurately relative to the light source.
12. A narrow-bandwidth red light source (full width at half maximum <20 nm) adjustable to different fluence rates (e.g., high power red (λ_{max} 660 nm) LEDs (Roithner Lasertechnik, Vienna)). The LEDs are wired in series to a power supply and mounted on an aluminum plate with a standard PC cooling fan. Thereby, fluence rates between 0.02 and 50 μmol m^{-2} s^{-1} can be obtained by adjusting the current. Lower fluence rates can be achieved by placing grey filters between LEDs and Petri dishes. More elegantly, a series of fluence rates can be produced from a single light source with a so-called threshold box, which constitutes a series of chambers connected by semitransparent mirrors oriented in a 45° angle in respect to the light beam so that half of the light passes through and the other half is reflected on the sample.
13. Fluence rates in the visible (photosynthetically active) region 400–700 nm can be measured with a PAR meter (SKP 215, Skye Instruments, UK or LI-190R, LiCor, Lincoln, USA) or a spectroradiometer (USB4000 spectrometer with fiber-optic cable and cosine-corrected head, Ocean Optics, Largo, USA).
14. Plastic polarization filters (Edmund Industrial Optics, Barrington, USA).
15. Dark room/chamber/cabinet at 21 °C.
16. Green (540 nm) safelight. A single LED typically driven by two 1.2 V rechargeable cells provides more light than necessary for the dark-adapted human eye. Do not use a filament

lamp with a green filter; this represents a rather powerful far-red light source.

17. Zoom macroscope with light source (e.g., Macro Z16, Leica Microsystems, Wetzlar, Germany) and CCD camera (e.g., DFX 500, Leica).
18. PC with image processing software for angle measurement (e.g., ImageJ, National Institutes of Health, USA).
19. Software for calculation and graphical representation of growth angle data (e.g., Microsoft Excel and/or Origin).

3 Methods

3.1 Preparation of Moss Cultures

All procedures are carried out in a sterile environment using presterilized equipment. Moss media are based on Ashton and Cove [32].

1. Inoculate ca. 20–40 ml liquid BCE225 medium with about 50 mg fresh weight of filaments using a flamed and cooled forceps in a 100 ml sterile flask equipped with a cellulose plug and glass pipette (*see* **item 4** in Subheading 2). Homogenize the culture with a sterilized shredder (Ultraturrax at 18,000 rpm for 5–10 s), flame the bottleneck and the aluminum foil around the plug and carefully close the flask so that the pipette submerses in the culture medium. Replace the aluminum foil at the top of the hose by a sterile filter and connect it to the aeration system in the growth chamber. Thereby the moist air will lead to a constant circulation and aeration of the culture (*see* **Note 1**).
2. Let the filaments grow until the culture reaches a sufficient density (maximally 2–3 weeks, the color of the culture should still be lime-green and not yet olive-colored; *see* **Note 2**). In order to prevent the filaments from forming buds and gametophores, the culture must be homogenized weekly (1–2 min Ultraturrax at 18,000 rpm). Check for bacterial or fungal contamination of the culture by touching the used Ultraturrax stick on the surface of a contamination control plate. Contaminants should become visible after 1–2 days.
3. Homogenize as above, transfer to a sterile 50 ml tube and let the material sink to the bottom. Decant excess culture medium from the tube (if necessary repeat the sedimentation of material) leaving 5–10 ml.
4. Overlay round BCE225 Petri dishes with a single sterile hydrated cellophane foil using sterile forceps.
5. Transfer ca. 4–6 ml of the sedimented culture to one BCE225 Petri dish overlaid with cellophane, for example, using a dispos-

able 3 ml pipette (*see also* **Note 3**). Disperse filaments by careful swirling of the petri dish to distribute the filaments evenly and to obtain a coherent patch of moss material. Pipet off excess liquid, seal the dish with permeable tape and place in the growth chamber for 7–10 days.

3.2 Preparation of Petri Dishes for Phototropism and Polarotropism Assays

Seven to ten days before the irradiation is planned, filaments are arranged on Petri dishes in order to generate standardized, gravitropically aligned, dark-adapted protonemata for appropriate treatments. In darkness, only one of two protonemal cell types can grow, namely, the so-called caulonemata. Caulonemal cells are very long, have oblique cell walls and only few small chloroplasts. In contrast to that, chloronemata have shorter cells and many large chloroplasts. To generate uniform caulonemata, strips of protonemata grown on cellophane (*see* Subheading 3.1; plates should carry both cell types) are cut out, placed on square Petri dishes previously overlaid with cellophane to prevent filament growth into the medium. The Petri dishes are then placed vertically in darkness using appropriate stands (or black, lightproof boxes wrapped in foil). This dark incubation allows the preexisting caulonemal filaments to branch and grow negatively gravitropically. After 7–10 days this material can be used for phototropism, gravitropism, or polarotropism experiments. Note that growth in darkness might be altered in mutants.

1. Prepare square BCE225 Petri dishes (one petri dish per light treatment; also include a dark control). Mark the desired positions of the filament strips on the back of the petri dish (thus the instruction is referring to the back view). For phototropism, draw four to five parallel sloped lines in a column starting at the left edge of the Petri dish and pointing to the upper right side with an inclination of ca. 35° from the horizontal (*see* Fig. 1a). Take care that the strips will not shade each other. For polarotropism experiments, draw horizontal lines with a distance of ca. 2 cm (*see* Fig. 1b).
2. Overlay the solid medium with sterile hydrated cellophane (as described in Subheading 3.1, *see* **Note 4**), trying to minimize transfer of liquid from the cellophane container. Appropriately placed small squares of cellophane can be easier to handle than large ones covering the whole area of the dish. When using small squares, take care that the cellophane edges do not interfere with the positions of the drawn lines (otherwise, they may interfere with negative gravitropic growth of filaments).
3. Cut strips (ca. 1–2 mm wide) from the 7–10 day-old lawn using a sharp, sterile scalpel (*see* Subheading 3.1, [33]). First cut the long edges with a sawing movement, then slice them to the right length. Transfer the strips to the square petri dish using sterile forceps. Seal the dish with Parafilm and place it vertically in total darkness at 21 °C for 7–10 days.

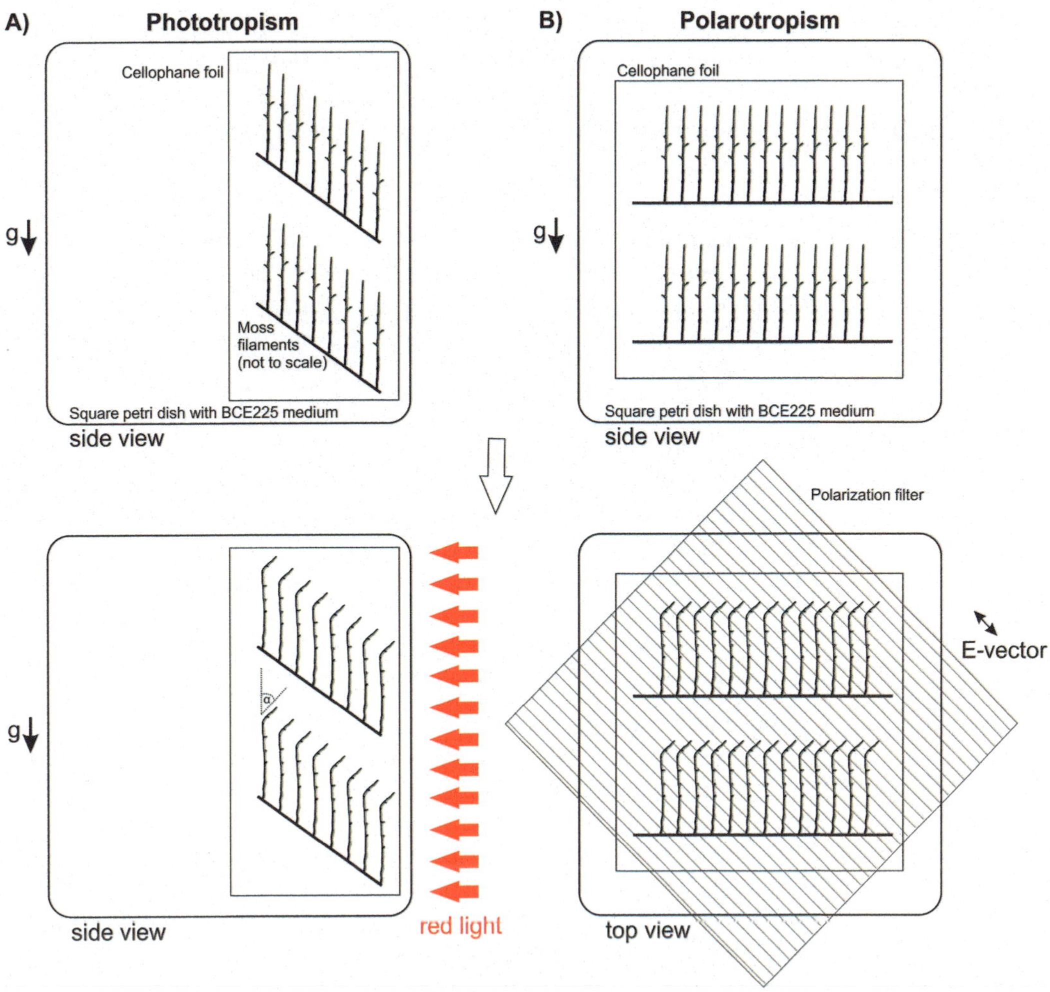

Fig. 1 Schematic representation of phototropism and polarotropism experiments. To investigate phototropic (**a**) and polarotropic responses (**b**), 7–10 day-old negatively gravitropically grown (= dark-adapted) filaments are transferred to continuous unilateral red light and to polarized red light, respectively. Upper panel depicts preparation of cellophane-overlaid BCE225 petri dishes with cut cellophane strips carrying the moss filaments (For clarity's sake, only two strips are shown, but four to five strips can be placed onto one plate.). Note that for phototropism, cellophane strips are placed with an inclination of 35° from the horizontal axis in order to prevent shading effects. Moss strips for polarotropism can be oriented horizontally on the square Petri dish. Petri dishes are placed vertically in darkness for 7–10 days to allow negative gravitropic growth in both cases. Lower panel shows irradiation setups in phototropism (**a**) and polarotropism assays (**b**). For phototropism (lateral view), vertically oriented/upright petri dishes are unilaterally irradiated for 16–17 h with 1.2–1.5 μmol m^{-2} s^{-1} continuous red light coming from the right. For polarotropism (top view), petri dishes are placed horizontally and irradiated from above for ca. 24 h with 0.03 μmol m^{-2} s^{-1} continuous linearly polarized red light. After irradiation, Petri dishes are photographed and the light-mediated reorientation of filaments is quantified by measuring the angle α between the filament and the vertical (dashed lines in panel A, bottom)

3.3 Irradiation of Dark-Grown Moss Cultures for Phototropism and Polarotropism Assays

1. After 7–10 days check the progress of negative gravitropic growth in a darkroom with a dim green safelight. The caulonemata should be more than 2 mm long. If not, prolong the dark incubation for up to three more days. Also pay attention to possible contamination.
2. Plan the start of the irradiation so that you are able to document the phototropic response 16–17 h or the polarotropic response >24 h after the beginning of the treatment.
3. *Phototropism*: set up a red light field to irradiate the *vertically* oriented petri dishes from the side. The field should be as uniform as possible (10% variation over the irradiated area is permissible). Set the fluence rate so that the front edge of the dish will receive 1.2–1.5 μmol m^{-2} s^{-1}. Remove the Parafilm from the side of the dish through which the moss filaments will be irradiated. Place the dish vertically so that the light impinges on the surface of the agar at an angle of ~10°, preferably with the help of an appropriate stand (*see* Fig. 1 and **Note 5**). Alternatively, enclose the dish in a black box with a slit for the light beam. To irradiate several dishes simultaneously, pack them side-by-side separated with black paper sheets. Incubate for 16–17 h.
4. *Polarotropism*: set up a red light field to irradiate the *horizontally* oriented petri dishes with light from above. Set the fluence rate to 0.03 μmol m^{-2} s^{-1} underneath the polarization filter. As the plastic lid of the Petri dish is (weakly) dichroic, replace it with a thin glass plate. Place a polarizing filter on top with the plane of polarization 45° to the vertical axis of the dish (*see* **Notes 6–10**). Place under the red light source on a black background. Incubate for >24 h.
5. Document the phototropic and polarotropic growth responses using a macroscope equipped with a digital CCD camera directly after the irradiation. Place the dish so that its original vertical axis corresponds exactly to that of the image (*see* Subheading 3.4).

3.4 Data Analysis and Representation of Growth Angle Data

In order to quantify the light-mediated reorientation of growth, the angle between the initial (i.e., vertical) and the new orientation of the filaments is measured using image-processing software such as ImageJ (https://imagej.nih.gov/ij/).

1. Start ImageJ and open your image (File>Open).
2. Convert the image to eight-bit greyscale (Image>Type>8-bit).
3. Select the line tool in the Tools menu.
4. Click on the point of the filament where the growth direction seems to deviate from the vertical, then click on the filament tip. Press "m" to add the measured value to the results table.

5. Repeat the step above for each measurable filament, ignoring those that have grown too densely and therefore interfere with each other. At least 50 measurements will be necessary to assay the response accurately.
6. Once finished with one filament line, copy-paste the results table into a spreadsheet. It may be necessary to convert your angle data as such, that the top of the vertical axis corresponds to 0° (in ImageJ this position by default corresponds to 90°). Likewise, a filament which shows a perfectly horizontal growth to or away from the light should result in a value of +90° or −90°, respectively. Group angles into classes of 5° or 10° (the modal class should contain at least 10 measurements). Transform the number of filaments in each class to % of total, then illustrate the angle class data as circular histogram. For this, either Microsoft Excel or Origin can be used (*see* **Note 11** and [34]).

4 Notes

1. Instead of the actively aerated cultivation system with a system of hoses and a pump, it is also possible to use sterile baffled flasks and a shaker. After inoculation and homogenization of the culture in a sterile 50 ml centrifuge tube, transfer the culture to the flask, close it with an air-permeable plug and place on a horizontal shaker (at ca. 150 rpm).
2. If a larger volume of liquid culture is desired, incubate the small culture until it is clearly lime green (approximately after 7–10 days), homogenize and transfer the entire culture to a 500 ml flask and add fresh medium up to ca. 100–150 ml.
3. Moss material on BCE225 media turns olive-brownish after 8–10 days. In order to synchronize all plated cultures, first disperse filaments on cellophane-overlaid solid BCE205 medium (BCE225 without glucose) and transfer to BCE225 medium at least 3 days before the preparation of moss strips.
4. Cellophane foil should not be autoclaved dry because it will wrinkle when placed on solid medium, disturbing negative gravitropic, phototropic or polarotropic growth.
5. The exact inclination of the light beam to the surface of the agar seems to have little effect on the result. However, the filaments should not be irradiated with light "through" the agar, so it is better to use a shallow angle of 10° to exclude this possibility.
6. Some petri dish lids may depolarize linearly polarized light. This can be checked by looking through the lid between two polarizing filters oriented so that the planes of polarization are perpendicular to each other. If the area is uniformly black it

should be OK to use the plastic lid in the polarotropism experiment. Another source of error can arise when the lid itself is dichroic (has polarizing characters). In this case, the lid will interfere with the polarizing filter so that the fluence rate over the agar surface varies. Thus we always use glass plates.

7. The axis along which a polarizing filter transmits the electromagnetic *E*-vector can easily be determined by observing a shallow reflection from a flat, non-metalic surface (such as water). Reflected light is strongly polarized with the *E*-vector parallel to the surface; consequently, the reflection will be blocked entirely when the polarizing filter is orientated perpendicularly to the surface.
8. Take care that the Petri dish is not closely sealed or water will condense on the lid or the glass plate. Water droplets will interfere with the light treatment, especially in the case of polarotropism. In our hands, using porous surgical tape instead of Parafilm in concert with a ventilated light cabinet is adequate.
9. Although it might appear to be the obvious procedure, *do not* place the polarizing filter with the *E*-vector perpendicular to the vertical axis of the Petri dish. In that case filaments must change their direction of growth either to the left or to the right to be perpendicular to the vertical axis and the *E*-vector of the light. Not only will they then grow into one another, the angle data will also comprise positive and negative angles, complicating analysis.
10. Sometimes in polarotropism experiments filaments grow upwards, away from the agar. To prevent this, cover the filament strips with sterile microscope coverslips. This has not been necessary in our hands.
11. Be aware that under some conditions the distribution of bending angles is bimodal (positive and negative phototropism with few aphototropic filaments), in which case the calculation of the mean value can be misleading.

Acknowledgments

This work was supported by DFG grant Hu702/5 to JH. We thank Mathias Zeidler for critical reading of the manuscript.

References

1. Li F-W, Melkonian M, Rothfels CJ, Villarreal JC, Stevenson DW, Graham SW, Wong GK-S, Pryer KM, Mathews S (2015) Phytochrome diversity in green plants and the origin of canonical plant phytochromes. Nat Commun 6:7852

2. Flint LH, McAlister ED (1935) Wave lengths of radiation in the visible spectrum inhibiting the germination of light-sensitive lettuce seed. Smiths Misc Coll 94:1–11
3. Borthwick HA, Hendricks SB, Parker MW, Toole EH, Toole VK (1952) A reversible photoreaction controlling seed germination. Proc Natl Acad Sci U S A 38:662–666
4. Nagatani A, Kay SA, Deak M, Chua NH, Furuya M (1991) Rice type I phytochrome regulates hypocotyl elongation in transgenic tobacco seedlings. Proc Natl Acad Sci U S A 88:5207–5211
5. Chen M, Chory J (2011) Phytochrome signaling mechanisms and the control of plant development. Trends Cell Biol 21:664–671
6. Godnev TN, Akulovich NK, Orlovskaia KI, Domash VI (1966) The influence of the phytochrome system on the formation of pigments in carrot tissue culture. Dokl Akad Nauk SSSR 169:692–694
7. Somers DE, Devlin PF, Kay SA (1998) Phytochromes and cryptochromes in the entrainment of the *Arabidopsis* circadian clock. Science 282:1488–1490
8. Searle I, Coupland G (2004) Induction of flowering by seasonal changes in photoperiod. EMBO J 23:1217–1222
9. Sakamoto K, Nagatani A (1996) Nuclear localization activity of phytochrome B. Plant J 10:859–868
10. Kircher S, Kozma-Bognar L, Kim L, Adam E, Harter K, Schäfer E, Nagy F (1999) Light quality-dependent nuclear import of the plant photoreceptors phytochrome A and B. Plant Cell 11:1445–1456
11. Nagy F, Kircher S, Schäfer E (2000) Nucleo-cytoplasmic partitioning of the plant photoreceptors phytochromes. Semin Cell Dev Biol 11:505–510
12. Kircher S, Gil P, Kozma-Bognár L, Fejes E, Speth V, Husselstein-Müller T, Bauer D, Ádám É, Schäfer E, Nagy F (2002) Nucleocytoplasmic partitioning of the plant photoreceptors phytochrome A, B, C, D, and E is regulated differentially by light and exhibits a diurnal rhythm. Plant Cell 14:1541–1555
13. Ma L, Li J, Qu L, Hager J, Chen Z, Zhao H, Deng XW (2001) Light control of *Arabidopsis* development entails coordinated regulation of genome expression and cellular pathways. Plant Cell 13:2589–2607
14. Hiltbrunner A, Viczián A, Bury E, Tscheuschler A, Kircher S, Tóth R, Honsberger A, Nagy F, Fankhauser C, Schäfer E (2005) Nuclear accumulation of the phytochrome A photoreceptor requires FHY1. Curr Biol 15:2125–2130
15. Quail PH (2010) Phytochromes. Curr Biol 20:R504–R507
16. Tepperman JM, Zhu T, Chang HS, Wang X, Quail PH (2001) Multiple transcription-factor genes are early targets of phytochrome A signaling. Proc Natl Acad Sci U S A 98:9437–9442
17. Tepperman JM, Hwang Y-S, Quail PH (2006) phyA dominates in transduction of red-light signals to rapidly responding genes at the initiation of Arabidopsis seedling de-etiolation. Plant J 48:728–742
18. Monte E, Tepperman JM, Al-Sady B, Kaczorowski KA, Alonso JM, Ecker JR, Li X, Zhang Y, Quail PH (2004) The phytochrome-interacting transcription factor, PIF3, acts early, selectively, and positively in light-induced chloroplast development. Proc Natl Acad Sci U S A 101:16091–16098
19. Osterlund MT, Hardtke CS, Wei N, Deng XW (2000) Targeted destabilization of HY5 during light-regulated development of *Arabidopsis*. Nature 405:462–466
20. Endo M, Tanigawa Y, Murakami T, Araki T, Nagatani A (2013) PHYTOCHROME-DEPENDENT LATE-FLOWERING accelerates flowering through physical interactions with phytochrome B and CONSTANS. Proc Natl Acad Sci U S A 110:18017–18022
21. Jenkins GI, Cove DJ (1983) Phototropism and polarotropism of primary chloronemata of the moss *Physcomitrella patens*: responses of the wild-type. Planta 158:357–364
22. Cove DJ, Schild A, Ashton NW, Hartmann E (1978) Genetic and physiological studies of the effect of light on the development of the moss, *Physcomitrella patens*. Photochem Photobiol 27:249–254
23. Cove D, Knight C (1987) Gravitropism and phototropism in the moss, *Physcomitrella patens*. Cambridge University Press, London
24. Hughes J (2013) Phytochrome cytoplasmic signaling. Annu Rev Plant Biol 64:377–402
25. Mittmann F, Brücker G, Zeidler M, Repp A, Abts T, Hartmann E, Hughes J (2004) Targeted knockout in *Physcomitrella* reveals direct actions of phytochrome in the cytoplasm. Proc Natl Acad Sci U S A 101:13939–13944
26. Jaedicke K, Lichtenthäler AL, Meyberg R, Zeidler M, Hughes J (2012) A phytochrome-phototropin light signaling complex at the plasma membrane. Proc Natl Acad Sci U S A 109:12231–12236
27. Jenkins GI, Cove DJ (1983) Phototropism and polarotropism of primary chloronemata of the moss *Physcomitrella patens*: responses of mutant strains. Planta 159:432–438

28. Kasahara M, Kagawa T, Sato Y, Kiyosue T, Wada M (2004) Phototropins mediate blue and red light-induced chloroplast movements in *Physcomitrella patens*. Plant Physiol 135:1388–1397
29. Ermert AL, Mailliet K, Hughes J (2016) Holophytochrome-interacting proteins in *Physcomitrella*: Putative actors in phytochrome cytoplasmic signaling. Front Plant Sci 7:613
30. Meske V, Ruppert V, Hartmann E (1996) Structural basis for the red light induced repolarization of tip growth in caulonema cells of *Ceratodon purpureus*. Protoplasma 192:189–198
31. Meske V, Hartmann E (1995) Reorganization of microfilaments in protonemal tip cells of the moss *Ceratodon purpureus* during the phototropic response. Protoplasma 188:59–69
32. Ashton NW, Cove DJ (1977) The isolation and preliminary characterisation of auxotrophic and analogue resistant mutants of the moss, *Physcomitrella patens*. Mol Gen Genet 154:87–95
33. Lamparter T, Esch H, Cove D, Hughes J, Hartmann E (1996) Aphototropic mutants of the moss *Ceratodon purpureus* with spectrally normal and with spectrally dysfunctional phytochrome. Plant Cell Environ 19:560–568
34. Zeidler M (2016) Physiological analysis of phototropic responses in *Arabidopsis*. Methods Mol Biol 1398:21–28

Chapter 20

CRISPR/Cas9-Mediated Knockout of *Physcomitrella patens* Phytochromes

Anna Lena Ermert, Fabien Nogué, Fabian Stahl, Tanja Gans, and Jon Hughes

Abstract

Here we describe procedures for gene disruption and excision in *Physcomitrella* using CRISPR/Cas9 (clustered regularly interspaced short palindromic repeat/CRISPR-associated 9) methods, exemplarily targeting phytochrome (*PHY*) gene loci. Thereby double-strand breaks (DSBs) are induced using a single guide RNA (sgRNA) with the Cas9 nuclease, leading to insertions or deletions (indels) due to incorrect repair by the nonhomologous-end joining (NHEJ) mechanism. We also include protocols for excision of smaller genomic fragments or whole genes either with or without homologous recombination-assisted repair. The protocol can be adapted to target several loci simultaneously, thereby allowing the physiological analysis of phenotypes that would be masked by functional redundancy. In our particular case, multiple *PHY* gene knockouts would likely be valuable in understanding phytochrome functions in mosses and, perhaps, higher plants too. Target sites for site-directed induction of DSBs are predicted with the CRISPOR online-tool and are inserted in silico into sequence matrices for the design of sgRNA expression cassettes. The resulting DNAs are cloned into Gateway DONOR vectors and the respective expression plasmids used for moss cotransformation with a Cas9 expression plasmid and a selectable marker (either on a separate plasmid or on one of the other plasmids). After the selection process, genomic DNA is extracted and transformants are analyzed by PCR fingerprinting.

Key words *Physcomitrella patens*, Phytochrome, Gene knockout, CRISPR/Cas9, Homologous recombination, Nonhomologous end joining

1 Introduction

Because of their developmental and morphological simplicity and the fact that the dominant phase of the life cycle is haploid, mosses represent useful model organisms for the investigation of developmental processes. As primitive land plants they are also interesting in the evolutionary context. *Physcomitrella patens* has in the past 25 years been used successfully in these contexts, including studies of phytohormone signaling and physiology, cell polarity, tip cell growth, gravitropism and light effects such as phototropism and polarotropism [1–6]. Notably, homologous recombination (HR)

Andreas Hiltbrunner (ed.), *Phytochromes: Methods and Protocols*, Methods in Molecular Biology, vol. 2026, https://doi.org/10.1007/978-1-4939-9612-4_20,

although almost absent in higher plants is highly efficient in mosses [7]. Thus mosses were until recently unique in the plant kingdom in providing access to genetic analysis via targeting methods. Although this situation has changed with the development of CRISPR/Cas9 which allows genome editing in higher plants, that system in conjunction with haploidy, rapid and reliable transfection, regeneration and selection methods [8–13], strong HR, and the complete genome sequence [14] render *Physcomitrella* well suited for gene function studies by reverse genetics.

Phytochrome effects in mosses are diverse and complex, often paralleling those in higher plants. Phytochromes use red light (RL) to control gravitropic growth, spore germination, regeneration of protonemata from protoplasts, and other responses [2, 6, 15, 16]. Key roles of phytochromes in *Physcomitrella* have been demonstrated using the *pubs hy2* chromophore biosynthesis mutant which revealed several developmental defects such as defective growth in RL and to a weaker extent also in white light (WL), reduced chlorophyll content, elongated growth of gametophores in WL, dwarf growth of gametophores in RL, and reduced phyllid development [17]. Comparative transcriptome analysis of dark (D)- and RL-grown WT and *pubs hy2* mutant plants corroborated a function of *Physcomitrella* phytochromes in the regulation of RL-dependent gene expression. In contrast to the well-characterized phytochrome-controlled transcription cascade in higher plants, however, transcription factors do not seem to be overrepresented among the RL-affected and thus phytochrome-regulated genes [17]. On the other hand, likely homologues of higher plant phytochrome signalling components have been identified in the *Physcomitrella* genome, such as *HY5*-like genes and *PIF*s [14, 18]. Molecular functionality of members of both families could recently be verified as well [18–20]. The role of phytochrome in steering directional responses such as phototropism and polarotropism (= growth in respect to the orientation of the electrical vector of light) and chloroplast photorelocation movements [4, 5, 21–25] is of particular interest as there is no possibility that this can occur through gene regulation [26]—although phytochromes are generally thought to act exclusively through gene regulation.

The *Physcomitrella* genome encodes seven phytochrome genes: *PHY1* to *PHY4* and *PHY5a* to *PHY5c*, which are evolutionarily (and therefore perhaps functionally) distinct from higher plant phytochromes since they form a distinct clade together with other moss phytochromes [27, 28]. However, their domain architecture, their autocatalytic binding of PΦB as a chromophore as well as their spectral and photoconversion properties all show great similarity to those of higher plant phytochromes [22]. Might then the action mechanism of phytochrome in mediating directional responses in mosses be represented in some way in higher plants?

Extensive knockout (KO) experiments in *Physcomitrella* implied that primarily *PHY4* is responsible for vectorial responses [22], whereas *PHY1* may predominate in regulating spore germination [16]. The physiological functions of the *PHY5a-c* subclade are still unknown. Although key roles of phytochromes in *Physcomitrella* development are well established, knowledge of the specific and/or overlapping functions of different phytochromes is still limited. Indeed, no cohesive study of phytochrome subfunctions has been carried out to date, the main limitation to this type of experiments being the difficulty to produce multiplex phytochrome knockouts. Thus it is of interest to create single and multiple knockouts of *Physcomitrella* phytochromes in different combinations. A systematic analysis would contribute substantially to our understanding of phytochrome functions in *Physcomitrella patens* and beyond.

Genome editing has recently been revolutionized by CRISPR/Cas9 methods. The type II CRISPR system utilizes a chimeric single guide RNA (sgRNA) instead of a CRISPR RNA (crRNA)-transactivating RNA (tracrRNA)-duplex [29]. It has already been used successfully in numerous diverse species—including plants [30, 31]. The sgRNA harbors a scaffold for binding by the Cas9 nuclease and a 20 bp target (= protospacer) sequence. This 20 bp target conveys sequence specificity for the induction of site-specific double strand breaks (DSBs) three nucleotides upstream of the so-called protospacer adjacent motif (PAM) which resides in the host genome sequence and thus is not part of the sgRNA. Once a DSB is formed, host-specific DSB repair mechanisms such as nonhomologous end joining (NHEJ) and homology-directed repair (HDR) are induced. In case of incorrect repair, nucleotides are inserted, deleted (indels) or replaced, often leading to frameshift mutations and thereby knockout of the desired gene. In conjunction with the availability of efficient HR in *Physcomitrella*, the system already proved useful in efficiently creating multiplex knockouts or deletions of up to 3 kbp [32–35]. In the following section we describe a protocol for the creation of phytochrome knockouts using different strategies: on the one hand, single DSBs (and subsequent NHEJ-repair) are induced with the help of a single sgRNA construct which is cotransformed in *Physcomitrella* protoplasts together with Cas9 and selectable marker plasmids for transient selection of transformants. On the other hand, smaller (~500 bp) and bigger (~6 kbp) gene fragments are removed by using two sgRNAs together with Cas9 and selectable marker plasmids either with or without a construct harboring two homologous regions flanking the DSB sites. It is also possible to replace the excised gene by a selectable marker gene cloned between the two homologous regions in the HR construct. This chapter describes the design and cloning of sgRNA expression constructs for the creation of phytochrome knockouts, moss protoplast transformation/selection, and finally the molecular analysis of transformants by PCR fingerprinting.

2 Materials

1. Genomic sequences of phytochrome genes:
   (a) PHY1 Pp3c25_2610V1.1
   (b) PHY2 Pp3c16_20280V1.1
   (c) PHY3 Pp3c16_18760V3.1
   (d) PHY4 Pp3c27_7830V3.3
   (e) PHY5a Pp3c3_23790V1.1
   (f) PHY5b Pp3c12_9240V3.1
   (g) PHY5c Pp3c4_15350V3.1
2. Plasmids (in high purity and concentration, >1 μg/μl):
   (a) pAct-Cas9 plasmid (contains the rice *ACTIN 1* promoter and a codon-optimized version of Cas9 from *Streptococcus pyogenes* fused to a SV40 nuclear localization signal [34], *see* **Note 1**).
   (b) pBHRF or pBNRF (plasmids containing a hygromycin or neomycin resistance cassette driven by the Cauliflower mosaic virus (CaMV) 35S promoter, allowing transient selection with hygromycin or G418, respectively [32, 34].
   (c) A donor vector for Gateway cloning (e.g., pDONR207, Invitrogen).
3. Gateway BP-Clonase kit (Invitrogen).
4. Chemically or electrocompetent *E. coli* cells (e.g., XL1-blue, Invitrogen).
5. Standard laboratory equipment and antibiotic-supplemented LB media for cloning and incubation of *E. coli* (incubators, shakers, agarose gel chambers etc.).
6. *Physcomitrella patens* wild type (WT).
7. Growth chamber (long-day; 16 h white light, ca. 50–80 μmol m^{-2} s^{-1} photosynthetically active radiation (PAR, 400–700 nm), 8 h D; at 21 °C).
8. Sterile bench and equipment for sterile handling of cultures (sterile 50 ml polypropylene tubes, disposable 3 ml plastic pipettes, etc.).
9. Facility for aeration of liquid cultures:
   (a) Sterile 100 ml and 500 ml flasks with suitable air-permeable cellulose plugs equipped with a 5 ml glass pipette so that—after assembly of flask and plug—the end of the pipette nearly touches the bottom. The flask–plug–pipette junction is wrapped in aluminium foil and the conical end of the pipette is connected to a 5–8 cm piece of a flexible

hose. The covering of the latter is closed with aluminium foil for autoclaving.

(b) Syringe sterile filter with cellulose mixed ester membrane (0.22 μm pore size, Roth) which are connected to each culture flask after inoculation of the moss culture.

(c) Network of hoses connected by T-fittings allows the individual flasks to be aerated via sterile filters.

(d) The network of hoses is connected with a glass olive filled with absorbent cotton and then with an impinger/wash bottle filled with autoclaved H_2O. The H_2O in the wash bottle is aerated with the help of an electrical air pump.

10. Mechanical shredder (e.g., T18 basic Ultra-Turrax, IKA, Staufen, Germany).
11. Micropore surgical tape.
12. BCE225 (BCE, 0.2% (w/v) glucose, 2 mM $CaCl_2$, 5 mM ammonium tartrate) liquid or solid medium [36]: 1× Solution B, 1× Solution C, 1× Solution E, 1× TES, 0.2% (w/v) glucose, where required 0.8% (w/v) agar for plant cultivation (Applichem). Autoclave, cool to ca. 50 °C and add 1× Solution N and 1× Solution Ca through a sterile filter. If required, add the appropriate antibiotic(s).

(a) Solution B (100×): 0.1 M $MgSO_4$. Autoclave and store at 4 °C.

(b) Solution C (100×): 184 mM KH_2PO_4 (pH adjusted to 6.5 with 4 M KOH). Autoclave and store at 4 °C.

(c) Solution E (100×): 1 M KNO_3, 1 mM Fe(III)$C_6H_5O_7$ (ferric citrate). Autoclave and store at 4 °C.

(d) Hoagland's trace element solution (1000× TES): 9.9 mM H_3BO_3, 0.12 mM $Al_2(SO_4)_3 \bullet K_2SO_4$, 0.22 mM $CuSO_4$, 0.24 mM KBr or $KBrO_3$, 0.66 mM LiCl, 0.10 mM Na_2MoO_4 (precipitates), 1.97 mM $MnCl_2$, 0.23 mM $CoCl_2$, 0.19 mM $ZnSO_4$, 0.17 mM KI, 0.15 mM $SnCl_2$.

(e) Solution Ca (100×): 200 mM $CaCl_2$. Autoclave separately from other BCE225 components and store at 4 °C.

(f) Solution N (100×): 500 mM $C_4H_{12}N_2O_6$ (ammonium tartrate). Autoclave separately from other BCE225 components and store at 4 °C.

(g) Hygromycin or G418 stock solutions: 50 mg/ml hygromycin dissolved in PBS (Invitrogen) or 50 mg/ml G418 dissolved in H_2O. Sterilize by sterile filtration.

13. PPNH4 medium: 1× Solution A, 1× Solution B, 1× Solution C, 1× alternative TES, 2.7 mM $C_4H_{12}N_2O_6$ (ammonium tartrate), 0.5% (w/v) glucose, 0.7% (w/v) agar for plant cultivation.
    (a) Solution A (100×): 0.5 M $Ca(NO_3)_2$, 4.5 mM $FeSO_4$. Autoclave and store at 4 °C.
    (b) Alternative TES (1000×): 220 nM $CuSO_4$, 10 mM H_3BO_4, 2 mM $MnCl_2$, 230 nM $CoCl_2$, 190 nM $ZnSO_4$, 170 nM KI, 100 nM Na_2MoO_4. Autoclave and store at 4 °C.
    (c) Solution B (100×) and Solution C (100×): *see* **item 12**.
14. Contamination control medium: 1 mM KNO_3, 0.1 mM $CaCl_2$, 1 mM KH_2HPO_4, 40 μM $MgSO_4$, 10 μM Fe(III) $C_6H_5O_7$ (ferric citrate), 1× TES, 27 mM glucose, 0.3% (w/v) yeast extract, 1.2% (w/v) agar for plant cultivation. Autoclave and pour small Petri dishes.
15. Sterile cellophane foils: 8 cm diameter circles to match standard Petri dishes. Place dry loose foils separated by a filter paper each in an autoclavable container and autoclave. Alternatively, autoclave the cellophane foils in water: Place individual foils in an autoclavable container with H_2O, let them soak for ca. 30 min, refresh the water and autoclave directly on the day of use (this best avoids contamination).
16. 8.5% mannitol solution: 8.5% (w/v) D-mannitol. Autoclave and store at room temperature.
17. 2% (w/v) Driselase (Sigma): dissolve powder in 8.5% mannitol solution and incubate for 15 min at room temperature; then spin down the insoluble components (2500 × *g*, 5 min), sterilize the clear brown supernatant by filtration, dispense it in 5 ml aliquots and store at −20 °C.
18. MMM buffer: 0.1% (w/v) MES, 9.1% (w/v) D-mannitol, 15 mM $MgCl_2$. Adjust the pH to 5.6 with KOH, sterilize by sterile filtration and store at 4 °C for up to 1 year.
    (a) MES: 1% (w/v) 2-(N-morpholino)ethanesulfonic acid in distilled water. Adjust pH to 5.6 with 0.1 M KOH and sterilize in 1.2 ml aliquots by autoclaving.
    (b) $MgCl_2$ stock solution: 1 M $MgCl_2$ in H_2O.
19. PEG transformation solution: autoclave 2 g PEG6000 aliquots in glass vials and melt the PEG in a microwave on the day of use in a small water bath in a beaker. Add 5 ml sterile 8.5% mannitol solution supplemented with 0.1 M Ca(NO)3 and 10 mM Tris–HCl pH 8.0, and mix well by pipetting. Leave at room temperature for 2–3 h before use.

20. PRMB (protoplast regeneration medium, bottom layer): 1× Solution B, 1× Solution C, 1× Solution D, 1× TES, 5 mM ammonium tartrate, 6% (w/v) D-mannitol, 0.7% (w/v) agar for plant cultivation, 0.5% (w/v) glucose. Sterilize by autoclaving and let cool to 50 °C, add 1 M $CaCl_2$ solution through a sterile filter to a final concentration of 10 mM and pour in standard round Petri dishes.
    (a) Solution B (100×), Solution C (100×), and TES (1000×): *see* **item 12**.
    (b) Solution D (100×): 1 M KNO_3, 4.5 mM $FeSO_4$. Autoclave and store at 4 °C; dissolve any precipitate that may have formed before use.
21. PRMT (protoplast regeneration medium, top layer): 1× Solution B, 1× Solution C, 1× Solution D, 1× TES, 5 mM ammonium tartrate, 8% (w/v) D-mannitol, 0.4% (w/v) agar for plant cultivation. Sterilize by autoclaving and let cool to 50 °C. Add 1 M $CaCl_2$ solution through a sterile filter to a final concentration of 10 mM freshly on the day of use ($CaCl_2$ tends to precipitate).
22. Sterile plastic sieves with 100 × 100 μm and 50 × 50 μm mesh sizes. We use 50 ml disposable tubes cut in half and with a hole cut in the lid which then holds the plastic mesh.
23. Autoclaved standard plastic pipette tips as well as autoclaved, cut and flamed plastic pipette tips for 1 ml and 20–200 μl pipettes. By flaming the cut pipette tips, sharp edges can be smoothed to minimize damage of protoplasts.
24. Hemocytometer (e.g., a Fuchs-Rosenthal counting chamber).
25. Microscope and/or zoom macroscope (e.g., Macro Z16, Leica Microsystems, Wetzlar, Germany).
26. 45 °C water bath and/or heating block for 50 ml disposable tubes and 14 ml round-bottom culture tubes.
27. Sterile polypropylene 14 ml round bottom culture tubes with low density polyethylene cap (e.g., Greiner).
28. Sterile 24-well plates (e.g., Nunclon Delta Surface, Thermo Scientific).
29. 2× CTAB (Cetyl trimethylammonium bromide) buffer: 2% (w/v) CTAB, 100 mM Tris–HCl pH 8.0, 20 mM EDTA (pH 8.0), 1.4 M NaCl, 1% (w/v) polyvinylpyrrolidone (add in small portions as it is difficult to dissolve), freshly add 50 mM DTT.
30. Optional (*see* **Notes 2** and **3**): Tungsten carbide beads (3 mm, Qiagen) or glass beads (2.85–3.3 mm, Carl Roth, Karlsruhe,

Germany) and a mixer mill (e.g., MM300, Retsch, Haan, Germany).

31. Optional (*see* **Note 3**):
    (a) 1.2 ml deep well plates with tightly closing lid strips (e.g., Sarstedt)
    (b) Multistep- and eight-channel (50–1200 μl) pipettes (e.g., Multipette plus and Xplorer, both from Eppendorf) with corresponding 1 ml, 5 ml and 10 ml tips (Eppendorf) and extra long tips (e.g., Sarstedt, 1250 μl), respectively.
    (c) Centrifuge capable of centrifugation of multiwell plates (e.g., Eppendorf 5810 R with S-4-104 rotor, 2 DWP buckets).
32. PCR ingredients:
    (a) Taq DNA Polymerase (NEB) including appropriate buffer.
    (b) 10× dNTPs (2 mM dATP, dTTP, dCTP, and dGTP each)
    (c) Oligonucleotides (ordered with default settings at Invitrogen/Thermo Fisher Scientific); *see* Table 1 for sequences.
33. PCR cycler (e.g., T3 Thermocycler from Biometra GmbH, Göttingen, Germany).
34. Gel chambers for agarose and (if necessary) PAGE gels.
35. Solutions for gel electrophoresis:
    (a) 15% PAGE gels: 50% (v/v) 30% acrylamide–bisacrylamide (29:1) solution, 1× TBE buffer, 10% (w/v) ammonium persulfate (APS), 0.13% (v/v) tetramethylethylenediamine (TEMED).
    (b) 1× TBE buffer/PAGE running buffer: 200 mM Tris–HCl pH 8.0, 166 mM H_3BO_3, 20 mM EDTA
    (c) Agarose gel: 1% (w/v) agarose melted in a microwave in 0.5× TBE buffer.
    (d) 0.5× TBE buffer/agarose gel running buffer: 100 mM Tris–HCl pH 8.0, 83 mM H_3BO_3, 10 mM EDTA
    (e) 6× Loading dye: 60% (v/v) glycerol, 0.03% (w/v) xylene cyanol, 0.03% (w/v) bromophenol blue, 60 mM EDTA, 10 mM Tris–HCl pH 7.6
    (f) DNA staining solution (e.g., ethidium bromide bath: 0.5–1 μg/ml ethidium bromide in 0.5× TBE buffer).
36. PCR reaction cleanup kit (e.g., from Qiagen or NEB).

Table 1
Oligonucleotides for generation and characterization of CRISPR/Cas9 *PHY4*-KOs

Primer name	Primer sequence	Description
ALL45	TACTCATTCCATCCTGCCCTCT	*PHY4* locus FW primer upstream sgRNA#3 target site
ALL46	TGGACTGGAAAGTCCACACTCA	*PHY4* locus RV primer downstream sgRNA#4 target site
ALL48	GTCTGTCGAGAAGTTTCTGATCG	FW primer in hygromycin resistance cassette
ALL113	CTTGCGACTTGTGACTTGTTA	PCR screening FW primer for CRISPR/Cas9 *PHY4*-KO's with sgRNA#4
ALL114	ACTTCGCACTACAACACGGG	PCR screening RV primer for CRISPR/Cas9 *PHY4*-KO's with sgRNA#4
ALL115	TCCTCAACTAATCTCAATCAATTT	PCR screening RV primer for CRISPR/Cas9 *PHY4*-KO's with sgRNA#3
ALL116	ATTGCTCGACTACATAACGGAT	PCR screening FW primer for CRISPR/Cas9 *PHY4*-KO's with sgRNA#3
ALL117	GCGCCAGAGATGCTTGACC	PCR screening FW primer for CRISPR/Cas9 *PHY4*-KO's with sgRNA#2
ALL118	GGTGAACAAAGTTCTTGCATCG	PCR screening RV primer for CRISPR/Cas9 *PHY4*-KO's with sgRNA#2
DL47	ATGCGACGCAATCGTCCGATC	FW primer in hygromycin resistance cassette
DL50	TCTTCTGAGACTGTATCTTTGATAT	RV primer in hygromycin resistance cassette
EG34	CAACACATGAGCGAAACC	RV primer in hygromycin resistance cassette
FST14	GAGTTCATGTTGCCCATGTAT	*PHY4*-specific RV primer downstream the sgRNA#2 target site and for WT-situation check
KJ31	CCCATGTATTGCGCATGGCA	*PHY4*-specific RV primer for PCR spanning sgRNA#2 target site for sequence analysis
KJ127	ATGTCGACCACCAAGTTGGCA	Sequencing (FW) primer for PCR products from clones resulting from transformation 1

3 Methods

3.1 Design of sgRNAs

The first step of sgRNA design is the prediction of putative target sites in the desired region of the *Physcomitrella* genome with the help of the CRISPOR online database (*see* below). The selected target sequence is subsequently inserted in silico into a sgRNA sequence template which is flanked by *attB* sites for Gateway cloning. The resulting DNA construct is finally chemically synthesized by a suitable company.

1. Download the genomic sequence including flanking sequences upstream and downstream of the gene of interest from http://www.cosmoss.org/ or https://phytozome.jgi.doe.gov/pz/portal.html#!info?alias=Org_Ppatens.
2. Copy 1000 nucleotides containing the desired target region to clipboard (for the creation of *Physcomitrella PHY4* KOs we used 1000 bp upstream and downstream of the ATG and stop codons and the first 1000 bp of the CDS, respectively).
3. Paste the sequence in the search field of the CRISPOR online database at http://crispor.tefor.net/ [37]. Select "Physcomitrella patens – Spreading earthmoss – Phytozome V11" as genome and "20bp-NGG-Sp Cas9, SpCas9-HF1, eSpCas9 1.1" as PAM and submit the request. All putative 20 bp target sequences of the sequence (forward and reverse is possible) are given, ranked according to their specificity score (estimates the uniqueness of a given sequence in the genome [38]). Select a target sequence with highest specificity and out-of-frame scores and with a low number of off-target sequences (if not avoidable with at least four mismatches). For *PHY4* we selected the following three target sequences: 123–104 nt (forward) upstream of the ATG start codon (GGGCCATACGTATTGGCAAC for sgRNA#3), 336–355 nt (forward) downstream of the ATG start codon (GGTGCCCAGTGTGGGTCAGC for sgRNA#2), and 358–377 nt (forward) downstream of the stop codon (ACTGAATGAAACCTTCGGGC for sgRNA#4), respectively.
4. The U3 and U6 promoters prefer A and G as the first transcribed nucleotide, respectively. Thus, in order to optimize the transcription of the sgRNA, attention must be paid to the first nucleotide of the selected 20 bp target sequence. If the target sequence starts with an A, use the U3 backbone matrix (below) and replace the 19 Ns with the 19 nucleotides of the target sequence omitting the first nucleotide. Possible target locus and U3 backbone matrix are given below:
   (a) Target locus in the *P. patens* genome, the PAM is in lower case:

   5′-ANNNNNNNNNNNNNNNNNNNngg-3′.

   (b) 5′ to 3′ sequence of the U3 backbone matrix. The two *attB* sites are in italics, the U3 promoter is underlined and the tracrRNA is in lowercase. The 19 Ns must be replaced by the 19 nucleotides of the target sequence (bold in the target locus):

   GGGGACAAGTTTGTACAAAAAAGCAG
   *GCTTC*TGTCCATTGAAGCAGACGTGTTGCGA

CAGGTTAGCGACGATGGGTGTAGATGTGATG
TGATGTGATGGTGTGGTTCTTCCACGGCG
GCGTCCTTGCGGTGGCGGAGAAGGGGATATC
CCGAAGGAGCGGCAGCGGGAGAGCACAAGC
AGAAAGGGTGCAGTGAGTGAGTGGGTCCAG
CTGGGTGGCTGGCCGAGTGGACGCGACCG
GGTTTCGAGGGGGCGGGGGAGAAAAGGGAT
GGAGCGAGGGATATAACCCACATGGAATGG
AGGTGGGTGTGAAGGCGGGTATATAGGAAG
GTGGAGGACTTACAACCCATA**NNNNNN**
NNNNNNNNNNNNNgttttagagctag
aaatagcaagttaaaataaggctagtccgttatcaacttgaaa
aagtggcaccgagtcggtgcttttttt*GACCCAGCTTTCTTGTA-CAAAGTGGTCCCC*

- If the target sequence starts with a G use the U6 backbone matrix (below) and replace the 19 Ns with the 19 nucleotides of the target sequence. If the target sequence starts with a C or a T, replace this first nucleotide (C or T) by a G in the target sequence and use the sgRNA U6 backbone matrix. Possible target locus and U6 backbone matrix are given below:

(c) Target locus in the *P. patens* genome, the PAM is in lowercase:

5′-GNNNNNNNNNNNNNNNNNNNngg-3′

5′-CNNNNNNNNNNNNNNNNNNNngg-3′

5′-TNNNNNNNNNNNNNNNNNNNngg-3′.

(d) 5′ to 3′ sequence of the U6 backbone matrix. The two *attB* sites are in italics, U6 promoter is underlined and tracrRNA is in lowercase. The 19 Ns must be replaced with the 19 nucleotides of the target sequence (bold in the target locus).

GGGGACAAGTTTGTACAAAAAAGCA
*GGCTTC*TGTCCATTGAAGCAGACGTGTTGCGACAG
TAGCGACGATGGGTGTAGATGTGATGTGATGTG
ATGGTGTGGTTCTTCCACGGCGGCGTCCTTGCG
GTGGCGGAGAAGGGGATATCCCGAAGGAGCG
GCAGCGGGAGAGCACAAGCAGAAAGGGT
GCAGTGAGTGAGTGGGTCCAGCTGGGTG
GCTGGCCGAGTGGACGCGACCGGGTTTCGA
GGGGGCGGGGGAGAAAAGGGATGGAGCG
AGGGATATAACCCACATGGAATGGAGGTGG
GTGTGAAGGCGGGTATATAGGAAGGTGGAGGA
CTTACAACCCATG**NNNNNNNNNNNNNN**
NNNNNgttttagagctagaaatagcaagttaaaataaggc
tagtccgttatcaacttgaaaaagtggcaccgagtcggtgctt
ttttt*GACCCAGCTTTCTTGTACAAAGTGGTCCCC*

5. Order as "gBlocks gene fragments 125-500 bp" at Integrated DNA Technologies (IDT, https://www.idtdna.com/pages/products/genes/gblocks-gene-fragments) or any company providing this service.

3.2 Design of HR Constructs

1. When removing a *Physcomitrella* gene fragment or a whole gene using two sgRNAs, efficiency can be increased by cotransformation with a template construct comprising two homologous regions which exactly flank the two cutting sites [34]. Thereby the repair of the two induced DSBs will use HR-assisted rather than NHEJ repair. The Cas9 nuclease cuts three nucleotides upstream the PAM sequence (pay attention to the orientation of the target sequence), so copy a stretch of 216 and 217 bp upstream and downstream of the two cutting sites and paste them individually, if desired separated by an *Xba*I site (*see* **Note 4**). The two homologous regions are subsequently flanked by *attB* sites for Gateway cloning (*see* **Note 5**), resulting in an *attB1*-216 bp-*Xba*I-217 bp-*attB2* fragment that can be ordered as 125–500 bp gBlocks fragment.
2. Order as "gBlocks gene fragments 125-500 bp" at Integrated DNA Technologies (IDT, https://www.idtdna.com/pages/products/genes/gblocks-gene-fragments) or any company providing a similar service.

3.3 Cloning and Preparation of sgRNA- and HR-Constructs

1. Dissolve the synthetic DNA fragments from Subheadings 3.1 and 3.2 according to manufacturer's instructions to a final concentration of 10 ng/μl.
2. Recombine the synthetic DNA into a Gateway donor vector (e.g., pDONR207) by BP-reaction using 1 μl of the synthetic DNA fragment according manufacturer's instructions (*see also* **Note 6**).
3. Transform the cloning reaction in electrocompetent or chemically competent *E. coli* cells and screen transformants using standard procedures [39].
4. Select single colonies, grow overnight cultures, and miniprep plasmid DNA to check integrity by restriction digest and sequencing.
5. Perform a large scale plasmid extraction either by alkaline lysis and PEG/NaCl precipitation according to Sambrook et al. [39], or using a plasmid midi/maxi preparation kit (e.g., from Macherey-Nagel or Qiagen).
6. After the final washing step in 70% (v/v) ethanol, DNA must be handled in a sterile environment and dissolved in sterile TE to a final concentration of >1 μg/μl. Check the quality of the preparation via restriction digest and/or agarose gel electrophoresis.

3.4 Preparation of Moss Cultures for Protoplast Transformation

All procedures are carried out in a sterile environment using presterilized equipment. Moss media are based on Asthon and Cove [36].

1. Inoculate ca. 20–40 ml liquid BCE225 medium with approximately 50 mg fresh weight of protonemata using a flamed and cooled forceps in a 100 ml sterile flask equipped with a glass pipette which was plunged through the cellulose plug (*see* **item 9** in Subheading 2). Homogenize the culture with a sterilized shredder (Ultra-Turrax at 18,000 rpm for 5–10 s), flame the bottleneck and the aluminium foil around the plug and carefully close the flask so that the pipette submerses in the culture medium. Replace the aluminium foil at the top of the hose by a sterile filter and connect it to the aeration system in the growth chamber. Thereby the moist air which is fed into the liquid will lead to a constant circulation and aeration of the culture (*see* **Note 7**).
2. Let the filaments grow until the culture reaches a sufficient density (maximally 2–3 weeks, the color of the culture should still be lime-green and not yet olive-colored, *see* **Note 8**). In order to prevent protonemata from forming buds and gametophores, the culture has to be homogenized 1–2 times a week (2-1 min Ultra-Turrax at 18,000 rpm). Check for bacterial or fungal contamination of the culture by touching the used Ultra-Turrax stick on the surface of a contamination control plate. Incubate in the growth chamber together with the liquid cultures. In case of a contamination, growth should become visible on the contamination control plate after 1–2 days.
3. After 1–2 weeks incubation with regular dispersion, filaments are plated on cellophane-overlaid PPNH4 medium. Homogenize the culture as described above, transfer to a sterile 50 ml tube and let the material sink to the bottom. Decant excess culture medium from the tube (if necessary repeat the sedimentation of material) leaving 5–10 ml.
4. Overlay round PPNH4 Petri dishes with a single sterile hydrated or dry cellophane foil using sterile forceps. Prepare at least one plate per planned transformation reaction.
5. Transfer ca. 4–6 ml of the sedimented culture to one PPNH4 Petri dish overlaid with cellophane, for example, using a disposable 3 ml pipette. Disperse filaments by careful swirling of the Petri dish to distribute the filaments evenly. Pipet off excess liquid, seal the dish with permeable tape and incubate in the growth chamber for 6 days prior to moss transformation (*see* **Note 9**).

3.5 Moss Transformation and Selection

The following protocol for transformation of *Physcomitrella* protonema-derived protoplasts is based on Cove et al. [11, 12]. If not stated otherwise, all steps are performed at room temperature under sterile conditions. First, protoplasts are produced from moss filaments by digestion of cell walls and middle lamellae by the enzyme cocktail Driselase. Released protoplasts are washed from Driselase with an isotonic mannitol solution. Protoplasts must be handled gently; in particular, turbulence during pipetting and resuspension should be minimized. Subsequently, protoplasts are transfected with different plasmid mixtures using PEG and a heat shock. After this, protoplasts are plated on cellophane-overlaid isotonic medium and allowed to regenerate for 1 week in the climate chamber before transfer to antibiotic-supplemented medium for selection of transformants. After another week, regenerants are placed back on antibiotic-free medium for growth (release of selection pressure). In case a genomic integration of a resistance cassette is intended, a second round of antibiotic selection follows. After the selection process is completed, surviving regenerants are transferred to multiwell masterplates and grown for extraction of genomic DNA and subsequent PCR fingerprinting.

3.5.1 Production of Physcomitrella Protoplasts

1. Select one PPNH4 plate carrying the moss material (*see* Subheading 3.4) per transformation reaction. Use only vital, actively branching and lime-green material without gametophores (since the latter do not transform well).
2. Harvest moss material from four plates at a time (~800 mg) and place the patches in a Petri dish containing 15 ml 8.5% mannitol solution. Disperse the material with a flamed forceps.
3. Add 5 ml of 2% Driselase solution (thereby the Driselase will be diluted to a final concentration of 0.5% (w/v) in a total volume of 20 ml). Mix by gentle swirling. Incubate for 30–45 min; swirl occasionally; avoid bright light. Observe the release of protoplasts using a zoom macroscope.
4. The spherical protoplasts are separated from undigested filaments using a fine sterile sieve with a mesh size of 100 × 100 μm. Place a sieve on a sterile 50 ml centrifuge tube and transfer the digested material carefully on the filter using a sterile flamed 3 ml plastic pipette.
5. Use 2 ml of 8.5% mannitol solution to recover loose plant material/residual protoplasts from the Petri dish and transfer it to the sieve (as above).
6. Spin the protoplast suspension for 5 min at 150–200 × *g*. The pellet (protoplasts) should be dark green and the supernatant brownish (from the Driselase). Remove the supernatant carefully. It is better to leave a residual volume of supernatant than to lose protoplasts.
7. The protoplasts are washed from Driselase twice with sterile 8.5% mannitol solution. By rotating the tube, gently resuspend

the protoplast pellet in the same volume of sterile 8.5% mannitol as the total volume used for the Driselase digest (i.e., 20 ml).

8. Place a sieve with a mesh size of 50 × 50 μm on a fresh sterile 50 ml centrifuge tube. Transfer the protoplasts to the sieve using a sterile flamed 3 ml plastic pipette.
9. Centrifuge for 5 min at 150–200 × *g*. This time the supernatant should be clear. Carefully discard the supernatant.
10. Repeat **steps 7–9** (omitting the filtration step) in order to wash protoplasts a second time.
11. Resuspend the protoplast pellet in 10 ml 8.5% mannitol solution.
12. Take a 20 μl aliquot of the protoplast suspension using a cut, flamed tip in order to estimate the protoplast density with a hemocytometer (Fuchs-Rosenthal counting chamber). Count only round, vital protoplasts in at least 3 (of the 16) large squares. Calculate the total cell number in 10 ml (it is possible to obtain up to 3–4 × 106 protoplasts from 800 mg moss material used for the isolation) and then the volume of MMM (*see* below) needed for a final concentration of 1.3–1.7 × 106 protoplasts per ml. 4–5 × 105 protoplasts (suspended in 300 μl MMM) are needed for each transformation.
13. Pellet the protoplasts gently by centrifugation (5 min at 150–200 × *g*) and discard the supernatant.

3.5.2 Physcomitrella Protoplast Transformation

1. Prepare one 14 ml round-bottom culture tube per transformation reaction and pipet the plasmids (~1 μg/μl) to the bottom of the tube (*see* **Note 10**). The total DNA amount must not exceed 30 μg because above this the DNA precipitates and transformation efficiency drops. For knockout of *PHY4*, 8 μg of the Cas9 expression plasmid (pAct-Cas9) was transfected and the remaining amount was divided by the number of the other plasmids (i.e., plasmids carrying the selection marker and the sgRNAs) so that the total amount adds up to 20 μg (Table 2). Reducing the amount of pAct-Cas9 to 5 μg is also possible. It is sufficient if one of the plasmids contains a selection marker (e.g., conferring resistance to hygromycin or G418).
2. Resuspend the protoplast pellet from **step 13** of Subheading 3.5.1 in the calculated volume of MMM solution (in **step 12** of Subheading 3.5.1).
3. Add 300 μl of the protoplast/MMM suspension to the culture tubes containing the plasmid DNA using a cut and flamed pipette tip. Mix by gentle swirling/rotation of the tube.
4. Add 300 μl of PEG transformation solution and immediately but gently mix by swirling/rotation in order to prevent aggregation of the protoplasts. Do not vortex or pipet up and down!

Table 2
Overview of different genome editing approaches and respective sgRNA combinations for modifying the *Physcomitrella patens PHY4* locus using CRISPR/Cas9

Transformation	Description	sgRNA(s)	HR	Selection
1	Single DSB+NHEJ	sgRNA#2	–	Transient
2	Removal of short *PHY4* fragment + NHEJ	sgRNA#2 + sgRNA#3	–	Transient
3	Removal of whole *PHY4* locus + NHEJ	sgRNA#3 + sgRNA#4	–	Transient
4	Removal of whole *PHY4* locus + HR with insertion of a resistance cassette	sgRNA#3 + sgRNA#4	HR plasmid with resistance cassette	Stable
5	Removal of whole *PHY4* locus + HR	sgRNA#3 + sgRNA#4	HR plasmid without resistance cassette	Transient

Each transformation reaction includes pAct-Cas9 with pBHRF for transient selection on hygromycin (except in the case of transformation 4 in which the hygromycin resistance cassette is cloned between two homologous regions and thereby stably integrated into the moss genome). The sgRNA-expression plasmids used are given in column 3

5. Heat shock: incubate the tube(s) for 5 min at 45 °C.
6. Let the tube(s) cool to room temperature for 5–10 min.
7. Since the PEG solution is toxic to protoplasts, the transformation mixture is diluted step-wise. At 3 min intervals add five times 300 μl 8.5% mannitol solution followed by five times 1 ml 8.5% mannitol solution. Mix gently after each dilution step.
8. Let the tubes stand for 30 min to 2 h at room temperature in weak light.
9. Discard the supernatant. If protoplasts are not entirely sedimented, spin them down (5 min at 150–200 × *g*).
10. Add 2 ml of fresh 8.5% mannitol solution and mix gently.
11. Add 3 ml of PRMT (low-melting protoplast regeneration medium top agar, preprepared, supplemented with $CaCl_2$ and kept molten at 45 °C in a heating block or water bath) and mix well but carefully using a flamed 3 ml pipette. Immediately dispense 1.25 ml each on four PRMB plates overlaid with cellophane. Spread the viscous suspension by swirling the Petri dish. Protoplasts can be dispersed on up to four Petri dishes at a time (not more because the PRMT cools quickly). Thus plate one transformation reaction at a time.
12. Seal the plates with surgical tape. Try to avoid tilting the plates since the protoplast/PRMT suspension tends to slip on the cellophane disks. Let plates stand for 30–60 min before transporting them to the climate chamber.

3.5.3 Protoplast Regeneration and Selection

1. Place the Petri dishes under an UV-protecting foil in the growth chamber and incubate for 6 days in white light. Observe protoplast regeneration using a zoom macroscope. Also pay attention to possible bacterial or fungal contamination (*see* **Note 11**).
2. Transfer the cellophane disks carrying the moss regenerants from PRMB to BCE225 plates containing the appropriate antibiotic (e.g., 30 mg/l hygromycin or 50 mg/l G418) using two flamed forceps.
3. Incubate in the growth chamber for 1 week (as in **step 1**). During this step nontransformed regenerants will die. Do not prolong this selective growth period because this favors the maintenance of episomal transformants (in addition, some antibiotics are light- sensitive).
4. Transfer the cellophane disks carrying the regenerants to BCE225 plates without antibiotic (release from selection).
5. Incubate for 2 weeks in the growth chamber (as in **step 1**). During this selection release moss plants will grow rapidly. Thereby episomal DNA will be diluted and finally lost in the youngest cells (at the edges of the moss colony).
6. If the transformation included a construct intended to integrate into the *Physcomitrella* genome by HR, repeat **steps 2–4** (second selection). During this step, up to 90% of the remaining moss colonies will die and ideally only stable transformants will survive.
7. To pick transient (after **step 5**) or stable transformants (after **step 6**, i.e., after 6–8 days of second selection) pluck single filaments from the edges of individual moss colonies under the zoom macroscope with a flamed and cooled forceps (dip the tip of the forceps into a Petri dish containing sterile agar medium). Transfer each filament to a single well of a 24-well plate with each well filled more than half with BCE225 agar medium (*see* **Note 12**). Seal plates with surgical tape to maximize growth and incubate plates in the growth chamber (*see* **Note 13**).
8. Note that the medium in the multiwell plates dries slowly (1–2 weeks). Once the agar begins to shrink, replace the surgical tape with Parafilm to avoid further desiccation.

3.6 Extraction of Genomic DNA from Moss Plants

PCR analysis allows likely transformants to be identified prior to sequencing. A prerequisite for the molecular analysis is the extraction of genomic DNA (gDNA) which is of adequate purity and concentration for PCR fingerprinting. Various protocols are possible, but most either involve numerous purification steps, which complicates medium- or high-throughput approaches, or do not reproducibly yield DNA of sufficient purity or concentration for PCR analysis. We have successfully used the E.Z.N.A SP Plant DNA Kit from Omega Bio-tek (Norcross, USA) as well as the comparatively quick protocol of Cove et al. [8]. However, we here describe a simplified protocol for low- to medium-throughput extraction of gDNA, which is based on Driselase digest of cell walls and extraction of DNA by complexation with cetyl trimethylammonium bromide (CTAB). For the extraction, no sterile handling is required, but take care not to contaminate the master plates carrying the moss transformants.

1. Prepare microcentrifuge tubes (or better a flat container) with 1 ml 0.5% (w/v) Driselase in 8.5% mannitol solution (*see* **Notes 2** and **3**).
2. Harvest ~50 mg of fresh moss protonemata (not gametophores, since they are not digested by Driselase) from each putative transformant from the master plate. Pull the filaments apart with a forceps, then transfer them to the Driselase solution.
3. Incubate for >45 min at room temperature. Occasionally invert/flick or swirl the microcentrifuge tube or container, respectively.
4. If filaments were digested in a flat container, transfer material to a microcentrifuge tube. Centrifuge for 5 min at >500 × *g* at RT and discard the clear supernatant.
5. Add 1 volume of 2× CTAB buffer and mix by vortexing. Thereby extracted protoplasts should burst and the solution should become green.
6. Incubate for 5 min at 65 °C.
7. Clarify the mixture by centrifugation at 16,000 × *g* for 5 min at RT.
8. Transfer the supernatant to a fresh tube. Take care to avoid cotransfer of debris.
9. Add 1 volume of isopropanol to precipitate the DNA. Mix well by inversion of the tube.
10. Centrifuge for 10 min at 16,000 × *g* (or faster) at RT to pellet the precipitated DNA.
11. Remove supernatant and wash the pellet with 500 μl 70% (v/v) ethanol.
12. Centrifuge for 5 min at 16,000 × *g* at RT.
13. Remove the supernatant and allow the pellet to dry (if necessary, remove droplets at the wall of the tube using the tip of a tissue).
14. Dissolve the pellet in 20–30 μl TE buffer. Store the DNA extracts at −20 °C. Use 1 μl of DNA extracts (concentration of nucleic acids: 100–500 ng/μl) for a 10 μl PCR reaction.

3.7 PCR Fingerprinting of Transformants

For molecular analysis of transformants, different PCR reactions using fresh or frozen gDNAs (from Subheading 3.6) are set up (*see* Fig. 1 and Table 3). If only indels around a single targeted DSB site are expected (as in case of transformation 1, Table 2), small (~100 bp) PCR products flanking the target site are generated and separated on a high-resolution agarose or, preferably, polyacrylamide gel. Thereby even single-base indels can be detected. In the case of excisions of genomic fragments or whole genomic loci using two sgRNAs, potential mutants can be tested using primers flanking the two target sites. Consequently, in case of a correct excision, a PCR product of reasonable size (e.g., 1 kbp) should be obtained. Otherwise, the wild-type product should result in either a longer PCR product (e.g., 1.2 kbp, if the excised fragment is short) or no product at all (if the excised fragment is too long for the selected elongation time). In case a resistance cassette is intended to be inserted, both this insertion and the 5′ and 3′ junctions should be checked (*see* Table 3 for detailed information on PCRs).

1. Set up standard PCR reactions (final volume 10 μl) containing 6.7 μl ddH_2O, 1 μl 10× NEB *Taq* Buffer (final conc. 1×), 1 μl dNTPs (2 mM each; final conc. 0.2 mM each), 1 μl template (100–500 ng/μl gDNA-extract; final conc. 10–50 ng/μl), 0.1 μl forward primer (100 μM; final conc. 1 μM), 0.1 μl reverse primer (100 μM; final conc. 1 μM), and 0.1 μl NEB *Taq* DNA-Polymerase (5 U; 0.5 U per 10 μl PCR reaction).
2. Set up the following standard PCR program: 2 min initial denaturation at 94 °C; 30 cycles each with 30 s denaturation at 94 °C, 30 s hybridization at 54–62 °C, and 1 min/kb elongation at 68 °C; 2 min final elongation at 68 °C.
3. For the analysis of a CRISPR/Cas9-induced *single DSB using sgRNA#2* (*see* Table 2, transformation 1), primer combination ALL117/ALL118 is used, which results in a 96 bp PCR product if the genomic locus has not been modified (*see* Table 3 and Fig. 1b). In order to detect small indels, PCR products are separated on 15–20% polyacrylamide gels following standard procedures [39]. If a given transformant turns out to bear a modification around the sgRNA#2 target site, a longer segment spanning the modification is amplified with primers ALL116/FST14 (1.2 kbp product size) and used for sequence analysis. Subsequently, resulting PCR products are purified using a PCR reaction cleanup kit and sequenced, for example, with primer KJ127.
4. For the analysis of clones in which the *removal of a ~480 bp fragment covering the beginning of the PHY4 CDS* using sgRNAs #3 and #2 (*see* Table 2, transformation 2) is intended, a PCR with the flanking primers ALL45/FST14 is performed (*see* Table 3 and Fig. 1c). In the WT and KO PCR products are 1.7 and 1.2 kbp respectively (*see* **Notes 14** and **15**).

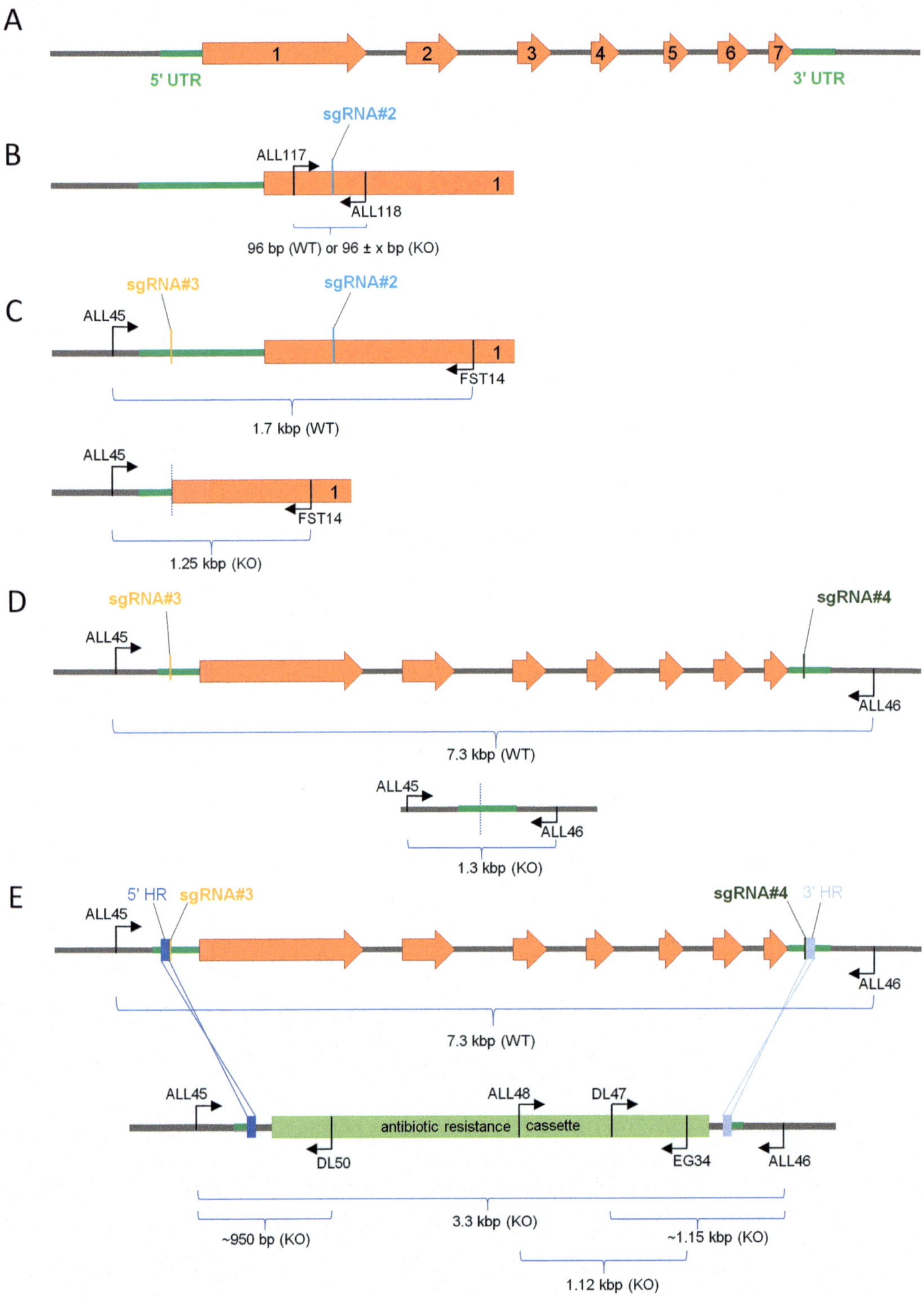

Fig. 1 The *PHY4* genomic locus of *Physcomitrella patens* with possible CRISPR/Cas9-mediated genomic alterations and associated primer binding sites for PCR fingerprinting of transformants. (**a**) Wild-type *PHY4* locus with seven exons (large arrows, numbered). (**b**) Close-up showing potential indels following targeting with sgRNA#2

5. For the analysis of clones in which the *whole PHY4 locus is intended to be removed* either with or without HR-mediated repair (*see* Table 2, transformations 5 and 3), the possible excision is probed with primers ALL45/ALL46. If the locus has been removed, this PCR yields a 1.3 kbp PCR product, whereas in the WT no product will be amplified because the elongation time (1 min 30 s) is too short to amplify the expected 7.3 kbp product (*see* Table 3 and Fig. 1d). Primers ALL45/FST14 are used for the WT positive control (*see* **Note 16**).
6. If the *whole PHY4 locus is intended to be replaced by a hygromycin resistance cassette* via HR (*see* Table 2, transformation 4), PCRs as described in **step 5** are performed. Additionally, the presence of the resistance cassette can be probed using primers ALL48/EG34, but note that this PCR does not tell anything about the site of the insertion. Finally, 5′ and 3′ junctions are checked with primers ALL45/DL50 or DL47/ALL46, respectively (*see* Table 3 and Fig. 1e).

4 Notes

1. A Cas9-encoding plasmid with resistance cassette could be used instead of two individual ones. Nomura et al. [33] utilized a fungal plant-codon-optimized Cas9 (fco Cas9) from *Streptococcus pyogenes* under the control of the *Scopelophila cataractae* (copper moss) EF1α promoter also containing an *NptII* expression cassette in the pSCOEI expression vector. We have now cloned the 35S::hygromycin resistance cassette from pBHRF in front of *Cas9* into pAct-Cas9, thereby reducing the number of plasmids necessary in the transformation mixture. This construct is available for testing.
2. An alternative method for extraction of moss tissue is mechanical disruption using a mixer mill. Either use fresh or prefrozen moss material (gametophores and filaments are adequate, however, in our hands gametophores tend to produce more

Fig. 1 (continued) (Table 2, transformation 1). Primer binding sites are shown as small arrows. The sizes of resulting PCR products in wild type and knockout are indicated. (**c**) Close-up showing sgRNA#2 and sgRNA#3 target sites. This sgRNA combination was used to excise a ~500 bp *PHY4* fragment (Table 2, transformation 2). Above, wild type; below, knockout. The sizes of resulting PCR products in wild type and knockout are indicated. (**d**) Removal of the entire ~6 kb *PHY4* locus using sgRNA#3 and sgRNA#4 (Table 2, transformations 3 and 5, but omitting homologous regions for HR-based repair in transformation 5 for clarity's sake). Above, wild type; below, knockout. Primer binding sites and sizes of PCR products are indicated (not to scale). (**e**) Replacement of the entire *PHY4* locus by an antibiotic (e.g., hygromycin) resistance cassette via HR-mediated repair of sgRNA#3 and sgRNA#4-induced DSBs (Table 2, transformation 4). Above, wild type; below, knockout. Primer binding sites and sizes of flanking and resistance cassette-specific PCRs as well as of PCRs spanning the 5′ and 3′ junctions in wild type and knockout are indicated (not to scale)

Table 3
Overview of PCR-fingerprinting approaches for molecular analysis of different putative CRISPR/Cas9-mediated *Physcomitrella PHY4*-KOs

Purpose of PCR	Appropriate for analysis of clones from transformation #	Primer combination	Annealing temperature (°C)	Elongation time	PCR product size of WT	PCR product size of KO
Control PCR for WT situation/gDNA quality	3–5	ALL45/FST14	56	1′40″	1.7 kbp	–
PCR spanning sgRNA#2 target site	1 (and 2 if desired)	ALL117/ ALL118	57	0–10″	96 bp	Different
PCR for sequence analysis of sgRNA#2-effects	1	ALL45/KJ31	60	1′40″	1.7 kbp	1.7 kbp
PCR spanning sgRNA#2 and sgRNA#3 target sites	2	ALL45/FST14	56	1′40″	1.7 kbp	~1.25 kbp
PCR spanning sgRNA#3 and sgRNA#4 target sites	3 and 5	ALL45/ALL46	60	1′30″	7.3 kbp	1.3 kbp
PCR spanning sgRNA#3 and sgRNA#4 target sites	4	ALL45/ALL46	60	3′20″	7.3 kbp	3.3 kbp
PCR for hygromycin resistance cassette	4	ALL48/EG34	53	1′10″	–	1.12 kbp
PCR spanning the 5′ junction (5′ → cassette)	4	ALL45/DL50	54	1′	–	~950 bp
PCR spanning the 3′ junction (cassette → 3′)	4	DL47/ALL46	60	1′10″	–	~1150 bp
PCR spanning sgRNA#3 target site	3–5 (if desired)	ALL116/ ALL115	53	0–10″	104 bp	Different
PCR spanning sgRNA#4 target site	3–5 (if desired)	ALL113/ ALL114	56	0–10″	101 bp	Different

The purpose of the respective PCR reaction, the transformation reaction corresponding to Table 2, and associated primer combinations are given in columns 1–3, respectively. The last four columns give characteristic numbers of the different PCR reactions, such as the annealing temperature, the elongation time and expected PCR product sizes for WT and KO, respectively.

reliable results compared to filaments, probably due to agar clumps adhering to filaments which are not easily disrupted when frozen). Add a tungsten or glass bead and (if applicable) shock-freeze in liquid nitrogen. Disrupt the still frozen moss material at 30 Hz for 1–2 min in a −20 °C frozen rack. Then add 2× CTAB buffer (as in **step 5**, Subheading 3.5.1) and subsequently follow the protocol from **step 6** on. Carbide beads can be reused after cleaning with 2 M HCl for 1 min and rinsing with an excess of H_2O.

3. Beyond that, one would prefer to perform many gDNA extractions at a time, ideally in 96-well plate format. The described procedure can readily be adjusted to such a format: put fresh (or prefrozen) moss material and a tungsten/glass bead into each well of a 96-deep-well plate and seal tightly with lid strips. Freeze the whole plate in liquid nitrogen and disrupt the material as described in **Note 2**. With a multistep pipette, add 1 volume 2× CTAB buffer and incubate at 65 °C for >10 min (the plate has to thaw first, so take care that the plate is indeed incubated at 65 °C for a while). Centrifuge for 10 min at >3200× *g* and transfer the clear supernatant to a fresh deep-well plate using a multichannel pipette with extra long (and slim) tips. Follow the protocol from **step 9** on but adapt centrifugation parameters of **steps 10** and **12** to 3200 × *g* (*see* Subheading 3.5.1) for 25 min and 10 min at RT, respectively.

4. In the HR construct, the two homologous regions may be separated by an *Xba*I site, which would enable the hygromycin or G418 resistance cassettes from pBHRF or pBNRF to be inserted. For this, digest the resistance plasmids with *Avr*II and *Spe*I, gel-purify the ~2 kb band, and ligate it into the opened HR-plasmid (= synthesized HR-fragment which is Gateway-cloned into a DONOR vector).

5. A HR construct can be directly fused to one of the sgRNAs flanked by *attB* sites in order to clone the whole cassette into a DONOR vector. In this case, synthesize a 1000 bp gBlocks fragment for sufficient capacity to prolong the HRs to 247 bp each. Sequence information on such a construct is available from Fabien Nogué.

6. The BP reaction can be down-scaled to half or quarter of the recommended volume.

7. Instead of the actively aerated cultivation system with a system of hoses and a pump, it is also possible to use sterile baffled flasks and a shaker. After inoculation and homogenization of the culture in a sterile 50 ml centrifuge tube, transfer the culture to the flask, close it with an air-permeable plug and place on a horizontal shaker (at ca. 150 rpm).

8. If a larger volume of liquid culture is desired, incubate the small culture until it is clearly lime green (approximately after 7–10 days), homogenize and transfer the entire culture to a 500 ml flask and add fresh medium up to ca. 100–150 ml.
9. Filaments can be kept on PPNH4 plates for up to 1 week. If fresh material is needed in the following week, harvest the entire material grown on cellophane on a PPNH4 (or a BCE225) plate using a flamed metal spatula. Place the filament patches in a 50 ml sterile centrifuge tube with ~25 ml sterile water and disperse it for ~1 min at 18,000 rpm with an Ultra-Turrax. Let the freshly dispersed material sink to the bottom of the tube, decant excess water and dispense on up to four fresh cellophane-overlaid PPNH4 plates as described above.
10. Typically, *Physcomitrella* transformations for stable DNA integration are performed with linearized DNA. Here, however, we recommend using circular plasmid DNA as this will reduce the risk of random integration of the plasmids (pAct-Cas9 and sgRNA-plasmids). Concerning the HR plasmid, recombination is equally efficient using a linear or a circular plasmid, thus the Nogué group by default uses circular DNA in this case as well.
11. To calculate the transformation efficiency, calculate the number of protoplasts regenerated prior to selection. This is the reference value for comparison with the number of transformants (following selection) and mutants (following PCR and sequencing).
12. Transformants grow into one another when using conventional Petri dishes with BCE225 medium as master plates. Thus we recommend instead using 24-well plates for maintaining isolated transformants. Thereby, it is unimportant if the colonies grow at different rates. The risk of cross-contamination is minimized too.
13. Fungal contamination of plates is sometimes a problem. Most filamentous fungi and yeasts can be killed without injuring the moss by adding nystatin and/or cefotaxime to a final concentration of 50 μg/ml and 100 μg/ml, respectively. For this, prepare sterile 1000× nystatin and cefotaxime stock solutions in 100% ethanol and H_2O, respectively. Some fungi are resistant to these treatments, but yeasts at least are unable to penetrate agar, so sometimes contaminants of this kind can be removed physically. Invert the plate so that the bottom faces upward. Then irradiate from above with red light at *ca.* 10 μmol m^{-2} s^{-1} for 1–2 weeks during which the filaments will grow upward (i.e., into the agar) phototropically. Once they have reached the "bottom" of the plate, turn it over and carefully cut out a section of the agar with a sterile scalpel. Invert the agar block and use a fresh scalpel to slice off a few mm of the agar containing the now-purified filaments.
14. If desired, further analysis of clones from transformations which are intended to produce genomic excisions (using two

sgRNAs) and which produced a WT band in the respective PCR (*see* Table 3) is also possible: in order to check the CRISPR efficiency in a double transformation, examine if indels were induced around the individual sgRNA target sites without an excision of the fragment in between. If so, refer to Table 3 for possible primer combinations around sgRNA target sites #2, #3, and #4, respectively.

15. If it proves necessary to design alternative primers, note that the *Physcomitrella* phytochrome sequences are very similar especially in the region of the first CDS exon. Thus care should be taken to select primers that are as specific as possible. It is advisable to ensure that the 3′ terminal residues are unique so that the DNA polymerase cannot amplify sibling gene fragments. When designing primers for PCRs spanning single sgRNA target sites, regions >15 bp upstream and downstream of the DSB site, respectively, should be selected. On the other hand, PCR products should not be too long in order to ensure an adequate resolution on PAGE gels.
16. If the PCR using primers ALL45/ALL46 is negative, this does not necessarily mean that the *PHY4* locus of that particular clone has not been modified: it might be that the gDNA sample itself is of poor quality. In this case it is advisable to check either with a primer combination specific for the locus-of-interest, which reliably provides a WT signal (e.g., using primers ALL45/FST14) or with primers for an independent gene (e.g., the *LHCB2* gene which is reliably amplified with EG15 (AATCCTTGCTCAAGCCTGTCAATG) and EG16 (CAAGTGGTCGTTCAAGTTCTCCAAT) to give a 900 bp PCR product).

Acknowledgments

This work was supported by DFG grant Hu702/5 to JH and by French ANR grant ANR11-BTBR-0001-GENIUS to NF. The IJPB benefits from the support of the LabEx Saclay Plant Sciences-SPS (ANR-10-LABX-0040-SPS). We thank Pierre François Perroud for his expertise and advice in the context of moss transformation.

References

1. Cove DJ (2000) The generation and modification of cell polarity. J Exp Bot 51:831–838
2. Jenkins GI, Cove DJ (1983) Light requirements for regeneration of protoplasts of the moss *Physcomitrella patens*. Planta 157: 39–45
3. Ashton NW, Schulze A, Hall P, Bandurski RS (1985) Estimation of indole-3-acetic acid in gametophytes of the moss, *Physcomitrella patens*. Planta 164:142–144
4. Jenkins GI, Cove DJ (1983) Phototropism and polarotropism of primary chloronemata of the moss *Physcomitrella patens*: responses of mutant strains. Planta 159:432–438
5. Jenkins GI, Cove DJ (1983) Phototropism and polarotropism of primary chloronemata of the moss *Physcomitrella patens*. responses of the wild-type. Planta 158:357–364

5. Jenkins GI, Courtice GR, Cove DJ (1986) Gravitropic responses of wild-type and mutant strains of the moss *Physcomitrella patens*. Plant Cell Environ 9:637–644
6. Schaefer DG, Zrÿd JP (1997) Efficient gene targeting in the moss *Physcomitrella patens*. Plant J 11:1195–1206
7. Cove DJ, Perroud P-F, Charron AJ, McDaniel SF, Khandelwal A, Quatrano RS (2009) Isolation of DNA, RNA, and protein from the moss *Physcomitrella patens* gametophytes. Cold Spring Harb Protoc 2009(2):pdb.prot5146. https://doi.org/10.1101/pdb.prot5146
8. Cove DJ, Perroud P-F, Charron AJ, McDaniel SF, Khandelwal A, Quatrano RS (2009) Transformation of moss *Physcomitrella patens* gametophytes using a biolistic projectile delivery system. Cold Spring Harb Protoc 2009(2):pdb.prot5145. https://doi.org/10.1101/pdb.prot5145
9. Cove DJ, Perroud P-F, Charron AJ, McDaniel SF, Khandelwal A, Quatrano RS (2009) Transformation of the moss *Physcomitrella patens* using T-DNA mutagenesis. Cold Spring Harb Protoc 2009(2):pdb.prot5144. https://doi.org/10.1101/pdb.prot5144
10. Cove DJ, Perroud P-F, Charron AJ, McDaniel SF, Khandelwal A, Quatrano RS (2009) Transformation of the moss *Physcomitrella patens* using direct DNA uptake by protoplasts. Cold Spring Harb Protoc 2009(2):pdb.prot5143. https://doi.org/10.1101/pdb.prot5143
11. Cove DJ, Perroud P-F, Charron AJ, McDaniel SF, Khandelwal A, Quatrano RS (2009) Isolation and regeneration of protoplasts of the moss *Physcomitrella patens*. Cold Spring Harb Protoc 2009(2):pdb.prot5140. https://doi.org/10.1101/pdb.prot5140
12. Cove DJ, Perroud P-F, Charron AJ, McDaniel SF, Khandelwal A, Quatrano RS (2009) Culturing the moss *Physcomitrella patens*. Cold Spring Harb Protoc 2009(2):pdb.prot5136. https://doi.org/10.1101/pdb.prot5136
13. Rensing SA, Lang D, Zimmer AD et al (2008) The *Physcomitrella* genome reveals evolutionary insights into the conquest of land by plants. Science 319:64–69
14. Cove DJ, Schild A, Ashton NW, Hartmann E (1978) Genetic and physiological studies of the effect of light on the development of the moss, *Physcomitrella patens*. Photochem Photobiol 27:249–254
15. Possart A, Hiltbrunner A (2013) An evolutionarily conserved signaling mechanism mediates far-red light responses in land plants. Plant Cell 25:102–114
16. Chen Y-R, Su Y, Tu S-L (2012) Distinct phytochrome actions in nonvascular plants revealed by targeted inactivation of phytobilin biosynthesis. Proc Natl Acad Sci U S A 109:8310–8315
17. Yamawaki S, Yamashino T, Nakanishi H, Mizuno T (2011) Functional characterization of *HY5* homolog genes involved in early light-signaling in *Physcomitrella patens*. Biosci Biotechnol Biochem 75:1533–1539
18. Possart A, Xu T, Paik I et al (2017) Characterization of phytochrome interacting factors from the moss *Physcomitrella patens* illustrates conservation of phytochrome signaling modules in land plants. Plant Cell 29:310–330
19. Xu T, Hiltbrunner A (2017) PHYTOCHROME INTERACTING FACTORs from *Physcomitrella patens* are active in *Arabidopsis* and complement the *pif* quadruple mutant. Plant Signal Behav 12:e1388975
20. Cove D, Knight C (1987) Gravitropism and phototropism in the moss, *Physcomitrella patens*. In: Developmental mutants in higher plants. Cambridge University Press, London, pp 181–196
21. Mittmann F, Brücker G, Zeidler M, Repp A, Abts T, Hartmann E, Hughes J (2004) Targeted knockout in *Physcomitrella* reveals direct actions of phytochrome in the cytoplasm. Proc Natl Acad Sci U S A 101:13939–13944
22. Kadota A, Sato Y, Wada M (2000) Intracellular chloroplast photorelocation in the moss *Physcomitrella patens* is mediated by phytochrome as well as by a blue-light receptor. Planta 210:932–937
23. Sato Y, Wada M, Kadota A (2001) Choice of tracks, microtubules and/or actin filaments for chloroplast photo-movement is differentially controlled by phytochrome and a blue light receptor. J Cell Sci 114:269–279
24. Uenaka H, Kadota A (2007) Functional analyses of the *Physcomitrella patens* phytochromes in regulating chloroplast avoidance movement. Plant J 51:1050–1061
25. Hughes J (2013) Phytochrome cytoplasmic signaling. Annu Rev Plant Biol 64:377–402
26. Mittmann F, Dienstbach S, Weisert A, Forreiter C (2009) Analysis of the phytochrome gene family in *Ceratodon purpureus* by gene targeting reveals the primary phytochrome responsible for photo- and polarotropism. Planta 230:27–37
27. Li F-W, Melkonian M, Rothfels CJ, Villarreal JC, Stevenson DW, Graham SW, Wong GK-S, Pryer KM, Mathews S (2015) Phytochrome diversity in green plants and the origin of

canonical plant phytochromes. Nat Commun 6:7852

28. Jinek M, Chylinski K, Fonfara I, Hauer M, Doudna JA, Charpentier E (2012) A programmable dual-RNA-guided DNA endonuclease in adaptive bacterial immunity. Science 337:816–821
29. Barrangou R, Doudna JA (2016) Applications of CRISPR technologies in research and beyond. Nat Biotechnol 34:933–941
30. Puchta H (2017) Applying CRISPR/Cas for genome engineering in plants: the best is yet to come. Curr Opin Plant Biol 36:1–8
31. Lopez-Obando M, Hoffmann B, Géry C, Guyon-Debast A, Téoulé E, Rameau C, Bonhomme S, Nogué F (2016) Simple and efficient targeting of multiple genes through CRISPR-Cas9 in *Physcomitrella patens*. G3 (Bethesda) 6:3647–3653
32. Nomura T, Sakurai T, Osakabe Y, Osakabe K, Sakakibara H (2016) Efficient and heritable targeted mutagenesis in mosses using the CRISPR/Cas9 system. Plant Cell Physiol 57:2600–2610
33. Collonnier C, Epert A, Mara K, Maclot F, Guyon-Debast A, Charlot F, White C, Schaefer DG, Nogué F (2017) CRISPR-Cas9-mediated efficient directed mutagenesis and RAD51-dependent and RAD51-independent gene targeting in the moss *Physcomitrella patens*. Plant Biotechnol J 15:122–131
34. Collonnier C, Guyon-Debast A, Maclot F, Mara K, Charlot F, Nogué F (2017) Towards mastering CRISPR-induced gene knock-in in plants: survey of key features and focus on the model *Physcomitrella patens*. Methods 121–122:103–117
35. Ashton NW, Cove DJ (1977) The isolation and preliminary characterisation of auxotrophic and analogue resistant mutants of the moss, *Physcomitrella patens*. Mol Gen Genet 154:87–95
36. Haeussler M, Schönig K, Eckert H et al (2016) Evaluation of off-target and on-target scoring algorithms and integration into the guide RNA selection tool CRISPOR. Genome Biol 17:148
37. Hsu PD, Scott DA, Weinstein JA et al (2013) DNA targeting specificity of RNA-guided Cas9 nucleases. Nat Biotechnol 31:827–832
38. Sambrook J, Fritsch E, Maniatis T (1989) Molecular cloning: a laboratory manual, 2nd edn. Cold Spring Harbor Laboratory Press, Cold Spring Harbor, NY

Chapter 21

Phylogenetic Methods to Study Light Signaling

Fay-Wei Li and Sarah Mathews

Abstract

Phylogenetic comparative methods (PCM) represent a rigorous approach for inferring functional evolution. To infer the origin and evolution of a function, PCM use a phylogenetic tree of the species in which the function has evolved and functional data from those species. These data enable reconstruction of ancestral states and inference of how the function evolved along the branches of the species tree. PCM can be applied to understand any aspect of light signaling, from early events in photoactivation, to interactions with signaling partners, to physiological responses. Integrating evolutionary histories of individual aspects of light signaling obtained through PCM with network modeling of protein–protein interactions for light signaling would enable a deep understanding of the evolution in light signaling pathways and their roles in helping plants adapt to changing environments. Here we describe the steps for using PCM to infer functional evolution using a species tree and trait data.

Key words Functional evolution, Light signaling, Phytochromes, Phylogenetic comparative methods

1 Introduction

Phytochromes are the major plant photoreceptors for light in the red (R) and far-red (FR) range of the spectrum, and the magnitude of genetic networks that they regulate is substantial, with many details remaining to be characterized [1]. Understanding the evolution of phytochrome-mediated light signaling is, nonetheless, of central importance in efforts to understand how plants have adapted to changing environments through time.

This chapter outlines methods for reconstructing the evolution of a trait or function in order to infer the origin of a trait and its degree of conservation. For example, we could map the presence of a sequence motif across the tree of life, or ask how the signaling components were assembled and evolved through time. To understand the evolution of a trait, we use a phylogenetic tree that describes the pattern of descent from a common ancestor [2]. The tree and the trait data from the living species allow us to reconstruct the steps by which the trait evolved along the branches of

Andreas Hiltbrunner (ed.), *Phytochromes: Methods and Protocols*, Methods in Molecular Biology, vol. 2026,
https://doi.org/10.1007/978-1-4939-9612-4_21,

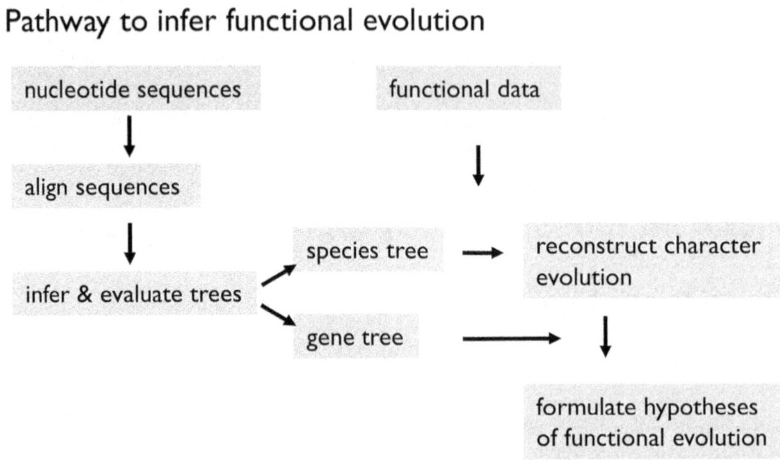

Fig. 1 Overview of the materials and steps for inferring functional evolution

the tree. Below we describe the approaches for obtaining a species tree and reconstructing the history of discrete traits (*see* Fig. 1 for an overview). The same analyses can be conducted for continuous traits by invoking alternative models of character evolution than are presented here.

We also include the steps for inferring a gene phylogeny, since the genes that encode signaling proteins have their own complex evolutionary histories. Most genes that function in light signaling pathways are members of gene families, with highly dynamic evolutionary histories. Taking phytochromes as an example, the data indicate that duplication and divergence occurred independently in three of the major clades of land plants (mosses, ferns, seed plants) starting in each case from a single ancestral phytochrome [3]. This implies that duplicates are generated and then lost during the course of evolution, and in this specific case, it means that similar functions in these three clades may reflect convergent patterns of functional diversification rather than conservation of an ancient function within a single gene lineage. We do not know whether the ancestral phytochromes for mosses, ferns, and seed plants were similar to one another in combining both phyA- and phyB-type functions, or whether the ancestors were more phyB-like, with phyA-type functions independently evolved in each clade (or vice versa). If the answer is that the ancestors were each more phyB-like, then we could consider phyB-like functions homologous, and conserved, across land plants. phyA-like functions would not be homologous but instead the results of independent, parallel evolution following gene duplication and divergence.

Gene phylogenies can therefore provide critical insights directly relevant to evaluating hypotheses of functional evolution, although they are not required for the steps we describe here for inferring functional evolution.

Despite the rigor of available methods for reconstructing functional evolution, the analyses can only be as good as the underlying data. The limiting factor is often the availability of functional data from an appropriate phylogenetic sample of the species in which the trait has evolved. Thus, experimental design is as important in evolutionary biology as it is in any other area of science. Additionally, for genes with similar patterns of evolution, statements about functional homology, or conservation, should be made with care. Some uncertainty will always remain in these studies, particularly in cases where functional information is available from a limited number of model systems rather than from an adequate sampling of phylogenetic diversity. The use of a small number of model organisms from very disparate clades (e.g., liverworts, mosses, and angiosperms) has provided insights into the endpoints in the evolution of some aspects of phytochrome signaling. Retracing the histories to those endpoints is still not possible, however, because functional data from major land plant clades remain unavailable. An example from seed plants below helps explain why this is the case.

2 Materials

1. Species tree. It is important that this tree captures a representative sample of phylogenetic diversity across the whole clade in which the function has evolved.
2. Gene phylogeny. Patterns of functional evolution can be inferred without knowledge of the underlying evolutionary history of the genes, but if the function has been linked with a gene in one or more model systems, understanding relationships among gene family members from different species may help identify which gene from a previously unstudied species might be most likely to share that function.
3. Functional data. It is critical that functional data are collected from species representing the phylogenetic diversity of the clade in which one wishes to understand functional evolution.

3 Methods

3.1 Obtaining a Species Tree

1. Obtain data from GenBank to infer a species tree (*see* **Note 1**): Use a pipeline or search the Nucleotide database of GenBank (https://www.ncbi.nlm.nih.gov/) to obtain sequences of several loci commonly used in plant phylogenetics, such as *mat*K, *rbc*L, *atp*B, *ndh*F, and the internal transcribed spacers (ITS) of the nuclear ribosomal repeats (*see* **Note 2**). Available pipelines greatly streamline the process of assembling a phylogenetic matrix from GenBank data (*see* **Note 3**).

2. Assemble a nucleotide alignment: Assemble nucleotide sequences from each gene into separate phylogenetic data matrices using a tool for multiple sequence alignment such as MACSE [4] (*see* **Note 4**). The alignment step can be performed through an online server (e.g., http://mbb.univ-montp2.fr/macse) or run locally using an open source alignment tool. Inspect the alignments for regions of high variability across sequences that are difficult to align. If these exist, remove them using Gblocks [5], (http://molevol.cmima.csic.es/castresana/Gblocks_server.html) because only nucleotides of clear homology should be included. Be sure to choose "Codon" as the sequence type. Inspect the individual gene trees to detect sequences that have been wrongly attributed to a species (i.e., species that are clearly in the wrong place in the tree based on other sources of information about species relationships). After removal of any problematic sequences, concatenate the gene sequences into a multigene alignment (*see* **Note 5**). This can be done with FASconCAT [6], or in Geneious [7], or abioscripts (http://ormbunkar.se/phylogeny/abioscripts/). FASconCAT is useful because it allows the output of coordinates of the individual gene boundaries. These will allow the best fitting model of nucleotide evolution to be indicated for each gene partition in the alignment in subsequent phylogenetic analyses.
3. Infer the species tree: Infer a tree using an approach that rigorously explores the possible tree space, that is, one that evaluates the fit of the data to as many of the possible tree topologies for a set of sequences as is computationally tractable, such as parsimony, maximum likelihood, or Bayesian inference, using open source software or an online server (*see* **Note 6**). Follow the guidelines for both inferring the tree and producing support values (e.g., bootstrap values or posterior probabilities) that are found in the relevant software documentation.
4. Determine the degree of support for the tree topology: Evaluate support in the data for nodes in the inferred phylogeny by inspecting the support values at each node. The degree of support is influenced by total phylogenetic information content of the alignment, the balance of information and noise, the amount of conflicting information among loci or nucleotide sites, and the number of both sequences in the alignment and tips in the tree. In thorough analyses of species relationships, additional tests are used to determine how well nodes are supported, but for the purposes of this chapter, it is important to note that bootstrap and posterior probability values below 0.90 and 0.95, respectively indicate that considerable uncertainty remains regarding the presence of that node in the tree.

3.2 Obtaining a Gene Phylogeny

Most genes that function in light signaling pathways are members of gene families. Gene phylogenies are not required to reconstruct functional evolution using PCM, but they may be important in identifying candidate functional homologs, and as in the phytochrome example, in providing information that may be important in distinguishing among evolutionary hypotheses. There now exist many resources for inferring the detailed gene trees that are needed to distinguish orthologs from paralogs.

In most cases, databases that do not provide annotation data will be included in the mining exercise in order to infer a tree that is as detailed as possible. Such databases include the transcriptome assemblies of the 1000 Plants (https://sites.google.com/a/ualberta.ca/onekp/) and many other projects, as well as databases that will be created by the 10 K Genomes Project [8] for plants (being initiated by the China National Gene Bank) and the Open Green Genomes Initiative (https://jgi.doe.gov/csp-2018-leebens-mack-open-green-genomes-initiative/). GenBank and Phytozome are the best sources of annotated sequences. *See* **steps 1–3** for instructions on mining databases that do or do not provide annotation data.

1. Obtain unannotated sequences: Search the desired database for sequences similar to the target using an approach such as tBLASTn [9] or HMMER [10], being sure to use a low threshold (E-value $<1e^{-5}$) to capture all the putative homologs. Filter these candidates by taking advantage of information on the target's well-defined functional domains by using the NCBI Conserved Domain Database to select sequences that share the same structure of core domains. An example of the approach can be found in a recent large-scale phylogenetic study on phytochromes [3].
2. Obtain annotated sequences from Phytozome: In the current version of Phytozome (https://phytozome.jgi.doe.gov/pz/portal.html#), the most efficient approach to obtaining data from a single locus or gene family is to use the Keyword Search option to provide the transcript name for *Arabidopsis thaliana* (e.g., for *PHYB*, AT2G18790) in the keyword field, choose *Arabidopsis thaliana* from the species display, and click on "GO" to search. On the search results page, choose gene view. On the gene view page, choose "Find Related, By sequence similarity" from the Actions menu. This brings up the BLAST menu. Choose the target clade from the species display and increase the number in the "alignments to show" field as needed. On the BLAST results page, select the appropriate hits (based on E value, length, and/or annotation) and add them to your cart using the My Data menu. Then choose "View Cart." Select all from the list of transcripts that is loaded and "Quick Download" from the My Data menu. Then choose "Sequence: CDS". The sequences will be downloaded in a fasta file.

3. Obtain annotated sequences from GenBank: Search the Nucleotide database of GenBank (https://www.ncbi.nlm.nih.gov/) to obtain sequences of the target signaling locus. On the NCBI Welcome page, choose Nucleotide from the database dropdown menu. In the search field, type a query such as "gene name AND taxon." For example, to retrieve records for phytochrome B from angiosperms, type "phyb AND Magnoliophyta" (without quotes). The taxon name must be a name in the NCBI taxonomy tree, which can be checked in the Taxonomy Browser on the NCBI webpage. Apply any relevant filters to the items obtained in the search, such as setting the length range to omit fragmentary sequences, and maximize the number of items displayed per page. There are two options for downloading the sequences in a fasta file. Using the "Send to:" option allows just coding sequences to be downloaded and items from all pages are sent to the file. Using the "Summary" option allows sequences to be downloaded as "FASTA (text)" files, but downloads only the items displayed on the current page. The chief advantage of the "Send to:" option is that untranslated regions can be avoided by choosing the coding sequence option and this substantially simplifies preparation of the final alignment. A disadvantage is that taxon names are not included in the fasta headers. To obtain the taxon names, use the ESummary or EFetch tool at the NCBI site to retrieve taxon names using the accession numbers from the headers. The headers in the fasta text files downloaded using the "Summary" option are much simpler and include the taxon name. However, depending on the gene of interest, a significant amount of alignment editing may be required to exclude untranslated regions from the alignment. In those cases, there will be a clear advantage to using the "Send to:" option to obtain coding sequences only.
4. Infer the gene tree: Infer the gene from the downloaded data by following **steps 2–4** of Subheading 3.1 Obtaining species tree.

3.3 Reconstructing Functional Evolution on the Species Tree

The evolutionary history of a particular trait or function is inferred by combining an estimate of relatedness (the species tree) and the data on traits from contemporary species. An optimality criterion such as parsimony, maximum likelihood, or Bayes values is used to reconstruct the evolution of the trait on the species tree by finding the best fit between the observed trait data and the tree topology. In the case of phytochrome-mediated light signaling, there is just one published example of an evolutionary analysis to reconstruct functional evolution. Mathews and Tremonte [11] reconstructed the evolutionary history of greening in the dark and the far-red high irradiance response in seed plants. We use this example to outline the steps required to infer origin and change of a function.

1. Collect functional data: The data should be collected from a representative sample of diversity in the target species clade. For example, in order to understand the evolution of phytochrome-mediated responses in seed plants, Mathews and Tremonte [11] collected data on greening of seedlings in the dark and in continuous red, far-red, and white light from multiple species of each gymnosperm clade (conifers, gnetophytes, cycads, *Gingko*) and from species in 17 major clades of angiosperms. When functional data are available from fewer species, the power of the analysis is reduced, because it increases the amount of uncertainty in the reconstructions at some or all the nodes in the tree (e.g., Fig. 2).
2. Map the origins and changes of function on the species tree: To follow the example of Mathews and Tremonte [11], use Mesquite [12] to find the character state assignment at each node that maximizes the probability of arriving at the observed states in the terminal taxa, given the data and a specific model of character evolution (*see* **Note 7**). Specifically, check that the input files containing the character data (i.e., greening of seedlings in different light conditions) and the species tree are in the format required by Mesquite. Open the data matrix, and import the tree file using the "Make New Trees Block from" option on the Trees & Taxa dropdown menu. Alter the tree for reconstruction of ancestral states by choosing "Alter/Transform Branch Lengths" and then "Arbitrarily Ultrametricize" on the Tree dropdown menus (*see* **Note 8**). Use the "Model Settings" option on the Characters menu to choose the Markov k-state 2-parameter model (AsymmMk) model, which has two parameters, one each for the change from 0 > 1 and 1 > 0. Use the "Trace Character History" module on the Analysis:Tree menu, then "Likelihood Ancestral States," then "Stored Probability Model," then the "Asymmetrical 2-param. Markov-k Model". If there are multiple columns (characters for which there are state data) in the data matrix, make sure that the appropriate character is selected for tracing. Choose "Tree Form," "Balls & Sticks" on the Display menu to see the marginal probabilities for the character data at each node in the tree (*see* **Note 9**).

3.4 Formulating Hypotheses of Functional Evolution

Factors that affect the accuracy in reconstruction of ancestral states include the completeness and phylogenetic distribution of the functional data, and the lability of the function during evolution. One cannot, for example, infer with any confidence the origin and conservation of a function in land plants by analyzing data from a handful of model organisms. To see why this is the case, compare the trees in Fig. 2a, b.

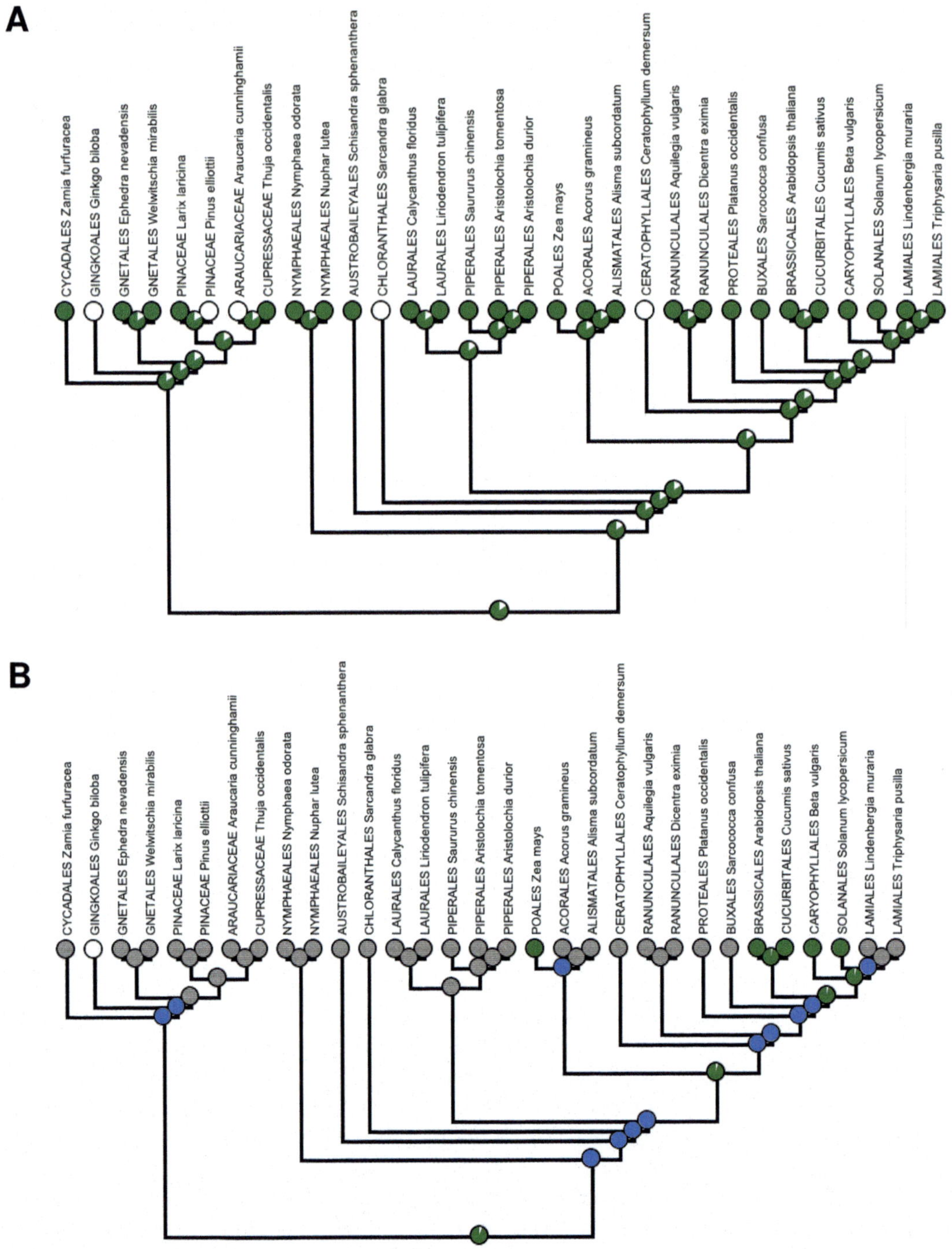

Fig. 2 Reconstruction of the elongation response to continuous far-red light (cFR) during seedling development on a phylogenetic tree of seed plants. (**a**) Reconstruction on the same tree with the same data used by Mathews and Tremonte [11]. Hypocotyls (or relevant organ) are short or long in cFR in species indicated by green-filled and open circles, respectively. (**b**) Reconstruction on the same tree, but using only the data available before the study by Mathews and Tremonte [11]. Circles with grey hashed fill represent species and clades without data. Circles with blue fill indicate nodes in the tree where the probability of an ancestral state cannot be determined

The results obtained by Mathews and Tremonte [11] provided a reasonably sound basis for inferring that being green in the dark is a derived condition in seedling development within a few seed plant clades, and furthermore, that the far-red high radiance response is ancestral in seed plants (Fig. 2a). The proportional probabilities for the presence or absence of the FR-HIR in ancestral seeds plants are 0.84 and 0.16 respectively. If they had, however, made conclusions based only on the functional data available before their study, their confidence in the state at the ancestral node would have been inflated, due to the fact that data from species that do not display inhibition of elongation in cFR were not accounted for in the functional data set for the tree in Fig. 2b. They would have been unable to infer the probabilities of ancestral states at key nodes in the phylogeny, including those at the base of gymnosperm, angiosperm, monocot, and eudicot crown groups (Fig. 2b). This paradox of a high probability supporting the ancestral condition of "inhibition in cFR" together with the uncertainty at major nodes in the backbone would have alerted them to the inadequacy of the data for inference of functional evolution.

4 Notes

1. In some cases, a published treefile will be available. A treefile in Newick Standard format (computer-readable treefiles that are widely used in evolutionary analyses) will be needed for the reconstruction of character evolution. For studies seeking to understand functional evolution across a wide phylogenetic breadth (e.g., across all land, vascular, seed, or flowering plants), good starting points are the trees from Wickett et al. [13], available at http://datacommons.cyverse.org/browse/iplant/home/shared/onekp_pilot/PNAS_alignments_trees/species_level/trees and from Ruhfel et al. [14], available at https://doi.org/10.5061/dryad.k1t1f. Open Tree of Life, a portal that is aggregating and synthesizing published phylogenetic trees, is an additional, potentially useful source of treefiles [15]. Example filenames for useful trees from Wickett et al. [13] include FNA2AA.trim50genes50sites.no3rd.partitioned.final.tre and astral.FNA2AA.trim50genes33taxa.no3rd.final.tre. A clade of interest, e.g., all angiosperms or all land plants, can be extracted from the tree and output into a new treefile by using Dendroscope [16] (dendroscope.org). Be sure to extract the complete clade.
2. Species trees are inferred from molecular sequence data. Analysis of nucleotide rather than amino acid sequences increases the number of characters in the matrix, which can increase phylogenetic power and accuracy. When studying

functional evolution across very deep nodes in a species tree, it may be more appropriate to analyze amino acid sequences because at certain nucleotide sites, multiple substitutions will lead to saturation, that is, sites that are changing will have changed before. Since gene and species phylogenies may differ, data from multiple genes are needed, ideally representing both the nuclear and plastid genomes. For nuclear data, it is important that orthologous sequences are compared. For genes that are members of gene families, determining orthology can be challenging due to incomplete sampling of all paralogs in a genome. Orthology may also be difficult to determine for ITS when they are sampled from genomes that have variable ribosomal repeats. Conserved ortholog sets (COS) are available and have been used effectively for some clades [17, 18]. These sets might include loci that may be more broadly useful and could be targeted when mining GenBank.

3. Open source pipeline for assembling phylogenetic matrices to infer a species tree include PHLAWD (https://github.com/FePhyFoFum/pyphlawd [19]) and SUMAC (https://github.com/wf8/sumac [20]).

4. Other powerful tools for multiple sequence alignment include MAFFT [21] and MUSCLE [22], but MACSE has the advantage of accommodating frameshifts that result from sequencing or are biologically induced without disrupting codon structure. This reduces the possibility of mistakenly aligning nonhomologous nucleotide positions. The default options of MAFFT and MUSCLE do not take codon position into account when aligning nucleotide sequences of protein-coding genes.

5. In some cases, the trees inferred from genes to be used in inferring the species tree may have conflicting topologies. If the conflicting nodes in the species tree inferred from the concatenated data are well supported by the data (high bootstrap or posterior probability values), consider omitting the conflicting individual locus alignment(s) before performing the concatenation, particularly if the conflict is at deeper nodes in the tree. Conflicting topologies at the tree's tips are often not well supported and also are less likely to impact downstream analyses. If removal of a gene alignment results in a plastid-only matrix to infer the species tree, be aware that the tree represents the organellar phylogeny and may differ from the species phylogeny.

6. A phylogenetic hypothesis of species relationships is the starting point for the study of trait evolution. To obtain the most accurate species tree possible, it is thus important to understand the basic differences among methods of phylogenetic inference to choose the most appropriate method for the data being used. An accessible and useful introduction to approaches to inferring evolutionary trees is found in Chapter 6 of Page and Holmes

[23], "Inferring molecular phylogenies." Several excellent online servers for tree inference are available. These include the servers for RAxML (https://embnet.vital-it.ch/raxml-bb/), IQ-TREE (http://www.iqtree.org), trex (http://www.trex.uqam.ca), PhyML (http://www.atgc-montpellier.fr/phyml/), and the CIPRES web portal (https://www.phylo.org).

7. BayesTraits is an alternative package for reconstructing ancestral states. A description and downloads are available at http://www.evolution.rdg.ac.uk/BayesTraitsV3/BayesTraitsV3.html. An online workflow is available at ARBOR (http://www.arborworkflows.com).
8. This will equalize all the root-to-tip path-lengths in the tree, a necessary step for trees inferred by maximum likelihood or Bayesian inference, which have branch lengths that are proportional to the amount of change along the branches.
9. These options are available in the current version of Mesquite.

References

1. Wang H, Wang H (2015) Phytochrome signaling: time to tighten up the loose ends. Mol Plant 8:540–551
2. Pagel M (1999) The Maximum Likelihood approach to reconstructing ancestral character states of discrete characters on phylogenies. Syst Biol 48:612–622
3. Li F-W, Melkonian M, Rothfels CJ, Villarreal JC, Stevenson DW, Graham SW, Wong GK-S, Pryer KM, Mathews S (2015) Phytochrome diversity in green plants and the origin of canonical plant phytochromes. Nat Commun 6:7852
4. Ranwez V, Harispe S, Delsuc F, Douzery EJP (2011) MACSE: Multiple Alignment of Coding SEquences accounting for frameshifts and stop codons. PLoS One 6:e22594
5. Castresana J (2000) Selection of conserved blocks from multiple alignments for their use in phylogenetic analysis. Mol Biol Evol 17:540–552
6. Kück P, Meusemann K (2010) FASconCAT: convenient handling of data matrices. Mol Phylogenet Evol 56:1115–1118
7. Kearse M, Moir R, Wilson A et al (2012) Geneious Basic: an integrated and extendable desktop software platform for the organization and analysis of sequence data. Bioinformatics 28:1647–1649
8. Cheng S, Melkonian M, Smith SA, Brockington S, Archibald JM, Delaux PM et al (2018) 10KP: A phylodiverse genome sequencing plan. Gigascience 7:1–9
9. Camacho C, Coulouris G, Avagyan V, Ma N, Papadopoulos J, Bealer K, Madden TL (2009) BLAST+: architecture and applications. BMC Bioinformatics 10:421
10. Finn RD, Clements J, Eddy SR (2011) HMMER web server: interactive sequence similarity searching. Nucleic Acids Res 39:W29–W37
11. Mathews S, Tremonte D (2012) Tests of the link between functional innovation and positive selection at phytochrome A: the phylogenetic distribution of far-red high-irradiance responses in seedling development. Int J Plant Sci 173:662–672
12. Maddison WP, Maddison DR (2017) Mesquite: a modular system for evolutionary analysis. http://mesquiteproject.org
13. Wickett NJ, Mirarab S, Nguyen N et al (2014) Phylotranscriptomic analysis of the origin and early diversification of land plants. Proc Natl Acad Sci U S A 111:E4859–E4868
14. Ruhfel BR, Gitzendanner MA, Soltis PS, Soltis DE, Burleigh JG (2014) From algae to angiosperms–inferring the phylogeny of green plants (*Viridiplantae*) from 360 plastid genomes. BMC Evol Biol 14:23
15. Hinchliff CE, Smith SA, Allman JF et al (2015) Synthesis of phylogeny and taxonomy into a comprehensive tree of life. Proc Natl Acad Sci U S A 112:12764–12769
16. Huson DH, Richter DC, Rausch C, Dezulian T, Franz M, Rupp R (2007) Dendroscope: An interactive viewer for large phylogenetic trees. BMC Bioinformatics 8:460

17. Levin RA, Whelan A, Miller JS (2009) The utility of nuclear conserved ortholog set II (COSII) genomic regions for species-level phylogenetic inference in *Lycium* (Solanaceae). Mol Phylogenet Evol 53:881–890
18. Jeong Y-M, Chung W-H, Chung H, Kim N, Park B-S, Lim K-B, Yu H-J, Mun J-H (2014) Comparative analysis of the radish genome based on a conserved ortholog set (COS) of *Brassica*. Theor Appl Genet 127:1975–1989
19. Smith SA, Beaulieu JM, Donoghue MJ (2009) Mega-phylogeny approach for comparative biology: an alternative to supertree and supermatrix approaches. BMC Evol Biol 9:37
20. Freyman WA (2015) SUMAC: constructing phylogenetic supermatrices and assessing partially decisive taxon coverage. Evol Bioinformatics Online 11:263–266
21. Katoh K, Kuma K, Toh H, Miyata T (2005) MAFFT version 5: improvement in accuracy of multiple sequence alignment. Nucleic Acids Res 33:511–518
22. Edgar RC (2004) MUSCLE: multiple sequence alignment with high accuracy and high throughput. Nucleic Acids Res 32:1792–1797
23. Page RDM, Holmes EC (1998) Molecular evolution: a phylogenetic approach. Blackwell Science, Oxford/Malden, MA

Index

Andreas Hiltbrunner (ed.), *Phytochromes: Methods and Protocols*, Methods in Molecular Biology, vol. 2026,
https://doi.org/10.1007/978-1-4939-9612-4, © Springer Science+Business Media, LLC, part of Springer Nature 2019

O

P

R

S

T

U

V

Y

Printed by Printforce, the Netherlands